Methods in Molecular Biology

Series Editor
John M. Walker
School of Life and Medical Sciences
University of Hertfordshire
Hatfield, Hertfordshire, UK

For further volumes:
http://www.springer.com/series/7651

For over 35 years, biological scientists have come to rely on the research protocols and methodologies in the critically acclaimed *Methods in Molecular Biology* series. The series was the first to introduce the step-by-step protocols approach that has become the standard in all biomedical protocol publishing. Each protocol is provided in readily-reproducible step-by-step fashion, opening with an introductory overview, a list of the materials and reagents needed to complete the experiment, and followed by a detailed procedure that is supported with a helpful notes section offering tips and tricks of the trade as well as troubleshooting advice. These hallmark features were introduced by series editor Dr. John Walker and constitute the key ingredient in each and every volume of the *Methods in Molecular Biology* series. Tested and trusted, comprehensive and reliable, all protocols from the series are indexed in PubMed.

Astrocytes

Methods and Protocols

Second Edition

Edited by

Barbara Di Benedetto

Department of Psychiatry and Psychotherapy, University of Regensburg, Regensburg, Germany

Humana Press

Editor
Barbara Di Benedetto
Department of Psychiatry
and Psychotherapy
University of Regensburg
Regensburg, Germany

ISSN 1064-3745 ISSN 1940-6029 (electronic)
ISBN 978-1-0716-4365-5 ISBN 978-1-0716-4366-2 (eBook)
https://doi.org/10.1007/978-1-0716-4366-2

This Humana imprint is published by the registered company Springer Science+Business Media, LLC, part of Springer Nature.
The registered company address is: 1 New York Plaza, New York, NY 10004, U.S.A.

Preface

This detailed volume collects a great variety of methods essential to investigate the multifaceted roles of astrocytes in both physiology and pathological conditions. The book covers several techniques for the preparation of astrocytes from animal models, for the examination of astrocyte morphology and functions in vitro and ex vivo, as well as for investigating astrocyte pathologies in the central nervous system. Written for the highly successful Methods in Molecular Biology series, chapters include introductions to their respective topics, lists of the necessary materials and reagents, step-by-step and readily reproducible laboratory protocols, and tips on troubleshooting and avoiding known pitfalls.

Authoritative and practical, *Astrocytes: Methods and Protocols* serves as an ideal guide for both experienced and beginner scientists working toward unraveling the novel, fascinating roles of these highly versatile cells.

Regensburg, Germany ***Barbara Di Benedetto***

The original version of the book has been revised. A correction to this book can be found at https://doi.org/10.1007/978-1-0716-4366-2_22

Contents

Contributors

IRIS ÁLVAREZ-MERZ • *Institute of Neurobiology, Heinrich Heine University Düsseldorf, Düsseldorf, Germany*

KATIA AVILA-GUTIERREZ • *Center for Interdisciplinary Research in Biology (CIRB), Collège de France, CNRS, INSERM, Université PSL, Paris, France*

LAURA BOI • *Department of Behavioral and Molecular Biology, University of Regensburg, Regensburg, Germany*

ANTHONY BOSSON • *Centre de Recherche du Centre Hospitalier de l'Université de Montréal (CRCHUM), Montréal, QC, Canada*

ANNE-CÉCILE BOULAY • *Center for Interdisciplinary Research in Biology (CIRB), Collège de France, CNRS, INSERM, Université PSL, Paris, France*

YOUKYEONG GLORIA BYUN • *Department of Biological Sciences, Korea Advanced Institute of Science and Technology, Daejeon, Republic of Korea; Center for Vascular Research, Institute for Basic Science, Daejeon, Republic of Korea*

SABRINA CHIERZI • *Centre for Research in Neuroscience, Department of Neurology & Neurosurgery, Brain Repair and Integrative Neuroscience Program, Research Institute of the McGill University Health Centre, Montreal General Hospital, Montreal, QC, Canada*

WON-SUK CHUNG • *Department of Biological Sciences, Korea Advanced Institute of Science and Technology, Daejeon, Republic of Korea; Center for Vascular Research, Institute for Basic Science, Daejeon, Republic of Korea*

CATERINA CIANI • *Neuroscience Department, SISSA, Trieste, Italy*

MARTINE COHEN-SALMON • *Center for Interdisciplinary Research in Biology (CIRB), Collège de France, CNRS, INSERM, Université PSL, Paris, France*

PATRIZIA DELLA VECCHIA • *Chair of Physiological Genomics, Biomedical Center, Faculty of Medicine, LMU, Munich, Germany*

LEWIS DEPAAUW-HOLT • *Département de Neurosciences, Université de Montréal, Montréal, QC, Canada; Centre de Recherche du Centre Hospitalier de l'Université de Montréal (CRCHUM), Montréal, QC, Canada*

BARBARA DI BENEDETTO • *Department of Psychiatry and Psychotherapy, University of Regensburg, Regensburg, Germany; Regensburg Center of Neuroscience, University of Regensburg, Regensburg, Germany*

SARA EITELMANN • *Institute of Neurobiology, Heinrich Heine University Düsseldorf, Düsseldorf, Germany*

CARMEN FALCONE • *Neuroscience Department, SISSA, Trieste, Italy*

KATJA FINK • *Laboratory of Neuroendocrinology-Molecular Cell Physiology, Institute of Pathophysiology, Faculty of Medicine, University of Ljubljana, Ljubljana, Slovenia; Celica Biomedical, Ljubljana, Slovenia*

KIJANA GEORGE • *Fralin Biomedical Research Institute at Virginia Tech Carilion, Roanoke, VA, USA*

SAMANTHA GOLF • *Department of Cell, Developmental and Integrative Biology, University of Alabama at Birmingham, Birmingham, AL, USA*

BENJAMIN HEITHOFF • *Fralin Biomedical Research Institute at Virginia Tech Carilion, Roanoke, VA, USA*

Christian Henneberger • *Institute of Cellular Neurosciences I, Medical Faculty, University of Bonn, Bonn, Germany; German Center for Neurodegenerative Diseases, Bonn, Germany*
Matthew Holt • *i3S – Instituto de Investigação e Inovação em Saúde, Universidade do Porto, Porto, Portugal*
Karl W. Kafitz • *Institute of Neurobiology, Heinrich Heine University Düsseldorf, Düsseldorf, Germany*
Gyuri Kim • *Department of Biological Sciences, Korea Advanced Institute of Science and Technology, Daejeon, Republic of Korea; Center for Vascular Research, Institute for Basic Science, Daejeon, Republic of Korea*
Marko Kreft • *Laboratory of Neuroendocrinology-Molecular Cell Physiology, Institute of Pathophysiology, Faculty of Medicine, University of Ljubljana, Ljubljana, Slovenia; Celica Biomedical, Ljubljana, Slovenia; Department of Biology, Biotechnical Faculty, University of Ljubljana, Ljubljana, Slovenia*
Si-Si Lin • *International Joint Research Centre on Purinergic Signalling of Sichuan Province /Research Centre on TCM-Rehabilitation and Neural Circuit, School of Acupuncture and Tuina/Health and Rehabilitation, Chengdu University of Traditional Chinese Medicine, Chengdu, China*
Philippe Mailly • *Center for Interdisciplinary Research in Biology (CIRB), Collège de France, CNRS, INSERM, Université PSL, Paris, France*
Rohit Menon • *Department of Behavioral and Molecular Biology, University of Regensburg, Regensburg, Germany*
Jan Meyer • *Institute of Neurobiology, Heinrich Heine University Düsseldorf, Düsseldorf, Germany*
José J. Miguel-Hidalgo • *Department of Psychiatry and Human Behavior, University of Mississippi Medical Center, Jackson, MS, USA*
Heloïse Monnet • *Center for Interdisciplinary Research in Biology (CIRB), Collège de France, CNRS, INSERM, Université PSL, Paris, France*
Carmen Muñoz-Ballester • *Department of Cell, Developmental and Integrative Biology, University of Alabama at Birmingham, Birmingham, AL, USA*
Keith K. Murai • *Centre for Research in Neuroscience, Department of Neurology & Neurosurgery, Brain Repair and Integrative Neuroscience Program, Research Institute of the McGill University Health Centre, Montreal General Hospital, Montreal, QC, Canada; Quantitative Life Sciences Graduate Program, McGill University, Montreal, QC, Canada*
Ciaran Murphy-Royal • *Département de Neurosciences, Université de Montréal, Montréal, QC, Canada; Centre de Recherche du Centre Hospitalier de l'Université de Montréal (CRCHUM), Montréal, QC, Canada*
Anastasiia Nagumanova • *Department of Psychiatry and Psychotherapy, University of Regensburg, Regensburg, Germany*
Domenico Natale • *Laboratory of Neuroplasticity and Neuroproteomics, Department of Biology, KU Leuven, Leuven, Belgium; i3S – Instituto de Investigação e Inovação em Saúde, Universidade do Porto, Porto, Portugal*
Yi Pang • *Department of Pediatrics, University of Mississippi Medical Center, Jackson, MS, USA*
Stefan Passlick • *Institute of Cellular Neurosciences I, Medical Faculty, University of Bonn, Bonn, Germany*
Sarah Peyrard • *Centre de Recherche du Centre Hospitalier de l'Université de Montréal (CRCHUM), Montréal, QC, Canada*

SAMO PIRNAT • *Laboratory of Neuroendocrinology-Molecular Cell Physiology, Institute of Pathophysiology, Faculty of Medicine, University of Ljubljana, Ljubljana, Slovenia; Celica Biomedical, Ljubljana, Slovenia*
STEFANIE ROBEL • *Department of Cell, Developmental and Integrative Biology, University of Alabama at Birmingham, Birmingham, AL, USA*
CHRISTINE R. ROSE • *Institute of Neurobiology, Heinrich Heine University Düsseldorf, Düsseldorf, Germany*
LEA R. SEEHOLZER • *Department of Psychiatry and Psychotherapy, University of Regensburg, Regensburg, Germany*
GERALD SEIFERT • *Institute of Cellular Neurosciences I, Medical Faculty, University of Bonn, Bonn, Germany*
SWETLANA SIRKO • *Chair of Physiological Genomics, Biomedical Center, Faculty of Medicine, LMU, Munich, Germany; Institute of Stem Cell Research, Helmholtz Center München, Deutsches Forschungszentrum für Gesundheit und Umwelt (GmbH), Munich, Germany*
ERIK SOMMER • *Institute of Cellular Neurosciences I, Medical Faculty, University of Bonn, Bonn, Germany*
MATJAŽ STENOVEC • *Laboratory of Neuroendocrinology-Molecular Cell Physiology, Institute of Pathophysiology, Faculty of Medicine, University of Ljubljana, Ljubljana, Slovenia; Celica Biomedical, Ljubljana, Slovenia*
YONG TANG • *International Joint Research Centre on Purinergic Signalling of Sichuan Province /Research Centre on TCM-Rehabilitation and Neural Circuit, School of Acupuncture and Tuina/Health and Rehabilitation, Chengdu University of Traditional Chinese Medicine, Chengdu, China; Acupuncture and Chronobiology Key Laboratory of Sichuan Province, Chengdu, China*
ALEXEI VERKHRATSKY • *International Joint Research Centre on Purinergic Signalling of Sichuan Province /Research Centre on TCM-Rehabilitation and Neural Circuit, School of Acupuncture and Tuina/Health and Rehabilitation, Chengdu University of Traditional Chinese Medicine, Chengdu, China; Faculty of Biology, Medicine and Health, The University of Manchester, Manchester, UK; University of the Basque Country, CIBERNED, Bizkaia, Spain and IKERBASQUE, Basque Foundation for Science, Bilbao, Spain*
DIEGO JULIAN VESGA-JIMENEZ • *Department of Psychiatry and Psychotherapy, University of Regensburg, Regensburg, Germany*
EUGENIA VIVI • *Department of Psychiatry and Psychotherapy, University of Regensburg, Regensburg, Germany*
BIN ZHOU • *Department of Anesthesiology, West China Hospital, Sichuan University, Chengdu, China; Laboratory of Anesthesia and Critical Care Medicine, National-Local Joint Engineering Research Center of Translational Medicine of Anesthesiology, Department of Anesthesiology, West China Hospital, Sichuan University, Chengdu, China*
ROBERT ZOREC • *Laboratory of Neuroendocrinology-Molecular Cell Physiology, Institute of Pathophysiology, Faculty of Medicine, University of Ljubljana, Ljubljana, Slovenia; Celica Biomedical, Ljubljana, Slovenia*

Chapter 1

Immunofluorescence Protocol to Label Astrocytes from Diverse Model and Non-model Mammals

Caterina Ciani and Carmen Falcone

Abstract

Conventional immunostaining protocols often lack versatility, catering predominantly to model organisms and failing to accommodate diverse specimens, particularly those that are non-perfused and are over-fixed. This chapter introduces an innovative immunofluorescence protocol designed to localize protein targets in brain tissue from a wide spectrum of mammals, spanning both model and non-model species. This method widely applies to label astrocytes from diverse mammalian species; however, it is usable to immunostain other protein targets. We provide a comprehensive guide encompassing the preparation of fresh and fixed tissues, incorporating essential steps such as deparaffinization, fixation, and cryoprotection. The immunofluorescence procedures outlined cover crucial aspects including antigen retrieval, mitigation of autofluorescence, nuclear staining, and meticulous mounting techniques, culminating in detailed instructions for image collection and resolutions for possible problems that could be encountered. By offering adaptable methodologies tailored to the complexities of diverse mammalian brain tissues, this chapter aims to advance the application of immunofluorescence techniques in astrocyte research and neurobiological research at large.

Key words Astrocytes, Immunofluorescence, Immunostaining, Cerebral cortex, Mammals, Evolution

1 Introduction

In neurobiological research, the preparation of tissue samples is a crucial step that significantly influences the outcome of experiments. However, researchers often encounter challenges due to the lack of control over the preparation protocols for tissues obtained from various sources such as collaborators or brain banks. Tissues treated with different fixation and embedding methods can yield varied results and require tailored immunofluorescence protocols [1–3]. Consequently, this diversity in preparation methods may introduce inconsistencies in data comparison across specimens. In this section, we aim to address these challenges by providing detailed guidelines for the fixation and embedding of

Barbara Di Benedetto (ed.), *Astrocytes: Methods and Protocols*, Methods in Molecular Biology, vol. 2896,
https://doi.org/10.1007/978-1-0716-4366-2_1,

tissues previously processed in different ways, including fresh, formalin/paraformaldehyde-fixed, or paraffin-embedded samples. Our approach aims to standardize the tissue preparation process, ensuring consistency and reliability in subsequent immunostaining procedures. Specifically, we outline a protocol for embedding tissues into 40-μm-thick, free-floating slices while acknowledging the flexibility to adapt the protocol for thinner slices or sections attached to glass slides (*see* **Notes** section). Following these guidelines, researchers can optimize their sample preparation methods, facilitating robust and reproducible immunofluorescence experiments in neurobiological studies.

2 Materials

Key reagents and materials for this method are listed in Table 1.

2.1 Samples

1. The tissues tested for this protocol span various mammalian orders such as Rodentia, Primates, Chiroptera, Artiodactyla, and Carnivora and each of them previously embedded and treated with different methods.

2.2 Preparation of Samples for the Protocol

2. Deparaffination.
3. Scalpel.
4. Xylene.
5. Decrescent concentration of ethanol solutions (100%, 96%, 75%, 50%).
6. Phosphate-buffered saline (PBS, 10×): 137 mM NaCl, 27 mM KCl, 14 mM Na2HPO4, and 43 mM KH2PO4. Bring total volume to 1 L with deionized water. Adjust pH to 7.5. Sterilize by autoclaving.
7. Fixation.
8. 4% Paraformaldehyde (PFA) in PBS 1×, pH 7.5, or formalin (use a chemical hood during the process).
9. Embedding.
10. Cryoprotective solution: 30% sucrose in PBS 1×.
11. Optimal cutting temperature (OCT) cryo-embedding medium.
12. Cryo-mold.

2.3 Preparation of Slices

1. Anti-freezing medium: sucrose 30% and ethylene glycol 30%; bring to 500 mL volume in phosphate buffer saline (PBS) 1×. Store at room temperature for a short period or at 4 °C for a longer period. Ethylene glycol is toxic; therefore, use a chemical hood during the preparation.

Table 1

Key reagents or resources. All reagents and resources needed for this protocol are listed, together with their source and catalogue/identifier number. These reagents are grouped into two categories: (1) antibodies and (2) chemicals, peptides, and recombinant proteins

Reagent or resource	Source
Antibodies	
Rabbit pAb to ALDH1L1	Abcam
Mouse mAb clone:5C10 to GFAP	Abcam
Rabbit anti-S100 beta antibody (EP1576Y)	Abcam
Donkey anti-mouse IgG (H + L) highly cross-adsorbed secondary antibody, Alexa Fluor™ 488	Invitrogen
Donkey anti-mouse IgG (H + L) highly cross-adsorbed secondary antibody, Alexa Fluor™ 594	Invitrogen
Donkey anti-rabbit IgG (H + L) highly cross-adsorbed secondary antibody, Alexa Fluor™ 488	Invitrogen
Donkey anti-rat IgG (H + L) highly cross-adsorbed secondary antibody, Alexa Fluor™ 594	Invitrogen
Chemicals, peptides, and recombinant proteins	
PBS tablets	Thermo Fisher
Killik O.C.T. compound embedding medium for cryostat	Bio-Optica Milano S. p.A.
Donkey serum	Sigma Aldrich
Sudan Black B	Sigma Aldrich
Triton™ X-100	Sigma Aldrich
TWEEN® 20	Sigma Aldrich
Sodium citrate tribasic dihydrate	Sigma Aldrich
Fisher Chemical Permount Mounting Medium	Fisher Scientific
NeuroTrace™ 435/455 Blue Fluorescent Nissl Stain	Thermo Fisher
DAPI 4′,6-diamidino-2-phenylindole dihydrochloride	Sigma Aldrich

2. Six-well plate (for the free-floating protocol) or glass slides (for immunofluorescence on slides).
3. Cryostat.

2.4 Antigen Retrieval

1. Sodium citrate buffer 10 mM: 10 mM Trisodium citrate dehydrated and 0.05% Tween-20; bring to volume to 1 L in ddH_2O, and use a stirrer to dissolve it properly.

2.5 Reduction of Autofluorescence

1. Sudan black solution:
   (a) Stock solution: 1% Sudan Black powder and 70% ethanol.
   (b) Working solution: 0.1% stock solution and 70% ethanol. It is essential to filter the solution to remove any undissolved powder. The solution should be stored at room temperature and shielded from light. Sudan Black powder can be toxic if inhaled, so it is imperative to use a chemical hood during the preparation process.

2.6 Permeabilization and Blocking

1. Blocking solution: 10% Donkey serum and 0.3% Triton ×-100; bring to (10 ml total volume) with PBS 1×.

2.7 Primary Antibody Labeling

1. Rabbit polyclonal anti-ALDH1L1 antibody (1:500 Abcam) to label astrocytes positive for the protein ALDH1L1.
2. Mouse monoclonal anti-GFAP clone 5C10 antibody (1:500 Abcam) to label astrocyte positive for the glial fibrillary acidic protein.
3. Rabbit anti-S100beta (1:300 Abcam) to label astrocyte positive for the calcium-binding protein S100beta.

2.8 Secondary Antibody Labeling

1. Donkey anti-mouse IgG Alexa Fluor 488 (1:1000, green channel).
2. Donkey anti-mouse IgG Alexa Fluor 594 (1:1000, red channel).
3. Donkey anti-rabbit IgG Alexa Fluor 488 (1:1000, green channel).
4. Donkey anti-rabbit IgG Alexa Fluor 594 (1:1000, red channel).

3 Methods

The step-by-step method details for this protocol are as follows. The protocol is executed in a six-multiwell plate, with a minimum volume of 1 mL added per well. Tissue transfer between wells is facilitated using a glass Pasteur pipette with a modified, hook-shaped tip. While a brush can also be utilized, it is not preferred due to the potential for tissue tear or damage. All incubation steps, unless specified otherwise, are conducted on a shaker at 80 rpm. In case of glass slides, the tissues should be incubated by applying enough volume to cover all the samples; the slides should be stored in a humidified box (to avoid evaporation of solutions), avoiding shaking (*see* **Note 1**).

3.1 Preparation of Tissues Treated with Different Protocols

Choose the correct protocol according to the type of tissue you need.

3.1.1 Fresh Tissues

1. Fix the tissue overnight by incubating it at +4°C in either 4% paraformaldehyde in PBS or formalin (37% formaldehyde aqueous solution).
2. Embed the tissue by incubating it in 30% sucrose in PBS for at least 2 days at +4 °C (until the specimen sinks) for cryoprotection.
3. Place the tissue in a cryo-mold filled with OCT cryo-embedding medium, ensuring no bubbles are present (to avoid problems during the cryo-sectioning), and then, solidify by placing the cryo-mold on dry ice or at −80 °C for a few minutes. The tissue is then stored at −80 °C.
4. Tissue cryo-sectioning at the cryostat involves preparing a six-well plate with 2 mL of anti-freezing medium per well and allowing the cryo-molded tissue and the six-well plate with anti-freezing medium to acclimatize in the cryostat (at −20 ° C) for 30 min. Cut the tissue sections at 40 μm thickness, and transfer them in the six-well plate filled with anti-freezing medium, with the number of sections per well depending on their size, typically no more than four. For sections cut thinner than 20 μm, we recommend the usage of glass slides, instead of the free-floating method.
5. Store the six-well plate (or the glass slide) at −20 °C.

3.1.2 Formalin or Paraformaldehyde-Fixed Tissues

1. Proceed directly with embedding as described above.

3.1.3 Paraffin-Embedded Tissues

1. Remove the paraffin-embedded sample from the plastic support, and trim the excess of paraffin with a scalpel.
2. Incubate in abundant xylene in a chemical hood until all paraffin is melted, with the xylene potentially needing to be changed every hour.
3. Rehydrate the tissue by incubating it in decreasing concentrations of ethanol: (i) ethanol 100% for 15 min, (ii) ethanol 96% for 15 min, (iii) ethanol 75% for 15 min, (iv) ethanol 50% for 30 min, and (v) PBS for 30 min.
4. For complete paraffin elimination and a renewal of the antigen retrieval solution, heat the antigen retrieval buffer solution until boiling, insert the tissue into the solution, and maintain it at a high temperature (not inferior to 70 °C) for 5 min. Our

experience suggests that boiling can lead to tissue deterioration, while temperatures around 65–70 °C are deemed sufficient for melting residual paraffin.

5. Subsequently, allow the antigen retrieval buffer to cool down to room temperature before proceeding to the next steps. This step is crucial to avoid thermal shock.
6. Fix the tissue with PFA 4% or formalin and embed it as described above.

3.2 Antigen Retrieval

Antigen retrieval is critical, especially when monoclonal antibodies are employed, as tissue fixation may mask the protein epitopes. This process is designed to unmask the target by altering pH and temperature.

1. Briefly wash the tissue section in PBS 1× to remove the antifreezing solution, and acclimatize it to room temperature. It is possible to process multiple sections in one immunostaining session (*see* **Note 2**).
2. Heat the antigen retrieval buffer in a beaker until boiling, either using a microwave or a laboratory heating stove. Submerge the tissue slice in the boiling buffer, and maintain it for 5 min (*see* **Note 3**). It's critical to allow the antigen retrieval buffer to cool for approximately 30 min at room temperature before using the section to prevent thermal shock.
3. Incubate the slice in PBS 1× to restore physiological pH.

3.3 Reduction of Autofluorescence

1. Incubate the tissue section with Sudan Black buffer for 30 min.
2. Wash the slice to remove excess solution. It's critical to note that Sudan Black buffer can impart a strong dark coloration in the well, making it advisable to transfer the tissue to a new multiwell.
3. Rehydrate the slice by incubating it in a decrescent concentration of ethanol solutions:
   (a) Ethanol 70% (for 5 s).
   (b) Ethanol 50% (for 5 s).
   (c) PBS 1× (for 4 min).

3.4 Permeabilization and Blocking

1. Incubate the slice for 1 h in the blocking solution (*see* **Note 4**) to permeabilize the cells, facilitating better antibody penetration and reducing nonspecific antibody binding. Notably, if the antibody's target is bound to the outer side of the cell membrane, Triton can be omitted from the blocking solution.

3.5 Primary Antibody Labeling

1. Dilute the blocking solution 1:5 in PBS.
2. Dilute the antibody/antibodies in this new blocking solution (*see* **Note 5**). Multiple antibodies from different host species can be used simultaneously.
3. Incubation is carried out overnight on a shaker at 4 °C. If necessary, it is possible to incubate the primary antibody for more than one night, up to 3 nights. However, this step is not recommended, because long incubation times on the shaker can damage the tissue.

3.6 Secondary Antibody Labeling

1. Dilute the blocking solution 1:5 in PBS. It's essential to note that the dilution of the secondary antibody should be determined experimentally, although a common range is typically between 1:400 and 1:1000 (*see* **Note 6**).
2. Wash the sample three times in PBS 1× (10 min each).
3. Dilute the blocking solution 1:5 in PBS, and add the secondary antibody at the appropriate volume and dilution. Incubation with secondary antibody/antibodies occurs for 2 h on a shaker at room temperature. Note that secondary antibodies are light-sensitive, requiring the multiwell plate to be shielded with aluminum foil from this point forward.
4. Wash the sample twice in PBS 1×.

3.7 Nuclear Staining

Conventional agents like 4,6-diamidino-2-phenylindole (DAPI) or Hoechst may not be effective on hyper-fixed tissues, such as those from long-term fixation in brain banks. An alternative option is NeuroTrace, the fluorescent counterpart of Nissl staining. Both protocols are outlined here, and the appropriate one should be selected based on the tissue's fixation duration. While we provide recommended dilutions for nuclear staining, these may need adjustment based on experimental findings.

- For DAPI/Hoechst staining:
  - Incubate with DAPI/Hoechst at a 1:500 dilution in PBS 1×.
  - Wash the sample twice with PBS 1× on a shaker (10 min each).
  - Dilute DAPI or Hoechst in PBS 1×, and incubate with the sample for 20 min.
  - Wash the sample three times in PBS on a shaker (10 min each).
- For NeuroTrace staining:
  - Incubate with NeuroTrace at a 1:20 dilution in PBS 1×.
  - Wash the sample twice in PBS 1× on a shaker (10 min each).
  - Wash the sample in PBS 1× containing 0.3% triton on a shaker (for 10 min).

- Dilute NeuroTrace in PBS 1×, and incubate on a shaker at room temperature for 1 h.
- Wash the sample in PBS 1× containing 0.1% triton on a shaker for 10 min.
- Wash twice in PBS 1× on a shaker (5 min each).
- Incubate in abundant PBS 1× at 4°C. CRITICAL: To avoid tissue damage, refrain from using the shaker at this step.

3.8 Mounting and Preservation

Following staining, the section should be carefully *mounted* on a glass slide using a brush to prevent folding or bubble formation (*see* **Note 7**).

1. Seal the slide with a mounting medium, such as Permount or Mowiol, and allow it to dry overnight at room temperature. Avoid using a shaker during this step.
2. Store the slide in a dark box at +4 °C to prevent signal bleaching.

3.9 Expected Outcome and Image Acquisition

1. This protocol has undergone extensive testing on brain tissue, particularly from prefrontal and frontal cortical regions, across a diverse range of species spanning various mammalian orders such as Rodentia, Primates, Chiroptera, Artiodactyla, and Carnivora. The resulting immuno-fluorescence staining exhibits a high signal-to-noise ratio, even under challenging tissue conditions, including non-model organisms and non-perfused or hyper-fixed tissue. It enables the execution of single immunostaining as depicted in Fig. 1, as well as double or triple immunostaining as illustrated in Fig. 2 (tissues displayed in the figures are 40 μm thick).
2. Capture images using a Nikon A1/R confocal microscope equipped with a 60× oil objective. To encompass all cells throughout the entire tissue thickness, acquire images as Z-stacks; e.g., for images of 40 μm, use a minimum of 42 steps, each 1 μm apart, thereby achieving an optical section thickness that complements all the section.
3. Image processing should be conducted using the Z-stack MAX-projection modality in FIJI software (ImageJ).

4 Notes

1. In the event of encountering challenges during the implementation of this protocol, several troubleshooting strategies can be employed to address potential issues in the resulting staining:
   (a) If tissue sections exhibit excessive autofluorescence, which may obscure the antibody signal, an incubation step with

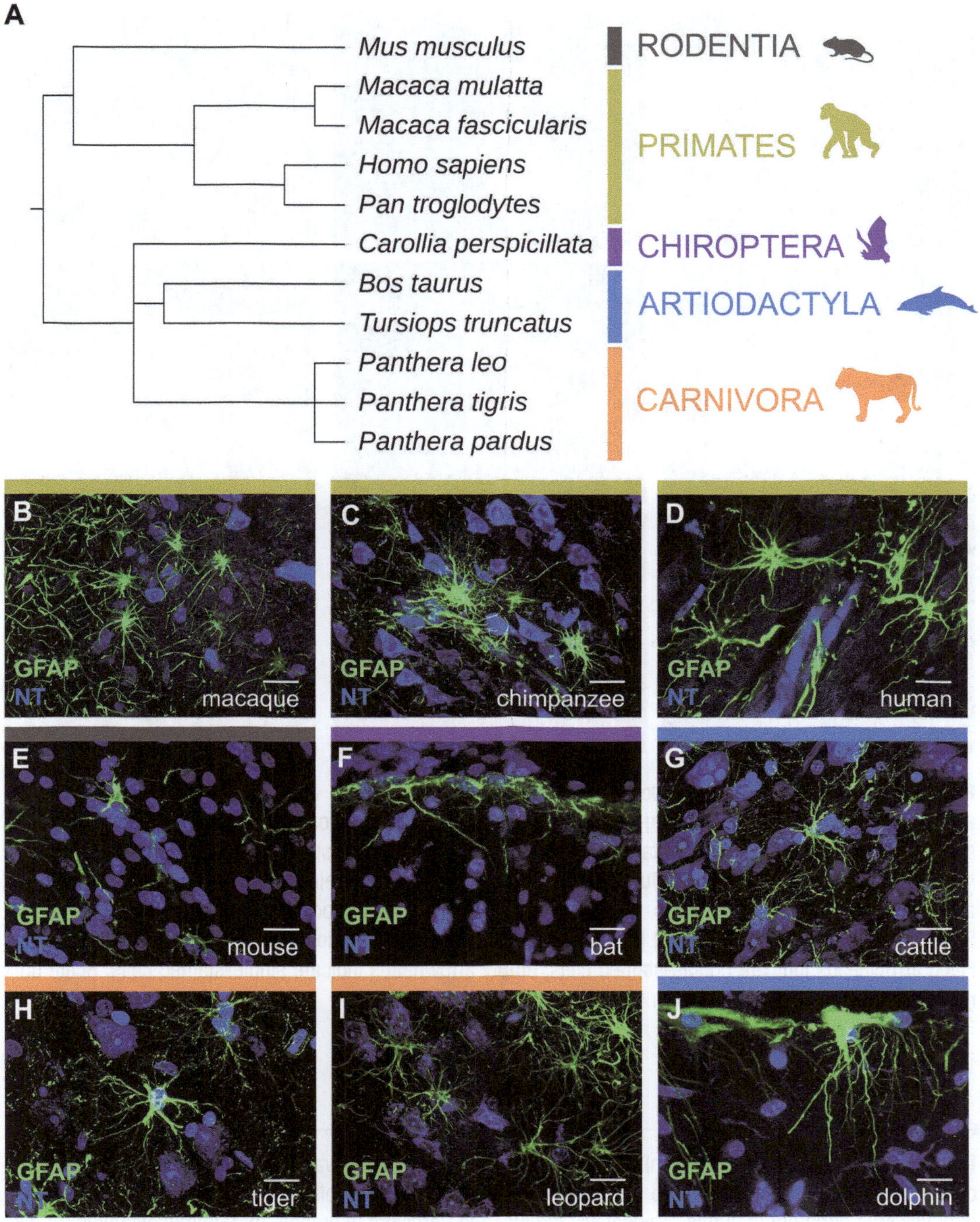

Fig. 1 Examples of outcome for single immunostainings performed on a wide range of species. The same immunostaining technique was evaluated across a diverse range of species. (**a**) On the left, an evolutionary tree depicting the species analyzed using this method. The tree was generated via https:/phylot.biobyte.de, and species taxonomy codes were sourced from the NCBI taxonomy database. The scientific names are matched with their common names as follows: *Mus musculus* (mouse), *Macaca mulatta* (Rhesus macaque), *Macaca fascicularis* (crab-eating macaque), *Homo sapiens* (human), *Pan troglodytes* (chimpanzee), *Carollia perspicillata* (Seba's short-tailed bat), *Bos taurus* (cattle), *Tursiops truncatus* (common bottlenose dolphin),

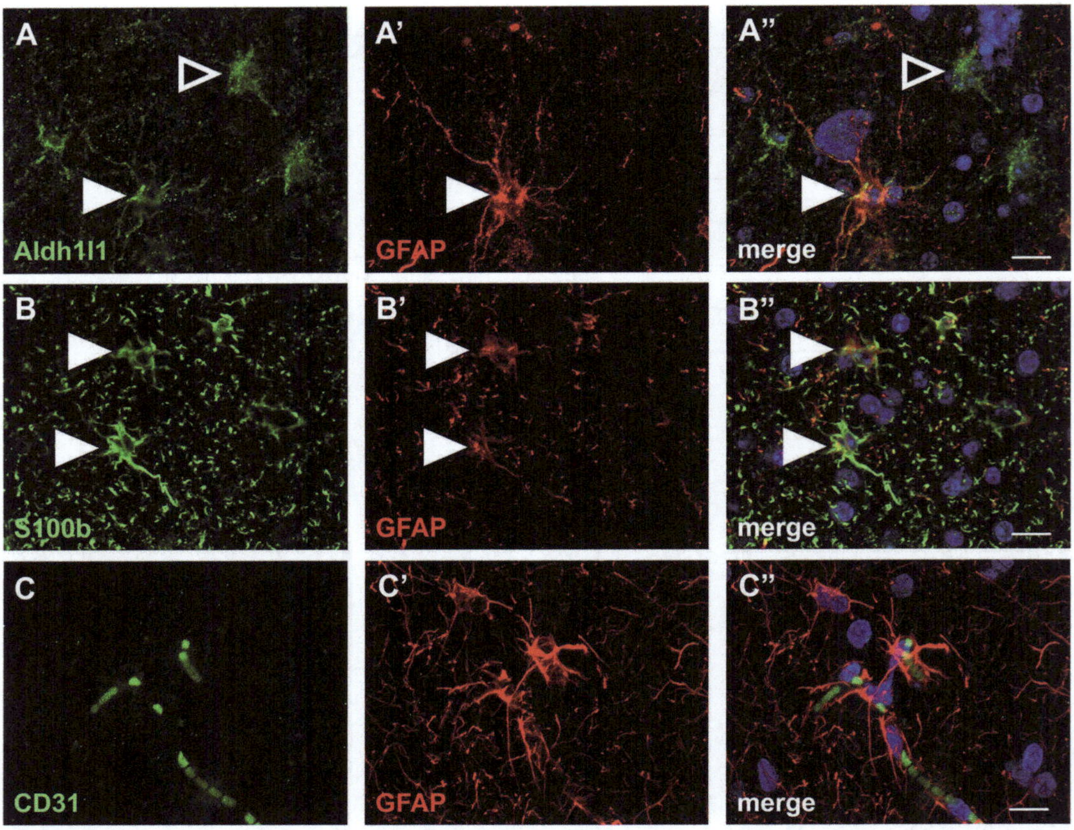

Fig. 2 Examples of outcome for double immunostaining tested on a wide range of species. Illustrative examples of double immunostaining on prefrontal cortex specimens from a selected species (*Bos taurus*, cattle). The same double immunostaining protocol was applied to the species listed in Fig. 1a (not shown here). (a–A″) NeuroTrace (NT) in blue; S100b (a calcium-binding protein expressed in a subpopulation of astrocytes) in green; GFAP in red. (**b**–B″) NT in blue; Aldh1l1 (a pan-astrocytic marker) in green; GFAP in red. Scale bars represent 20 μm. (c–C″) NT in blue; CD31 (a marker for endothelial cells) in green; GFAP in red. Filled arrowheads indicate double-stained cells; empty arrowheads indicate single-stained cells. Scale bars represent 20 μm. (Reproduced from Ciani et al. [4])

Fig. 1 (continued) Panthera leo (lion), *Panthera tigris* (tiger), and *Panthera pardus* (leopard). On the right, the corresponding taxonomic orders are indicated. The color coding is as follows: Rodentia in grey, Primates in dark yellow, Chiroptera in purple, Artiodactyla in blue, and Carnivora in orange. (**b**–**j**) Immunostaining for glial fibrillary acidic protein (GFAP) is shown in green, labeling astrocyte intermediate filament-III; nuclei are stained in blue with NeuroTrace. The species from which each brain section is derived are indicated for each micrograph. Each micrograph is color-coded according to the taxonomic order of the species. Scale bars = 20 μm. (Reproduced from Ciani et al. [4])

Sudan Black buffer can be incorporated to reduce background noise effectively. However, it's crucial to note that the Sudan Black buffer can impart a strong dark coloration to the well, necessitating caution to prevent potential interference with subsequent steps.

(b) If encountering difficulties with nuclear staining, particularly with hyper-fixed tissues, alternative staining agents such as NeuroTrace can be utilized as substitutes for conventional intercalants like DAPI or Hoechst. This alternative approach offers improved efficacy, especially in challenging tissue conditions.

(c) If improper antibody dilutions are observed, it's recommended to conduct a thorough evaluation of the experimental variables, including species specificity, tissue thickness, and fixation state, to determine the optimal antibody dilution for achieving desired staining outcomes.

(d) It's often debated whether to replace the buffer in the same well or transfer the tissue to a new one each time. To prevent tissue damage, it's generally preferable to replace the old solution with a new one in the same well, although critical steps may necessitate a fresh environment for the sample.

2. It is possible to process multiple sections in one immunostaining session, particularly when comparing multiple specimens that can be easily and macroscopically recognizable one from the other. However, it is advisable to perform antigen retrieval for each section individually. Once all sections have undergone antigen retrieval, subsequent steps can be carried out in parallel.
3. A mild buffer (pH 6) and a temperature not exceeding 100°C are recommended to prevent tissue damage while facilitating antigen retrieval.
4. When considering alternatives, *donkey serum* can be substituted with other serums such as fetal bovine serum or goat serum. However, if different types of serum are used, the host species of the secondary antibodies should match the serum used.
5. While this protocol has demonstrated efficacy with all tested antibodies listed in the key resources table, determining the appropriate antibody dilution necessitates consideration of factors such as species, tissue thickness, and fixation state of the sample. It is important to note that the protocol was exclusively tested on mammalian species. Nonetheless, its potential applicability to other vertebrates beyond mammals, with similar tissue conditions, appears promising.

6. Selecting an appropriate secondary antibody can also pose a challenge. Factors to consider include the host species of the primary antibody, the fluorophore color, and the serum used in the blocking solution. Matching these criteria ensures optimal antibody performance and compatibility with microscopy equipment.

7. Mounting slices onto glass slides can be a delicate process. We recommend immersing a glass slide in PBS 1× within a Petri dish, carefully transferring the slice into the dish, and gently unfolding it using a brush. As the slice assumes the correct position, carefully remove excess PBS, allowing the section to adhere to the slide securely.

References

1. Kogata N, Howard BA (2013) A whole-mount immunofluorescence protocol for three-dimensional imaging of the embryonic mammary primordium. J Mammary Gland Biol Neoplasia 18:227–231
2. Donaldson JG (2015) Immunofluorescence Staining. Curr Protoc Cell Biol 69:431–437
3. Aros CJ (2022) Indirect Immunofluorescence of Tissue Sections. Methods Mol Biol 2386:17–26
4. Ciani C, Pistorio G, Mearelli M, Falcone C (2023) Immunofluorescence protocol for localizing protein targets in brain tissue from diverse model and non-model mammals. STAR Protoc 4:102482

Chapter 2

Retro-Orbital Delivery of AAVs for CNS Wide Astrocyte Targeting

Domenico Natale and Matthew Holt

Abstract

Viral vector-mediated astrocyte targeting in live mice is a popular and valuable method to investigate astrocyte function in the context of intact neural circuits and complex brain physiology. Targeted genetic manipulation and functional investigation of this cell population can be accomplished by utilizing cell type–specific promoters to drive adeno-associated virus (AAV)-mediated transgene expression specifically in astrocytes. Here, we provide a comprehensive protocol for non-invasive retro-orbital (RO) administration of blood–brain barrier (BBB)-crossing AAVs in neonatal and adult mice, such as AAV-PHP.B, AAV-PHP.eB, and AAV.CAP-B22, which results in central nervous system (CNS)-wide transduction. Key procedures outlined include the preparation of AAV solutions for injection, a modified two-handed injection technique for precise and consistent RO injections, and a training strategy to practice mock RO injections using non-toxic dyes. This protocol serves as a valuable resource for researchers interested in exploring the roles of astrocytes in brain functions and neurological disorders.

Key words Retro-orbital injection (RO injection), Blood–brain barrier (BBB), Recombinant adeno-associated virus (rAAV)-based vector, BBB-crossing AAV vectors, Fast Green FCF, Astrocytes

1 Introduction

Recombinant adeno-associated virus (rAAV)-based vectors are widely used for transgene delivery in the central nervous system (CNS), due to their low immunogenicity, their ability to transduce post-mitotic cells, and the fact that they support long-term gene expression in various cell types [1, 2].

Several methods exist for the delivery of adeno-associated viruses (AAVs) into the CNS, including local intraparenchymal, cerebrospinal fluid (CSF), and systemic injection techniques [2–4]. Direct local CNS administration of rAAVs via intraparenchymal

Supplementary Information The online version contains supplementary material available at https://doi.org/10.1007/978-1-0716-4366-2_2.

Barbara Di Benedetto (ed.), *Astrocytes: Methods and Protocols*, Methods in Molecular Biology, vol. 2896, https://doi.org/10.1007/978-1-0716-4366-2_2,

stereotactic injections is effective and uses low vector amounts. However, it is generally limited to transduction of small populations of cells local to the injection site, and the process is invasive (effectively a stab wound injury) and may actually induce reactive gliosis along the needle tract. This alteration in the cellular environment may significantly impact experimental readouts and introduce artifacts into the experiment [3, 5]. Injection into the CSF surrounding the brain and spinal cord offers the advantage of widespread distribution of the viral vector throughout the CNS. However, this method typically requires a surgical procedure, such as intracisternal or intrathecal injection, to access the CSF space—posing surgical risks, such as infection, damage to surrounding structures (such as blood vessels or nerves), and cerebrospinal fluid leakage [3, 4]. Systemic injection involves delivering the viral vectors into the bloodstream, allowing them to circulate throughout the body and reach the CNS after crossing the blood–brain barrier (BBB). This method offers the advantage of non-invasive administration, as it does not require direct injection into the brain or CSF space. However, systemic injection may result in lower transduction efficiency in the CNS compared to local or CSF delivery methods, as the viral vector may transduce peripheral cells/organs (such as the liver) before crossing the blood–brain barrier (BBB) to reach the CNS [3, 6]. Following the initial discovery that AAV9 is able to cross the intact BBB in 2009 [7], there have been concerted efforts to generate capsids with improved BBB crossing and CNS tropism, which have resulted in the generation of engineered AAV capsids, such as AAV-PHP.B, AAV-PHP.eB, and AAV.CAP-B22 (*see* **Note 1**) [1, 6, 8–13]. Traditionally, systemic delivery of AAVs was achieved by injecting into the tail vein of the mouse [5]. However, retro-orbital (RO) injection provides a simpler and less invasive alternative for rAAV delivery, particularly in juvenile mice where tail vein injections can pose technical challenges [6, 14, 15]. Hence, RO is rapidly becoming the preferred route for systemic administration [6, 13].

As AAVs have a tendency to transduce multiple cell types within the CNS, various strategies have been developed to achieve cell type–specific expression [1, 8, 12, 16–19]. Specific promoters like GFAP and Aldh1l1 can be incorporated into the expression cassette to drive transgene expression selectively in astrocytes (*see* **Note 2**) [17, 20]. This approach has been used, for example, to express designer receptors exclusively activated by designer drugs (DREADDs) or Cre recombinase in astrocytes [21–24]. The use of specific promoters may further be complemented by the incorporation of microRNA (miRNA) target sequences enriched in non-astrocytic cells, such as miR-124 (neuron-specific), into the 3′ untranslated region (UTR) of the transgene [25]. This strategy helps to block "off-target" transgene expression, thereby ensuring that the transgene is predominantly expressed in astrocytes [5, 17,

25–27]. A further strategy to obtain astrocyte-specific transgene expression makes use of transgenic mouse lines expressing Cre in an astrocyte-specific manner (e.g., *Fgfr3*-iCreER^{T2}) to activate transgene expression [1, 28–32]. This generally occurs either through the removal of a Lox-STOP-Lox cassette downstream of the promoter allowing transcription, or through the use of incompatible Lox sites to produce a Cre-mediated transgene inversion that reorientates the transgene in relation to the promoter allowing transcription to take place - so-called FLEX (Flip-Excision) or DiO (double-floxed inverse orientation) systems [2, 30, 33, 34]. Ultimately, the combination of AAV capsid and expression cassette used depends on the specific experimental goals and the mouse lines available.

The combination of various strategies for astrocyte-specific targeting and CNS-wide transgene expression through systemic delivery of novel rAAVs allows for innovative and unique experimental designs, including the concurrent targeting of multiple brain areas with a single vector injection (which is essential for investigating and treating multi-system neuropsychiatric disorders) [1, 6, 12, 13, 19, 35–37]. However, the correct execution of RO injections is crucial to maximize solution delivery and minimize harm to the experimental animal. An issue commonly encountered with the traditional one-handed technique is the backflow of AAV solution during injection, due to excessive pressure and poor plunger control [14, 15]. Hence, in this protocol, we present a streamlined procedure including a modified two-handed retro-orbital injection technique in adult and neonatal mice, with tips to avoid common pitfalls and facilitate successful AAV delivery while minimizing harm to the experimental animal.

2 Materials

2.1 Mock Retro-Orbital Injections of Fast Green FCF

1. Sterile 1.5 mL Eppendorf/microcentrifuge tubes.
2. Tube rack, to hold the Eppendorf/microcentrifuge tubes.
3. 20, 200, and 1000 μL micropipettes, correctly calibrated.
4. Pipette tips with sterile filters for 20, 200, and 1000 μL micropipettes.
5. Sterile nylon, polyethersulfone (PES) or cellulose acetate (CA), membrane filters (0.22 μm pore size), compatible with disposable Luer lock or Luer slip tip syringes.
6. Sterile 5, 10, or 20 mL disposable syringes, preferably with Luer lock instead of Luer slip tip.
7. 70% ethanol.

8. 27-gauge (or smaller) single-use disposable insulin syringes with fixed needle design (low dead space syringes (LDSS) [38], *see* **Note 3**).
9. Sharps disposal container.
10. 2% w/v Fast Green FCF solution prepared in sterile saline (0.9% NaCl) or sterile phosphate-buffered saline (PBS).
11. Headband magnifying glasses (optional, *see* **Note 4**).
12. Digital laboratory scale (weight range 0.01–500 g) with empty pipette tip box for weighing the mouse (*see* **Note 5**).
13. Light source, preferably a LED light with flexible arm.
14. Recovery cage with nest made of soft gauze pads.
15. Heating pad to provide gentle warmth and prevent hypothermia.
16. Thin absorbent gauze pads, preferably triangle-shaped.
17. 0.5% proparacaine hydrochloride ophthalmic solution for topical eye anesthesia.
18. Isoflurane anesthesia setup. This setup may vary between institutions, but, at a minimum, it should include a calibrated vaporizer, an oxygen source, and a transparent anesthesia chamber (*see* **Note 6**). Using a face mask or funnel-shaped nose cone for the mouse's snout can aid in finer control of anesthesia maintenance (*see* **Note 7**). Additionally, a charcoal absorber device is strongly recommended for scavenging waste anesthetic gases to mitigate the health risks for personnel (e.g., CNS effects, reproductive issues, hepatic and renal effects, respiratory irritation, cancer risk, allergies, nausea, and occupational hazards) [39, 40]. Please consult local health, safety, and environment (HSE) committee guidelines and regulations regarding this matter.
19. Standard personal protective equipment (PPE) for working with mice in a laboratory setting.

2.2 Retro-Orbital Injections of BBB-Crossing rAAVs in Adult Mice

1. Here, use the same materials as for the mock retro-orbital injections, with the following additions or modifications:
   (a) Mouse strain appropriate for the AAV serotype used (*see* **Note 1**).
   (b) AAV solution (of defined titer) (*see* **Notes 8** and **9**).
   (c) 1% Virkon® disinfectant solution.
   (d) AAV waste container containing a 10% bleach solution (0.5% sodium hypochlorite).
   (e) Biological safety cabinet or laminar flow hood in an appropriate biosafety level environment for AAV work (depending on local regulations, usually BSL 2).

(f) Appropriate personal protective equipment (PPE) to ensure safety and prevent potential exposure to AAVs (lab coat or disposable gown, double gloves, safety goggles or face shield, respiratory protection, closed-toe shoes). Please consult local regulations and always follow institutional biosafety guidelines and procedures (which are generally informed by national legislation covering biosafety and the use of genetically modified animals).

2.3 Retro-Orbital Injections of BBB-Crossing rAAVs in Neonatal Mice

1. Besides the material listed in the previous sections, the following pieces of equipment are required for RO injections of neonatal mice:
   (a) 31-gauge (or higher gauge number) single-use disposable insulin syringes with fixed needle design (low dead space syringes (LDSS) [38]) or a 25 μL Hamilton syringe for life science applications with an appropriate needle length, gauge and point style.
   (b) Magnifying glasses or, preferably, a dissecting stereomicroscope (able to reach at least 6× magnification).
   (c) A clean, empty cage to be kept in the animal holding room, for separating the dam from the litter.
   (d) A small container for gathering the pups to be injected.
   (e) Cryoanesthesia setup (cooling pad or water containing crushed ice [41, 42]).

3 Methods

3.1 Increasing Dexterity and Precision of the Injection Technique Through Mock Retro-Orbital Injections

In most institutes, interventions in experimental animals will require permission from the local ethics committee, who enforce the relevant national legislation covering the use of animals in experiments. Obtaining this permission requires that experiments will be performed by approved personnel who have the required training and certification (e.g., FELASA B minimum or equivalent).

Injection into the retro-orbital (RO) venous sinus serves as a faster and less traumatic alternative to tail vein injection, albeit requiring a high level of training, manual dexterity, and precision [14, 15]. Due to these requirements, operators must undergo adequate training before injecting high amounts of costly BBB-crossing rAAVs, both to minimize the risk of wasting AAV and to prevent injury to the experimental animals. Therefore, it is crucial to allocate sufficient time before the actual procedure to practice mock RO injections using a harmless marker dye, which on entering the bloodstream will accumulate in the mouse's skin and provide an indirect indication that the injection procedure is being

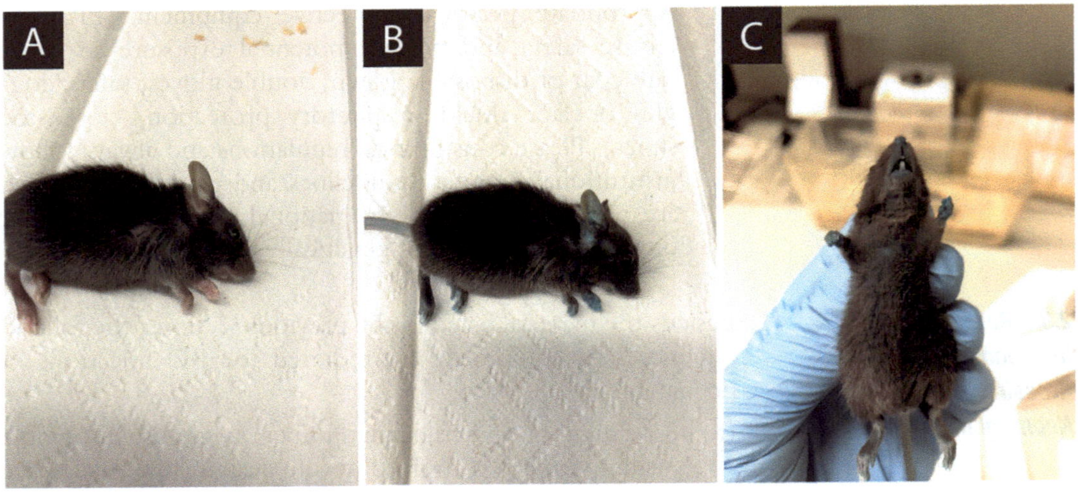

Fig. 1 Use of Fast Green FCF as a training aid for RO injections. Mouse appearance before (**a**), 5 s after (**b**), and 1 h (**c**) following retro-orbital injection of a 2% w/v Fast Green FCF solution in sterile saline (0.9% NaCl). Note the blue appearance of the skin indicating that the injection was successful and the Fast Green FCF injectate successfully entered into the systemic circulation

performed correctly and that the injectate is reaching the systemic circulation. We recommend utilizing Fast Green FCF as a marker dye (*see* **Note 10**).

Generally, a concentration of 1–5% Fast Green FCF in saline or phosphate-buffered saline (PBS) is appropriate for intravenous injection in mice [43]. In our experience, the best concentration in terms of combining quality of visualization and safety for the mouse is a concentration of 2% Fast Green FCF in saline. We recommend injecting a maximum of 100 μL of solution for a P21 mouse (*see* **Note 11**). The injection of 2% Fast Green FCF in saline has been reported to result in no adverse effects in young mice, except for a temporary greenish appearance of the skin lasting approximately 18 h, after which the substance is metabolized and the skin returns to its normal color [43] (Fig. 1).

1. Prior to starting, wipe off working surfaces with 70% ethanol.
2. Prepare a solution of Fast Green FCF at 2% (w/v) in 0.9% NaCl (or PBS as an alternative). Pour into a disposable sterile syringe, and filter with a 0.22 μm filter to remove any particulates. A large stock can be prepared and stored at room temperature for use over several months.
3. Tare the scale with the empty pipette tip box and weigh the mouse. The purpose here is merely to determine the maximum injectable volume of Fast Green FCF solution (*see* **Note 5**).
4. If necessary, weigh the charcoal canister before use and note its weight.

5. Place the animal in the induction chamber and seal/close the top.
6. Anesthetize the mouse with the calibrated vaporizer set to 5% isoflurane concentration flow. This should take approximately 5 min.
7. Ensure that the animal is deeply anesthetized using the toe pinch reflex.
8. Switch the system flow to the face mask/nose cone, transfer the animal from the anesthesia chamber to the heating pad, and position the mouse's snout into the face mask/nose cone.
9. Position the mouse in a recumbent posture on its left flank, ensuring that its right eye is accessible to the operator (for right-handed individuals) (*see* **Note 12**). IMPORTANT: Position the mouse's head close to the edge of the operation table, to have more range of motion for the left hand.
10. Adjust gas flow to 3–4% isoflurane concentration to keep a deep anesthesia (*see* **Note 13**). If the animal starts responding to a toe pinch, gently restrain in the face mask/nose cone until full anesthesia is reestablished.
11. Draw up the appropriate volume of Fast Green FCF solution into the LDSS. Expel any air bubbles. Perform this step slowly to prevent solution splattering, as Fast Green FCF stains. Wear headband magnifying glasses for a clearer view of the injection site (optional).
12. Apply one drop of proparacaine hydrochloride onto the right eye for topical eye anesthesia, and leave it to act for 1 min.
13. Position the light source to illuminate the mouse's head directly, ensuring that it does not cast shadows when hands are placed above the head.
14. Remove the drop of proparacaine hydrochloride using a triangle-shaped absorbent gauze pad.
15. Switch off the vaporizer flow, remove the face mask/nose cone from the mouse's snout, and move yourself into the right position to inject the Fast Green FCF solution using the two-handed RO injection technique (*see* **Note 14**; Electronic Supplementary Material (ESM) Video 1).
16. Hold the mouse's face, and gently displace the right eye from the socket with your left hand, bulging it between the index finger and thumb. Hold the syringe by its body with your right hand and advance the needle tip to the medial canthus (bevel down). IMPORTANT: place the syringe along an axis that passes through an idealized line connecting the snout tip and the injected eye, with an angle of ca. 30 degrees between the skin of the face and the needle.

17. Insert the needle into the medial canthus. IMPORTANT: insert the needle for a maximum of 4–5 mm into the retro-orbital space.
18. Once the needle has advanced without resistance and is positioned correctly, remove your left hand (and your hold on the mouse's head). This will make the snout fall slightly downward. When this happens, accommodate the change in angle by altering the position of your right hand (rotating the wrist slightly, if necessary).
19. Swap hands, so you are now holding the syringe body with your left hand and the plunger with your right hand. Stabilize the syringe firmly in place with your left hand (thumb below the syringe body, index finger above), while controlling the plunger tip with your right hand (pinched between thumb and index finger for a more controlled grip, if necessary).
20. Inject very slowly (ca. 20 μL per second) at a constant speed. Do not aspirate accidentally before injection, as this will collapse the blood vessels of the retro-orbital sinus. IMPORTANT: to ensure proper delivery of the injectate with minimal solution backflow, it is strongly advised to hold the syringe as illustrated in ESM Video 1.
21. After injecting the entire volume (100 μL), wait 3–4 s before retracting the needle. IMPORTANT: Retract the needle very slowly with both hands to prevent bleeding or backflow of injectate, and maintain visual control throughout the entire movement.
22. Apply mild pressure to the injection site with a soft gauze pad.
23. Place the mouse in the recovery cage and monitor it during the recovery process (*see* **Note 15**).
24. Dispose of the syringe into a suitable sharps container. Do not attempt to recap the syringe before disposal.
25. Weigh the charcoal canister after use and mark down its weight once all procedures have been performed (*see* **Note 16**).
26. Clean up and disinfect the working surfaces with 70% ethanol.

3.2 RO Injection of BBB Crossing AAVs

Here, the procedure mirrors that described for RO injections of Fast Green FCF, except that a solution of BBB-crossing rAAVs is used [6, 13]. Determine the precise quantity of genome copies (GC) to administer and, based on the measured titer, prepare an injectate solution by diluting the AAV stock solution into sterile saline or PBS. Use properly calibrated micropipettes and pipette tips with filters at this stage (*see* **Notes 8** and **9**). Doses of rAAV for systemic injections can be very high, typically ranging from 1.0×10^{11} to 1.0×10^{14} GC (absolute dose per animal) [6, 13, 44–47]. The optimal dose to use depends on factors such as the specific rAAV serotype, intended application, mouse age, body size, tolerability, and the starting titer of the AAV stock solution (which limits the maximum deliverable dose due to restrictions on the volume that can be injected at one time) [2, 6, 13, 26, 44–47]. It is important to note that high rAAV titers may increase the risk of immune responses and toxicity [2, 48]. Careful dose optimization, including use of viral vector dose escalations (VVDEs), is necessary to achieve the desired level of transgene expression while minimizing potential side effects [35] (Fig. 2).

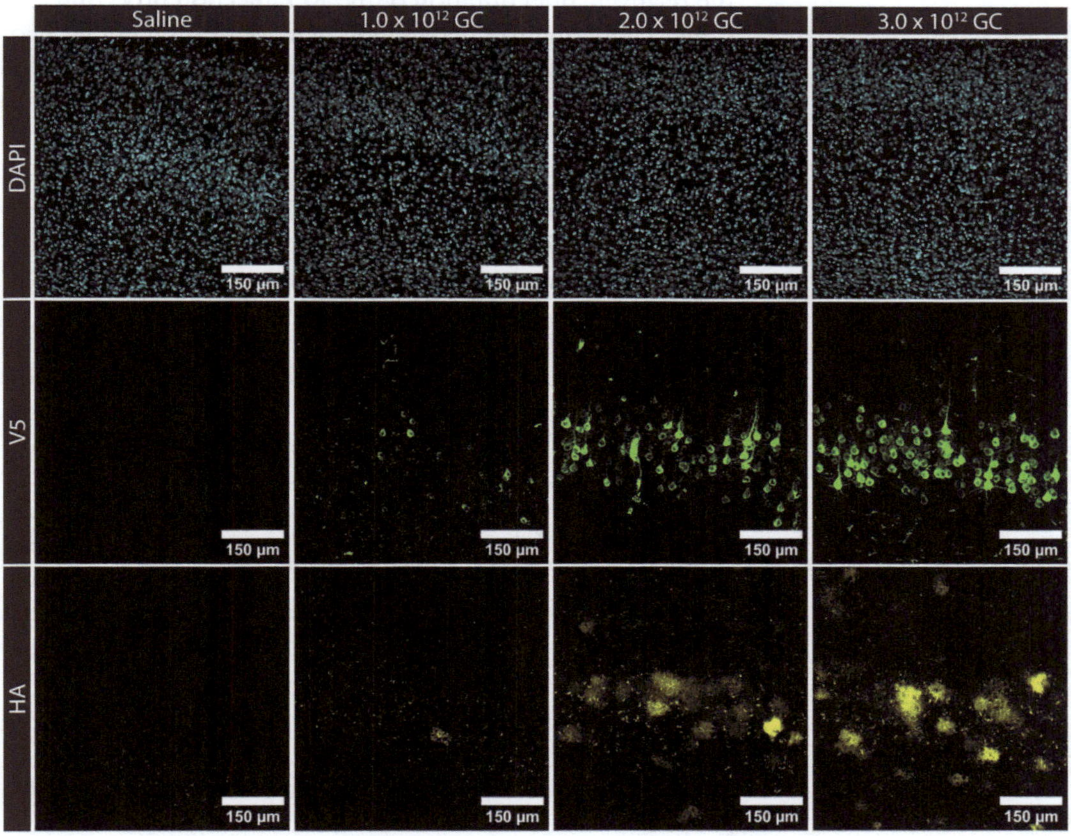

Fig. 2 Determining the optimal dosing regimen for transgene expression using a viral vector dose escalation (VVDE) experiment. Immunohistochemistry (IHC) staining of cortical brain tissue from *Fgfr3*-iCreER^{T2} mice injected with a 1:1 mixture of viral vectors encoding transgenes fused, respectively, to V5 and hemagglutinin (HA) epitope tags to allow detection. Expression of the V5-tagged construct specifically in neurons was ensured using the α-CaMKII promoter. Expression of the HA-tagged construct in astrocytes was achieved using astrocyte-specific Cre expression to activate a FLEX type system. Mice were injected with increasing amounts of viral vectors (with total vector dose in genome copies (GC) indicated). Optimal expression levels were observed starting from a dose of 2.0×10^{12} GC. Staining with anti-V5 (green) and anti-HA (yellow) confirms specific construct expression in pyramidal neurons and astrocytes. Cell types are identified by their stereotypical morphologies

3.3 Details for Injecting Neonatal Mice (P1–P2) via the Retro-Orbital Space

Here, the procedure includes special points that should be considered due to the specifics of injecting neonates (at postnatal day 1 or 2; P1-P2) rather than adult mice [14, 49]:

1. Magnification requirement: The use of magnifying glasses, or a dissecting stereomicroscope (at least 6× magnification), is necessary to ensure the injection site and procedure can be clearly visualized.
2. Anesthesia: While it is possible to manually restrain the pups for this procedure instead of using sedatives or anesthetics, we recommend using hypothermic cooling (cryoanesthesia) prior to the injection to minimize discomfort to the animal. Given the relatively brief duration of the RO injection procedure, we suggest adhering to the common practice of submerging neonates in crushed ice or using simple cooling pads. For detailed instructions on performing cryoanesthesia in mouse pups up to 7 days of age, please refer to published protocols [41, 42].
3. Needle size: Ideally, use a 31-gauge 0.3 mL insulin syringe with a fixed very short needle of 6 mm (such as the BD Ultra-Fine™ system), which is suitable for most applications. If higher accuracy is required, we recommend investing in a specialist Hamilton syringe, such as the 25 μL Neuros Syringe, Model 1702 RN, 33 gauge, Point Style 4, with an adjustable 0–20 mm needle length. Another possible option is the 25 μL Model 702 RN (removable needle) Syringe, compatible with Small Hub Removable (RN) needles available in various custom lengths, gauges, and needle point styles. Alternatively, consider the Intraocular Injector Syringe. This syringe is ideal for delicate and precise injections, with the plunger operated by pressing a lever while the syringe is held like a pen. This system benefits from a needle designed for quick replacement but has the drawback that the maximum syringe volume is approximately 6.5 μL.
4. Limited injectate volume: Ensure that the volume does not exceed 20 μL, since a newborn mouse pup weighs approximately 1–1.5 grams and the circulating blood volume is limited [14, 50].
5. Mouse restraint: Since the eye is covered by the skin at this age, instead of applying proparacaine hydrochloride for topical anesthesia, simply clean the area with sterile saline (0.9% NaCl). Gently restrain the head between the tip of the thumb and forefinger of the non-dominant hand. Using the dominant hand, inject into the spot that will develop into the medial canthus with a single smooth, yet resolute, push on the plunger.
6. Maternal separation: Prior to injections, transfer the dam to a separate cage and leave this in the animal holding room.

Holding the mother in a separate location while injecting the pups (in a specialist procedure room) ensures that the mother has limited exposure to potentially distressing audible and ultrasonic vocalizations that the pups may emit during manipulation.

7. Nest preservation: Place all the pups to be injected from a litter into a small container and cover them with bedding from the original nest to provide warmth and maintain the original scent. After a pup has received the RO injection, return it to the home cage, ensuring it is coated with the original nest bedding to facilitate reacceptance by the dam. Reintroduce the dam into the home cage to rejoin the litter.

3.4 Results

1. Validation of cell type specificity. The injections should result in the successful expression of transgenes in astrocytes. Expression and cell type specificity can most simply be assayed using standard techniques such as immunohistochemistry (IHC), with antibodies specific to the transgene of interest and astrocyte markers (e.g., S100β) (Fig. 3); transgenic animals expressing fluorescent proteins in astrocytes (e.g., Aldh1l1-eGFP line) are also useful in this respect [21]. If high-quality antibodies are not available against transgenes of interest fluorescence in situ hybridization is an alternative to detect the mRNAs encoding transgenes [51].

2. Increasing AAV bioavailability and transgene expression. In some circumstances, particularly with systemic delivery of rAAVs, the challenge lies not in achieving cell type specificity

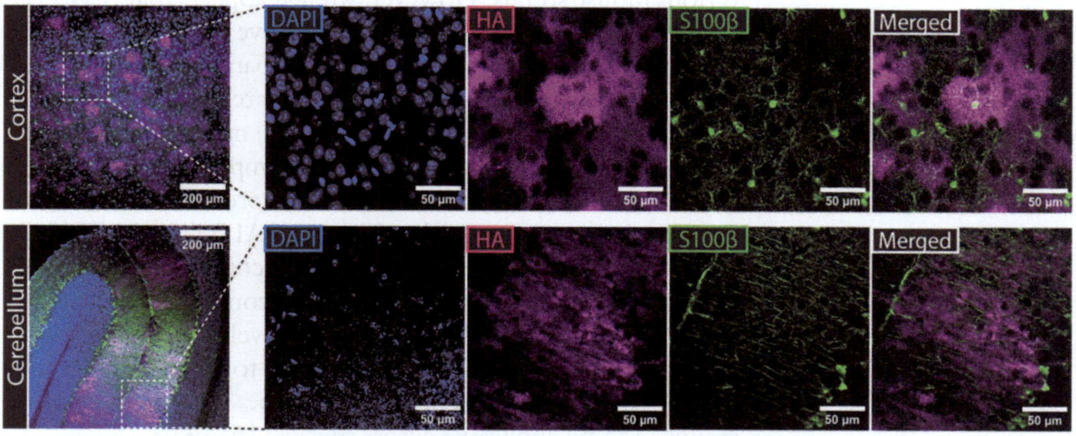

Fig. 3 Confirming cell type–specific transgene expression in astrocytes using antibody counterstaining. Immunohistochemistry (IHC) staining of cortical and cerebellar brain tissue from a *Fgfr3*-iCreER^{T2} mouse injected with 1.0×10^{12} GC of rAAVs. Expression of the HA-tagged transgene in astrocytes was achieved using astrocyte-specific Cre expression to activate a FLEX type system. Both protoplasmic astrocytes of the cortex and specialized astrocytes of the cerebellum (Bergmann glia) are identified by counterstaining against S100β, a common astrocyte marker

but rather in achieving optimal levels of transgene expression in vivo. This depends on various factors such as total amount of AAVs penetrating the CNS, total numbers of AAVs transducing the cells of interest, viral genome structure (single- versus double-stranded), and permissivity—the capacity of the host cell to express the viral genome once it receives it [2, 26, 35]. Therefore, optimizing the delivery and expression system becomes crucial [2, 26]. This can involve strategies like VVDEs (Fig. 2), testing different AAV capsids, and evaluating different expression cassettes to drive optimal transgene expression. Often, combining multiple approaches can synergistically enhance AAV bioavailability and transgene expression, leading to more effective targeting of astrocytes [2, 18].

3. Timing of transgene expression post-injection. The time frame for observing transgene expression following RO injection of rAAVs can vary depending on several factors, including the characteristics of the viral vector and the target cell type [1, 2]. However, transgene expression can typically be observed within 2–4 weeks post-injection, and stable and sustained transgene expression may be maintained for several months to years post-injection [2]. The use of techniques such as IHC or quantitative PCR can provide valuable insights into the kinetics and dynamics of transgene expression.

4 Notes

1. The retro-orbital injection procedure is generally applicable to various mouse strains. However, while it appears that AAV9 can be used in systemic injections to achieve CNS transduction across mouse strains (albeit at comparatively low rates), recently developed AAV capsids (which are generally based on engineering of the original AAV9 capsid) may not display such a broad range of BBB crossing. For example, the dramatically enhanced CNS tropism of AAV-PHP.B present in C57BL/6J mice has been reported to be absent in BALB/cJ mice [52]. As such, researchers are recommended to consult the published literature and select the capsid most compatible with their experimental system. Note that capsid development is an ongoing process and future efforts are likely to develop more efficient BBB crossing AAVs which can be used across mouse strains (and be applicable across species) [6, 13].
2. The design of rAAV vectors must carefully consider the expression cassette size [2, 26]. A standard single-stranded adeno-associated virus (ssAAV) expression cassette has a packaging capacity of approximately 4.7 kb (including transgene of interest as well as additional regulatory elements). The onset of

transgene expression from ssAAVs tends to be generally less efficient due to obligate second strand DNA synthesis in the transduced cell, which is rate-limiting. In contrast, a self-complementary adeno-associated virus (scAAV) expression cassette overcomes this limitation because it is designed as an intramolecular double-stranded DNA template. The drawback is that scAAV vectors are limited by a smaller packaging capacity (approximately 2.4 kb). Hence, transgene size will influence vector design and expression kinetics [35, 53]. It will also influence the choice of regulatory elements. Astrocyte-specific promoters such as GFAP and Aldh1l1 tend to be larger and weaker compared to strong, constitutively active promoters such as CMV or CAG [2]. Finally, sequences such as the Woodchuck Hepatitis Virus Post-transcriptional Regulatory Element (WPRE), which is thought to promote mRNA stability and boost transgene expression, all take up space in the expression cassette [26, 54].

3. A 27-gauge (or smaller) needle on a 0.5 mL insulin syringe is recommended to reduce the potential for eye trauma. A sharp cutting needle is preferred over a blunt-tipped needle, as it leads to less tissue distortion and damage [50]. Due to animal welfare requirements, each mouse should be injected using a single disposable syringe. This raises the issue that valuable AAV solution can be retained in the "dead volume" between the syringe barrel and needle. This limitation can be overcome using so-called low dead space syringes (LDSS), such as standard insulin syringes with a fixed needle design [38]. In our lab, we use a 0.3 mL insulin syringe with a fixed short 30-gauge needle of 8 mm length (such as the BD Micro-Fine™ system). We find this syringe is the easiest to handle and allows better (visual) control over AAV delivery.
4. While visualizing the point of injection (medial canthus) is feasible with the naked eye in a P21 (or older) mouse, the optional use of magnifying glasses may enhance accuracy and efficacy of the procedure.
5. Weighing the animal serves as a guide to determine the maximum volume of AAV solution that can be injected intravenously. To mitigate the risk of hemodynamic changes and pulmonary edema associated with intravenous administration, in the case of RO injections, the injection volume per individual dose is restricted to a maximum of 1% of the animal's body weight. For example, for a P21 mouse, we routinely inject 100 μL of solution, since the body weight of an average P21 C57BL/6J mouse is approximately 10 grams [14, 50]. The weight of an individual mouse is most accurately determined by directly weighing the animal; alternatively, weight can be inferred by mouse age using a body weight chart, such as the

one provided by the JAX organization for the C57BL/6J mouse strain, which is freely available at https://www.jax.org/jax-mice-and-services/strain-data-sheet-pages/body-weight-chart-000664.

6. Besides isoflurane anesthesia, injectable anesthetics can also be a suitable option, as long as one uses doses within the therapeutic window and ensures a margin of safety [39, 40]. In fact, in circumstances where an isoflurane anesthesia setup was not available, we have effectively used a combination of Domitor® (medetomidine hydrochloride):Xylazine (1000 μg/kg: 75 mg/kg), delivered via intraperitoneal injection. We use intraperitoneal injection of the Domitor® antagonist, Antisedan® (atipamezole), given at five times the dose of Domitor®, to completely reverse anesthesia. In our hands, this reversal gives an improved margin of safety when using Domitor® as an injectable anesthetic in small rodents, and we prefer the Domitor®:Xylazine combination over the more conventional Ketamine:Xylazine cocktail. The advantages of using a Domitor®:Xylazine-based protocol are the consistent and predictable sedation and anesthesia, the precise control over the duration of anesthesia, and the added analgesic action of Domitor® [40, 55]. In our experience, the main practical drawback with using injectable anesthetics over inhalant anesthetics is the longer time the animal spends under anesthesia (about 10 min for induction and circa 30 min for full recovery after reversal).

7. A nose cone for anesthesia delivery can be fashioned from a cut-down plastic tube with a piece of latex glove stretched over the tube opening. A slit cut in the lower half of the latex membrane serves as the opening for the mouse's snout.

8. Highly purified AAV vectors are necessary for consistent and reliable in vivo transduction. Therefore, use only research-grade ultra-purified AAV preparations for in vivo work, due to the presence of contaminants in lower-grade preparations that can affect transduction efficiency from animal to animal and potentially induce an immune reaction [6]. We prefer to use AAV packaging services provided by an external company with recognized and robust quality control. Their enhanced recombinant AAV production protocols ensure guaranteed titer and purity, resulting in more consistent transduction profiles across batches, reducing the number of vector productions required. Outsourcing AAV production can also prove cost-effective in the long run, as there is no need to maintain complex production facilities with tight biosafety regulations on site. Companies that offer such services are VectorBuilder (which offers a simple vector design interface), PackGene Biotech, GenScript, OriGene, GENEWIZ from Azenta, and Charles River, among

others. The choice of the company depends on issues such as the exact serotype of AAV needed and quantity and quality (GMP) of vector required. Alternatively, use a viral vector core service at your local institution or refer to published AAV production protocols if you want to obtain your rAAVs in house [6, 56, 57].

9. Avoid freeze–thaw cycles for AAV stock solutions. On arrival, the AAV should be aliquoted for long-term storage at −80 °C, which is sufficient to maintain long-term AAV functionality [6, 58]. We generally use 20 μL aliquots, but the aliquot size can vary, depending on the initial GC/mL titer of the stock solution. When needed, the aliquots should be thawed on wet ice. Once thawed, individual aliquots can be stored at 4 °C for up to 2 months. When handling AAV vectors, ensure the use of appropriate personal protective equipment, and adhere to local regulations. Work in a standard biosafety cabinet or under a laminar flow hood for optimal safety. To prepare the AAV solution for injection, use sterile 1.5 mL Eppendorf/microcentrifuge tubes, as the 27-gauge (or smaller) single-use LDSS disposable insulin syringes that we recommend generally do not fit into the smaller 0.5 mL tubes. Use a tube rack to hold the tubes upright and minimize the risk of an AAV spill. In the case of a spill, clean up immediately with 10% bleach (0.5% sodium hypochlorite). Generally, we calculate the amount of AAV needed and add an additional 15–20% to the final volume to account for small inaccuracies inherent in solution preparation, e.g., pipetting. If multiple mice require injection with the same AAV solution, prepare a master mix which will cover all the animals to avoid unnecessary wastage of AAV. After the RO injection, do not retain any remaining AAV solution. Instead, dispose of it properly by discarding it into a designated waste container, containing 10% bleach (0.5% sodium hypochlorite). After the injections have been performed, clean up and disinfect working surfaces with 1% Virkon® solution.
10. The aim is to observe changes in skin color post-RO injection, indicating that the injection procedure has been performed correctly and the injectate has entered the systemic circulation. Training in this manner prevents wastage of AAVs by ensuring optimal injection technique.
11. The maximum volume of injectate for an adult mouse is limited to 200 μL, with an ideal volume of 100 μL [14, 50]. A mouse should receive no more than one injection per day. When more than one injection is required, use each eye alternately, and allow at least 24 h between injections [14].
12. Arrange the front paws and the left hind limb beneath the body while extending the right hind limb away from the body. For

left-handed individuals, the mouse position should be inverted.

13. The ideal respiratory rate under deep isoflurane anesthesia can vary, depending on factors such as the concentration of isoflurane being used and the individual mouse's response to anesthesia [39, 40]. However, in our experience, an appropriately deep anesthesia is reached when the respiratory rate stabilizes at around 27–30 breaths per minute: the mouse should perform a deep breath approximately every 2 s, but without forceful reflexive thrusts of the body.
14. We recommend that the operator kneel down, aligning their head with the level of the operating table. This positioning allows for closer proximity to the snout of the mouse, facilitating better hand placement and maneuvering during the two-handed RO injection technique.
15. The actual injection process should take approximately 30 s, and the mouse typically recovers within 2–3 min.
16. The charcoal canister should be replaced once it reaches the recommended maximum weight [39, 40].

Acknowledgments

DN is supported by a FWO PhD Fellowship (Grant 11M4323N) from the Research Foundation—Flanders (FWO). MGH acknowledges support from the Thierry Latran Foundation (SOD-VIP), FWO (Grant 1513616N), and European Research Council (ERC) (Starting Grant 281961, AstroFunc; Proof of Concept Grant 713755, AD-VIP). He is currently the ERA Chair (NCBio) at i3S Porto funded by the European Commission (H2020-WIDE-SPREAD-2018-2020-6; NCBio; 951923).

References

1. Bedbrook CN, Deverman BE, Gradinaru V (2018) Viral strategies for targeting the central and peripheral nervous systems. Annu Rev Neurosci 41:323–348. https://doi.org/10.1146/annurev-neuro-080317-062048
2. Nectow AR, Nestler EJ (2020) Viral tools for neuroscience. Nat Rev Neurosci 21:669–681. https://doi.org/10.1038/s41583-020-00382-z
3. Haery L, Deverman BE, Matho KS, Cetin A, Woodard K, Cepko C, Guerin KI, Rego MA, Ersing I, Bachle SM, Kamens J, Fan M (2019) Adeno-associated virus technologies and methods for targeted neuronal manipulation. Front Neuroanat 13:93. https://doi.org/10.3389/fnana.2019.00093
4. Glascock JJ, Osman EY, Coady TH, Rose FF, Shababi M, Lorson CL (2011) Delivery of therapeutic agents through intracerebroventricular (ICV) and intravenous (IV) injection in mice. J Vis Exp 56:2968. https://doi.org/10.3791/2968
5. Stoica L, Ahmed S, Gao G, Sena-Esteves M (2013) Gene transfer to the CNS using recombinant adeno-associated virus. Curr Protoc Microbiol 14:1451–1458. https://doi.org/10.1002/9780471729259.mc14d05s29
6. Challis RC, Ravindra Kumar S, Chan KY, Challis C, Beadle K, Jang MJ, Kim HM, Rajendran PS, Tompkins JD, Shivkumar K, Deverman BE, Gradinaru V (2019) Systemic AAV vectors for widespread and targeted gene

delivery in rodents. Nat Protoc 14:379–414. https://doi.org/10.1038/S41596-018-0097-3

7. Foust KD, Nurre E, Montgomery CL, Hernandez A, Chan CM, Kaspar BK (2009) Intravascular AAV9 preferentially targets neonatal neurons and adult astrocytes. Nat Biotechnol 27:59–65. https://doi.org/10.1038/nbt.1515
8. Chen W, Yao S, Wan J, Tian Y, Huang L, Wang S, Akter F, Wu Y, Yao Y, Zhang X (2021) BBB-crossing adeno-associated virus vector: an excellent gene delivery tool for CNS disease treatment. J Control Release 333:129–138. https://doi.org/10.1016/j.jconrel.2021.03.029
9. Gradinaru V (2020) Expanding the brain researcher's toolkit. Science 369:637. https://doi.org/10.1126/science.abd2660
10. Deverman BE, Pravdo PL, Simpson BP, Kumar SR, Chan KY, Banerjee A, Wu W-L, Yang B, Huber N, Pasca SP, Gradinaru V (2016) Cre-dependent selection yields AAV variants for widespread gene transfer to the adult brain. Nat Biotechnol 34:204–209. https://doi.org/10.1038/nbt.3440
11. Chan KY, Jang MJ, Yoo BB, Greenbaum A, Ravi N, Wu W-L, Sánchez-Guardado L, Lois C, Mazmanian SK, Deverman BE, Gradinaru V (2017) Engineered AAVs for efficient noninvasive gene delivery to the central and peripheral nervous systems. Nat Neurosci 20:1172–1179. https://doi.org/10.1038/nn.4593
12. Ravindra Kumar S, Miles TF, Chen X, Brown D, Dobreva T, Huang Q, Ding X, Luo Y, Einarsson PH, Greenbaum A, Jang MJ, Deverman BE, Gradinaru V (2020) Multiplexed Cre-dependent selection yields systemic AAVs for targeting distinct brain cell types. Nat Methods 17:541–550. https://doi.org/10.1038/s41592-020-0799-7
13. Goertsen D, Flytzanis NC, Goeden N, Chuapoco MR, Cummins A, Chen Y, Fan Y, Zhang Q, Sharma J, Duan Y, Wang L, Feng G, Chen Y, Ip NY, Pickel J, Gradinaru V (2022) AAV capsid variants with brain-wide transgene expression and decreased liver targeting after intravenous delivery in mouse and marmoset. Nat Neurosci 25:106–115. https://doi.org/10.1038/s41593-021-00969-4
14. Yardeni T, Eckhaus M, Morris HD, Huizing M, Hoogstraten-Miller S (2011) Retro-orbital injections in mice. Lab Anim (NY) 40:155–160. https://doi.org/10.1038/laban0511-155
15. Prabhakar S, Lule S, da Hora CC, Breakefield XO, Cheah PS (2021) AAV9 transduction mediated by systemic delivery of vector via retro-orbital injection in newborn, neonatal and juvenile mice. Exp Anim 70:450–458. https://doi.org/10.1538/expanim.20-0186
16. Jüttner J, Szabo A, Gross-Scherf B, Morikawa RK, Rompani SB, Hantz P, Szikra T, Esposti F, Cowan CS, Bharioke A, Patino-Alvarez CP, Keles Ö, Kusnyerik A, Azoulay T, Hartl D, Krebs AR, Schübeler D, Hajdu RI, Lukats A, Nemeth J, Nagy ZZ, Wu KC, Wu RH, Xiang L, Fang XL, Jin ZB, Goldblum D, Hasler PW, Scholl HPN, Krol J, Roska B (2019) Targeting neuronal and glial cell types with synthetic promoter AAVs in mice, non-human primates and humans. Nat Neurosci 22:1345–1356. https://doi.org/10.1038/s41593-019-0431-2
17. O'Carroll SJ, Cook WH, Young D (2021) AAV targeting of glial cell types in the central and peripheral nervous system and relevance to human gene therapy. Front Mol Neurosci 13:618020. https://doi.org/10.3389/fnmol.2020.618020
18. Merienne N, Le Douce J, Faivre E, Déglon N, Bonvento G (2013) Efficient gene delivery and selective transduction of astrocytes in the mammalian brain using viral vectors. Front Cell Neurosci 7:106. https://doi.org/10.3389/fncel.2013.00106
19. Yshii L, Pasciuto E, Bielefeld P, Mascali L, Lemaitre P, Marino M, Dooley J, Kouser L, Verschoren S, Lagou V, Kemps H, Gervois P, de Boer A, Burton OT, Wahis J, Verhaert J, Tareen SHK, Roca CP, Singh K, Whyte CE, Kerstens A, Callaerts-Vegh Z, Poovathingal S, Prezzemolo T, Wierda K, Dashwood A, Xie J, Van Wonterghem E, Creemers E, Aloulou M, Gsell W, Abiega O, Munck S, Vandenbroucke RE, Bronckaers A, Lemmens R, De Strooper B, Van Den Bosch L, Himmelreich U, Fitzsimons CP, Holt MG, Liston A (2022) Astrocyte-targeted gene delivery of interleukin 2 specifically increases brain-resident regulatory T cell numbers and protects against pathological neuroinflammation. Nat Immunol 23:878–891. https://doi.org/10.1038/s41590-022-01208-z
20. Borodinova AA, Balaban PM, Bezprozvanny IB, Salmina AB, Vlasova OL (2021) Genetic constructs for the control of astrocytes' activity. Cells 10:1600. https://doi.org/10.3390/cells10071600
21. Yu X, Nagai J, Khakh BS (2020) Improved tools to study astrocytes. Nat Rev Neurosci 21:121–138. https://doi.org/10.1038/s41583-020-0264-8

22. Luo R, Zhou B, Liao P, Zuo Y, Jiang R (2023) Disrupting cortical astrocyte Ca^{2+} signaling in developing brain induces social deficits and depressive-like behaviors. Glia 71:1592–1606. https://doi.org/10.1002/glia.24358
23. Requie LM, Gómez-Gonzalo M, Speggiorin M, Managò F, Melone M, Congiu M, Chiavegato A, Lia A, Zonta M, Losi G, Henriques VJ, Pugliese A, Pacinelli G, Marsicano G, Papaleo F, Muntoni AL, Conti F, Carmignoto G (2022) Astrocytes mediate long-lasting synaptic regulation of ventral tegmental area dopamine neurons. Nat Neurosci 25:1639–1650. https://doi.org/10.1038/s41593-022-01193-4
24. Nagai J, Rajbhandari AK, Gangwani MR, Hachisuka A, Coppola G, Masmanidis SC, Fanselow MS, Khakh BS (2019) Hyperactivity with disrupted attention by activation of an astrocyte synaptogenic cue. Cell 177:1280–1292. https://doi.org/10.1016/j.cell.2019.03.019
25. Taschenberger G, Tereshchenko J, Kügler S (2017) A microRNA124 target sequence restores astrocyte specificity of gfaABC$_1$D-driven transgene expression in AAV-mediated gene transfer. Mol Ther Nucleic Acids 8:13–25. https://doi.org/10.1016/j.omtn.2017.03.009
26. Domenger C, Grimm D (2019) Next-generation AAV vectors – do not judge a virus (only) by its cover. Hum Mol Genet 28:R3–R14. https://doi.org/10.1093/hmg/ddz148
27. Gleichman AJ, Kawaguchi R, Sofroniew MV, Carmichael ST (2023) A toolbox of astrocyte-specific, serotype-independent adeno-associated viral vectors using microRNA targeting sequences. Nat Commun 14:7426. https://doi.org/10.1038/s41467-023-42746-w
28. Endo F, Kasai A, Soto JS, Yu X, Qu Z, Hashimoto H, Gradinaru V, Kawaguchi R, Khakh BS (2022) Molecular basis of astrocyte diversity and morphology across the CNS in health and disease. Science 378:eadc9020. https://doi.org/10.1126/science.adc9020
29. Young KM, Mitsumori T, Pringle N, Grist M, Kessaris N, Richardson WD (2010) An *Fgfr3-iCreER*^{T2} transgenic mouse line for studies of neural stem cells and astrocytes. Glia 58:943–953. https://doi.org/10.1002/glia.20976
30. Gierut JJ, Jacks TE, Haigis KM (2014) Strategies to achieve conditional gene mutation in mice. Cold Spring Harb Protoc 2014:339–349. https://doi.org/10.1101/pdb.top069807
31. Hu NY, Chen YT, Wang Q, Jie W, Liu YS, You QL, Li ZL, Li XW, Reibel S, Pfrieger FW, Yang JM, Gao TM (2020) Expression patterns of inducible Cre recombinase driven by differential astrocyte-specific promoters in transgenic mouse lines. Neurosci Bull 36:530–544. https://doi.org/10.1007/s12264-019-00451-z
32. Srinivasan R, Lu TY, Chai H, Xu J, Huang BS, Golshani P, Coppola G, Khakh BS (2016) New transgenic mouse lines for selectively targeting astrocytes and studying calcium signals in astrocyte processes *in situ* and *in vivo*. Neuron 92:1181–1195. https://doi.org/10.1016/j.neuron.2016.11.030
33. Saunders A, Sabatini BL (2015) Cre activated and inactivated recombinant adeno-associated viral vectors for neuronal anatomical tracing or activity manipulation. Curr Protoc Neurosci 72:1.24.1–1.24.15. https://doi.org/10.1002/0471142301.ns0124s72
34. Octeau JC, Chai H, Jiang R, Bonanno SL, Martin KC, Khakh BS (2018) An optical neuron-astrocyte proximity assay at synaptic distance scales. Neuron 98:49–66.e9. https://doi.org/10.1016/j.neuron.2018.03.003
35. Rincon MY, de Vin F, Duqué SI, Fripont S, Castaldo SA, Bouhuijzen-Wenger J, Holt MG (2018) Widespread transduction of astrocytes and neurons in the mouse central nervous system after systemic delivery of a self-complementary AAV-PHP.B vector. Gene Ther 25:83–92. https://doi.org/10.1038/s41434-018-0005-z
36. Marino M, Holt MG (2022) AAV vector-mediated antibody delivery (A-MAD) in the central nervous system. Front Neurol 13:870799. https://doi.org/10.3389/fneur.2022.870799
37. Marino M, Zhou L, Rincon MY, Callaerts-Vegh Z, Verhaert J, Wahis J, Creemers E, Yshii L, Wierda K, Saito T, Marneffe C, Voytyuk I, Wouters Y, Dewilde M, Duqué SI, Vincke C, Levites Y, Golde TE, Saido TC, Muyldermans S, Liston A, De Strooper B, Holt MG (2022) AAV-mediated delivery of an anti-BACE1 VHH alleviates pathology in an Alzheimer's disease model. EMBO Mol Med 14:e09824. https://doi.org/10.15252/emmm.201809824
38. Kesten JM, Ayres R, Neale J, Clark J, Vickerman P, Hickman M, Redwood S (2017) Acceptability of low dead space syringes and implications for their introduction: a qualitative study in the west of England. Int J Drug Policy 39:99–108. https://doi.org/10.1016/j.drugpo.2016.09.005

39. Hoogstraten-Miller SL, Brown PA (2008) Techniques in aseptic rodent surgery. Curr Protoc Immunol 1:12–14. https://doi.org/10.1002/0471142735.im0112s82
40. Oh SS, Narver HL (2024) Mouse and rat anesthesia and analgesia. Curr Protoc 4:e995. https://doi.org/10.1002/cpz1.995
41. Danneman PJ, Mandrell TD (1997) Evaluation of five agents/methods for anesthesia of neonatal rats. Lab Anim Sci 47:386–395
42. Phifer CB, Terry LM (1986) Use of hypothermia for general anesthesia in preweanling rodents. Physiol Behav 38:887–890. https://doi.org/10.1016/0031-9384(86)90058-2
43. Wong HE, Qi W, Choi HM, Fernandez EJ, Kwon I (2011) A safe, blood-brain barrier permeable triphenylmethane dye inhibits amyloid-β neurotoxicity by generating nontoxic aggregates. ACS Chem Neurosci 2:645–657. https://doi.org/10.1021/cn200056g
44. Jackson KL, Dayton RD, Deverman BE, Klein RL (2016) Better targeting, better efficiency for wide-scale neuronal transduction with the synapsin promoter and AAV-PHP.B. Front Mol Neurosci 9:116. https://doi.org/10.3389/fnmol.2016.00116
45. Hu S, Yang T, Wang Y (2021) Widespread labeling and genomic editing of the fetal central nervous system by *in utero* CRISPR AAV9-PHP.eB administration. Development 148:dev195586. https://doi.org/10.1242/dev.195586
46. Radhiyanti PT, Konno A, Matsuzaki Y, Hirai H (2021) Comparative study of neuron-specific promoters in mouse brain transduced by intravenously administered AAV-PHP.eB. Neurosci Lett 756:135956. https://doi.org/10.1016/j.neulet.2021.135956
47. Dayton RD, Grames MS, Klein RL (2018) More expansive gene transfer to the rat CNS: AAV PHP.eB vector dose–response and comparison to AAV PHP.B. Gene Ther 25:392–400. https://doi.org/10.1038/s41434-018-0028-5
48. Peden CS, Burger C, Muzyczka N, Mandel RJ (2004) Circulating anti-wild-type adeno-associated virus type 2 (AAV2) antibodies inhibit recombinant AAV2 (rAAV2)-mediated, but not rAAV5-mediated, gene transfer in the brain. J Virol 78:6344–6359. https://doi.org/10.1128/JVI.78.12.6344-6359.2004
49. Lampe SEG, Kaspar BK, Foust KD (2014) Intravenous injections in neonatal mice. J Vis Exp 93:e52037. https://doi.org/10.3791/52037
50. Diehl K-H, Hull R, Morton D, Pfister R, Rabemampianina Y, Smith D, Vidal J-M, Van De Vorstenbosch C (2001) A good practice guide to the administration of substances and removal of blood, including routes and volumes. J Appl Toxicol 21:15–23. https://doi.org/10.1002/jat.727
51. Zhang Y, Chen K, Sloan SA, Bennett ML, Scholze AR, O'Keeffe S, Phatnani HP, Guarnieri P, Caneda C, Ruderisch N, Deng S, Liddelow SA, Zhang C, Daneman R, Maniatis T, Barres BA, Wu JQ (2014) An RNA-sequencing transcriptome and splicing database of glia, neurons, and vascular cells of the cerebral cortex. J Neurosci 34:11929–11947. https://doi.org/10.1523/JNEUROSCI.1860-14.2014
52. Hordeaux J, Wang Q, Katz N, Buza EL, Bell P, Wilson JM (2018) The neurotropic properties of AAV-PHP.B are limited to C57BL/6J mice. Mol Ther 26:664–668. https://doi.org/10.1016/j.ymthe.2018.01.018
53. McCarty DM (2008) Self-complementary AAV vectors; advances and applications. Mol Ther 16:1648–1656. https://doi.org/10.1038/mt.2008.171
54. Choi JH, Yu NK, Baek GC, Bakes J, Seo D, Nam HJ, Baek SH, Lim CS, Lee YS, Kaang BK (2014) Optimization of AAV expression cassettes to improve packaging capacity and transgene expression in neurons. Mol Brain 7:17. https://doi.org/10.1186/1756-6606-7-17
55. Roughan J, Ojeda O, Flecknell P (1999) The influence of pre-anaesthetic administration of buprenorphine on the anaesthetic effects of ketamine/medetomidine and pentobarbitone in rats and the consequences of repeated anaesthesia. Lab Anim 33:234–242. https://doi.org/10.1258/002367799780578183
56. Fripont S, Marneffe C, Marino M, Rincon MY, Holt MG (2019) Production, purification, and quality control for adeno-associated virus-based vectors. J Vis Exp 143. https://doi.org/10.3791/58960
57. Bilal AS, Parker SN, Murray VB, MacDonnell LF, Thuerauf DJ, Glembotski CC, Blackwood EA (2023) Optimization of large-scale adeno-associated virus (AAV) production. Curr Protoc 3:e757. https://doi.org/10.1002/cpz1.757
58. Bee JS, Zhang YZ, Phillippi MK, Finkner S, Mezghebe T, Webber K, Den CW, Marshall T (2022) Impact of time out of intended storage and freeze-thaw rates on the stability of adeno-associated virus 8 and 9. J Pharm Sci 111: 1346–1353. https://doi.org/10.1016/j.xphs.2022.01.002

Chapter 3

An Empirical–Theoretical Approach to Determine Astroglial Potassium Upon Ischemic Stress

Iris Álvarez-Merz and Sara Eitelmann

Abstract

Dynamic regulation of intracellular potassium is a key feature of astrocytes, as it will determine their capability to regulate extracellular potassium homeostasis during physiology and pathology. However, the study of intracellular potassium dynamics is hindered by the lack of adequate techniques. This chapter outlines a method for calculating astrocytic potassium concentration using the Goldman–Hodgkin–Katz equation. The method integrates experimental measurements of astroglial sodium concentration and membrane potential, along with extracellular potassium and sodium concentrations. We will show how this empirical-theoretical method reveals an initial loss of intracellular potassium and a subsequent transient potassium overshoot upon a protocol that mimics metabolic ischemic stress in acute brain slices.

Key words Intracellular potassium, Astrocytes, Goldman–Hodgkin–Katz equation, Acute brain slices, Ischemia, Metabolic inhibition

1 Introduction

Maintaining low extracellular K^+ levels is crucial for normal cellular function within the central nervous system [1]. Excessive accumulation of K^+ in the extracellular space can trigger neuronal depolarization and disrupt neurotransmitter homeostasis [2–4]. As far back as 1965 [5], Hertz proposed the significant involvement of astrocytes in clearing K^+ from the extracellular environment and modulating neuronal activity via manipulation of extracellular K^+ concentration ($[K^+]_e$). One of the primary mechanisms involves the spatial buffering of K^+ through gap junction–coupled networks, wherein K^+ is taken up at sites of high neuronal activity and redistributed to regions of lower activity [6]. This is facilitated by astrocytic membrane transporters and passive K^+ channels, such as the Na^+/K^+-ATPase (NKA) and inwardly rectifying potassium channels (K_{ir}) [7, 8]. Importantly, the astroglial intracellular potas-

Barbara Di Benedetto (ed.), *Astrocytes: Methods and Protocols*, Methods in Molecular Biology, vol. 2896,
https://doi.org/10.1007/978-1-0716-4366-2_3,

sium concentration ($[K^+]_i$) serves as a critical determinant in this regulatory process. Elevated extracellular K^+ levels enhance the driving force for astroglial K^+ uptake, facilitating efficient clearance of excess K^+. Conversely, increased astroglial K^+ levels attenuate K^+ uptake capacity, potentially leading to elevated $[K^+]_e$ and neuronal dysfunction [7]. Thus, astrocytes ensure precise regulation of $[K^+]_e$, crucial for sustaining optimal neuronal function and preventing excitotoxicity [9, 10].

Despite its high relevance, quantitative data of astroglial K^+ remain scarce [11] due to technical challenges inherent in its measurement. The current toolkit mainly comprises ion-sensitive microelectrodes and ion-sensitive fluorometric indicators. For instance, K^+-sensitive microelectrodes, although mainly used to quantify $[K^+]_e$ [12], have also been utilized for intracellular measurements of $[K^+]$ [13–15]. However, this method is cell-invasive and does not provide direct spatial information on changes in $[K^+]_i$. Conversely, only a few K^+-sensitive fluorescent dyes have allowed to detect changes in $[K^+]_i$, but these indicators typically have poor selectivity and bind to Na^+ with similar affinity [16, 17]. Moreover, genetically encoded K^+-sensitive indicators have not been effectively used in astrocytes in situ up to now [18, 19].

Previous attempts to determine astroglial K^+ have often relied on mathematical modelling, based on the predominant permeability of the astrocytic membrane to K^+ [20–26]. Here, we adopt an approach based on the Goldman–Hodgkin–Katz equation to gain insights into astroglial K^+. A combination of intracellular Na^+ imaging, extracellular K^+ and Na^+ recordings, and electrophysiological results in murine neocortical tissue slices is employed, coupled with the utilization of the Goldman–Hodgkin–Katz equation (Eq. 1) for calculating $[K^+]_i$.

$$E_M = \frac{RT}{F} \ln \frac{P_K[K^+]_e + P_{Na}[Na^+]_e}{P_K[K^+]_i + P_{Na}[Na^+]_i}, \tag{1}$$

where R is the universal gas constant, T is the absolute temperature, F is the Faraday constant, $[X^Z]_{e/i}$ are the respective extra- and intracellular ion concentrations, and P_X is the membrane permeability of the respective ion.

However, determining ion permeabilities can be challenging. To this end, the relative membrane permeability for Na^+ over K^+ (α) is used here. Equation (2) can be rearranged using $\alpha = P_{Na+}/P_{K+}$:

$$E_M = \frac{RT}{F} \ln \frac{[K^+]_e + \alpha[Na^+]_e}{[K^+]_i + \alpha[Na^+]_i}. \tag{2}$$

In this chapter, we will also show how the calculation of $[K^+]_i$ in astrocytes reflects dynamic K^+ transients upon specific protocols. During conditions such as brain ischemia, the effective regulation of $[K^+]_e$ by astrocytes is compromised due to the disruption of their energy metabolism and NKA activity [11, 27–29]. The ionic breakdown not only impedes extracellular K^+ clearance but also causes K^+ loss from astrocytes, intensifying the elevation of $[K^+]_e$ and promoting neuronal depolarization amid energy deprivation [20, 30]. Ischemic stress can be replicated in situ using various protocols. This chapter demonstrates how to induce transient energy restriction by inhibiting glycolysis and oxidative phosphorylation simultaneously. To achieve this, a glucose-free saline was used, to which the non-metabolizable glucose analogue 2-deoxyglucose and the cytochrome C oxidase (complex IV) antagonist sodium azide were added. This approach can be used to mimic mild to severe metabolic stress, depending on the duration (1–10 min) and/or the concentration of sodium azide (5–10 mM) [31, 32].

Understanding the intricate role of astrocytic K^+ regulation in the context of brain ischemia is essential for elucidating the mechanisms of neuronal injury and identifying potential therapeutic targets. Thus, this chapter presents an empirical–theoretical method for quantitatively determining the complete time course of ischemia-induced changes in astroglial K^+ in acutely isolated brain slices.

2 Materials

Chemicals were purchased from Merck KGaA (Darmstadt, Germany) or AppliChem (Darmstadt, Germany), unless otherwise stated.

2.1 Animals

1. Male and female Balb/C mice aged 14–21 days were used for this study. Mice were bred in a strictly controlled environment with a 12-h light/dark cycle and a constant temperature (20–24 °C). Food and water were provided *ad libitum*.

2.2 Solutions

1. Standard artificial cerebrospinal fluid (ACSF): 130 mM NaCl, 2.5 mM KCl, 1.25 mM NaH_2PO_4, 26 mM $NaHCO_3$, 2 mM $CaCl_2$, 1 mM $MgCl_2$, and 10 mM glucose (*see* **Note 1**); carbonated with 95% O_2 and 5% CO_2, resulting in a pH of 7.4 and an osmolarity of 305–310 mOsm (*see* **Note 2**).
2. Modified ACSF for dissection of brain slices: 130 mM NaCl, 2.5 mM KCl, 1.25 mM NaH_2PO_4, 26 mM $NaHCO_3$, 0.5 mM $CaCl_2$, 6 mM $MgCl_2$, and 10 mM glucose (*see* **Note 1**); carbonated with 95% O_2 and 5% CO_2, resulting in a pH of 7.4 and an osmolarity of ~300 mOsm (*see* **Note 2**).

3. HEPES-based ACSF: 125 mM NaCl, 3 mM KCl, 1.25 mM NaH_2PO_4, 25 mM N-(2-hydroxyethyl)piperazine-N′-2-ethanesulfonic acid (HEPES), 2 mM $CaCl_2$, 2 mM $MgSO_4$, and 10 mM glucose (*see* **Note 1**); pH 7.4 and 290 mOsm.
4. SR101 stock solution: 1.5 mM sulforhodamine 101 (SR101; CAS no.: 60311-02-6) in ultrapure double-processed water. Aliquot the solution in 50 μL reaction tubes and store at −20 ° C until use.
5. SBFI-AM solution: 150 μM sodium-binding benzofuran isophtalate-acetoxymethyl ester (SBFI-AM) diluted in HEPES-based ACSF (*see* **Note 3**). Vortex and microfilter the solution using a pore size of 0.22 μm. Store the solution at +4 ° C until use.
6. Patch pipette solution: 116 mM K-methanesulfonate, 32 mM KCl, 10 mM HEPES, 10 mM NaCl, 4 mM Mg-ATP, and 0.4 mM Na_2-GTP; pH 7.3 and 380 mOsm. Filter the solution using a 0.22 μm pore size, and store it at −20 °C for up to 1 month.
7. "Chemical ischemia" solution: glucose-free standard ACSF containing 5 mM sodium azide (NaN_3; CAS no.: 26628-22-8) and 2 mM 2-deoxy-D-glucose (2-DG; CAS no.: 154-17-6). Can be stored at +4°C for 1 week.

2.3 Equipment for the Preparation of Acute Brain Slices

1. CO_2 chamber (*see* **Note 4**).
2. Container with ice.
3. 250 mL glass beaker filled with ice-cold, carbonated modified ACSF.
4. Surgical tools for brain preparation and extraction, including glass Petri dishes, scissors of varying sizes, a scalpel, spatulas, filter paper, and superglue. Metal tools should be placed in a small beaker with ice-cold, carbonated modified ACSF. Before use, pre-cool the Petri dishes.
5. Thermostat–immersion circulator (C10; Haake Technik GmbH, Vreden, Germany) set at 34 °C.
6. Two glass beakers with self-made slice-holding nets (*see* Fig. 1e), one containing modified ACSF and the other containing standard ACSF. Both beakers are preheated to 34 °C.
7. Microtome (e.g., HM 650 V; Thermo Fisher Scientific, Walldorf, Germany).
8. Razor blades (e.g., Wilkinson Sword; Wilkinson Sword GmbH, Solingen, Germany) cleaned with isopropanol to remove any residues.
9. Pre-cooled buffer tray and cooling element (kept at −20 °C until use).

10. Cutting stage.
11. To transfer slices as gently as possible, use a pummel made of, e.g., the flame-polished, wide end of borosilicate glass Pasteur pipettes together with a suction bulb.

2.4 Experimental Materials and Devices

1. Borosilicate glass capillaries (e.g., 1405059; Hilgenberg GmbH, Malsfeld, Deutschland) (*see* **Note 5**).
2. Borosilicate glass capillaries with filament (e.g., GC100F-15, GC150F-15; Harvard Apparatus, Holliston, MA, USA).
3. Vertical micropipette puller (e.g., Narishige PP-830; Narishige, Tokyo, Japan).
4. Microloaders (e.g., 20 μL epT.I.P.S.; Eppendorf, Hamburg, Germany).
5. Binoculars.
6. Slice-holding grids made of a stainless steel frame with 1.5-mm-spaced, 100-μm-thick Lycra threads, finished with a plastic coating (e.g., SHD-42/15 or SHD-41/15; Warner Instruments, Holliston, MA, USA).
7. Peristaltic pump (e.g., Minipuls 3; Gilson, Middleton, USA).
8. Micromanipulator (e.g., PatchStar; Scientifica, Uckfield, UK).
9. Upright microscope (e.g., Eclipse FN1; Nikon, Tokyo, Japan) equipped with a 40× water immersion objective (e.g., Fluor 40×/0.8 W; Nikon) and a 60× water immersion objective (e.g., Fluor 60×/1.00 W, DIC H/N2, ∞/0 WD 2.0; Nikon).
10. Epi-fluorescent lamp (e.g., Intensilight C-HGFI; Nikon).
11. Transmission lamp (e.g., FN-LH Precentered Lamphouse; Nikon).
12. Monochromator (e.g., Polychrome V; Thermo Fisher Scientific/FEI, Planegg, Germany).
13. CMOS camera (ORCA FLASH 4.0LT; Hamamatsu Photonics, Herrsching, Germany).
14. Pressure application device for dye injection (e.g., PDES-02D; NPI Electronic GmbH, Tamm, Germany), attached to a pressure/vacuum generator (e.g., 2PK+; ALA Scientific-Instruments, Schwäbisch Gmünd, Germany).
15. Focus drive (e.g., NIKRFK; Prior Scientific Instruments Ltd., Cambridge, UK).
16. SBFI-AM filter set: 409 nm beam splitter and 510/84 nm emission filter (AHF Analysentechnik AG, Tübingen, Germany).

17. SR101 filter set: 585 nm beam splitter and 630/75 nm emission filter (AHF Analysentechnik AG).
18. Patch amplifier (e.g., EPC10 USB; Harvard Bioscience/HEKA Elektronik, Lambrecht, Germany).

2.5 Software

1. Nikon NIS-Elements v4.51 (Nikon).
2. pClamp 11 (Axon Instruments; CA, USA).
3. Patchmaster Next (Harvard Bioscience/HEKA Elektronik).
4. OriginPro 2021 (OriginLab Corporation; Northampton, MA, USA).
5. Excel (Microsoft; Washington, USA).

3 Methods

3.1 Preparation of Acutely Isolated Brain Slices

Animal procedures are conducted in strict accordance with the institutional guidelines of the Heinrich Heine University Düsseldorf as well as European Union Council Directive (2010/63/EU). These procedures are communicated to and approved by the Animal Welfare Office of the Animal Care and Use Facility of the Heinrich Heine University Düsseldorf (institutional act number: O52/05). According to the German Welfare Act (Tierschutzgesetz; Subheading 4, **step 3**), *post mortem* removal of brain tissue does not require additional approval.

The aim of this procedure is to obtain parasagittal tissue slices that include the somatosensory cortex. For that purpose, the slicing procedure is described in the following steps:

1. Following the recommendations of the European Commission [33], anesthetize mice older than postnatal day 14 with CO_2 and quickly decapitate.
2. Immediately after decapitation, transfer head into ice-cold (3 ± 1 °C) modified ACSF.
3. Cut the skin along the sagittal plane to expose the skull (*see* Fig. 1a).
4. Using smaller scissors, cut the skull from the back of the head to the snout, between the eye sockets and through the lambdoid suture (*see* Fig. 1a, **steps 1–3**).
5. Rapidly remove the brain using a spatula, and transfer to a Petri dish containing ice-cold modified ACSF.
6. Cut off the olfactory bulbi and the cerebellum using a sharp scalpel (*see* Fig. 1b).
7. Separate the hemispheres with a sagittal cut.

8. Perform a parasagittal cut at a 45° angle (*see* Fig. 1c).
9. Carefully dry the area surrounding the brain to ensure proper adhesion to the glue. For that, we recommend holding the hemisphere with a spatula and gently but rapidly drying the area with filter paper, avoiding tissue damage as much as possible.
10. Fix the hemisphere with the cut side downward onto the cutting stage of a microtome with superglue. Allow the superglue to harden by gently dropping some ice-cold modified ACSF over it.
11. Put the pre-cooled buffer tray inside the microtome. Place the cutting stage into the buffer tray, and add a cooling element (kept at −20 °C until use). Submerge the stage and the tissue in ice-cold modified ACSF (*see* Fig. 1d). To ensure sufficient tissue oxygenation and maintain a constant pH, continuously carbonate the ACSF with 95% O_2 and 5% CO_2.
12. Cut 250-μm-thick parasagittal slices (microtome settings: speed, 20; frequency, 60; amplitude, 1.0).
13. After slicing, transfer each brain slice onto a slice-holding net in a glass beaker filled with carbonated modified ACSF containing 0.5–1 μM SR101 at 34 °C for 20 min to selectively stain astrocytes [34] (*see* **Note 6**). Keep the slices in the dark, and avoid any unnecessary illumination from now on. The beaker can be covered with aluminum foil for this purpose.
14. Transfer the slices onto the slice-holding net of a second beaker containing carbonated standard ACSF at 34 °C for 10 additional min.
15. To ensure optimal results, it is recommended to keep the slices in standard ACSF at room temperature (20–22 °C) for at least 30 min before using them for experiments. It is important to note that acute brain slices should not be used for more than 6 h after their isolation.
16. Transfer one brain slice to the recording bath and secure it with a grid. Ensure that the grid threads do not damage the tissue, particularly the region of interest (ROI) (i.e., layers II/III of the somatosensory cortex, *see* Fig. 1f).
17. Place recording bath with the fixed brain slice onto the stage of an upright microscope. Perfuse it constantly with carbonated standard ACSF at a flow speed of 2–2.5 mL/min. Ensure that the recording bath is stable when installed on the microscope stage.

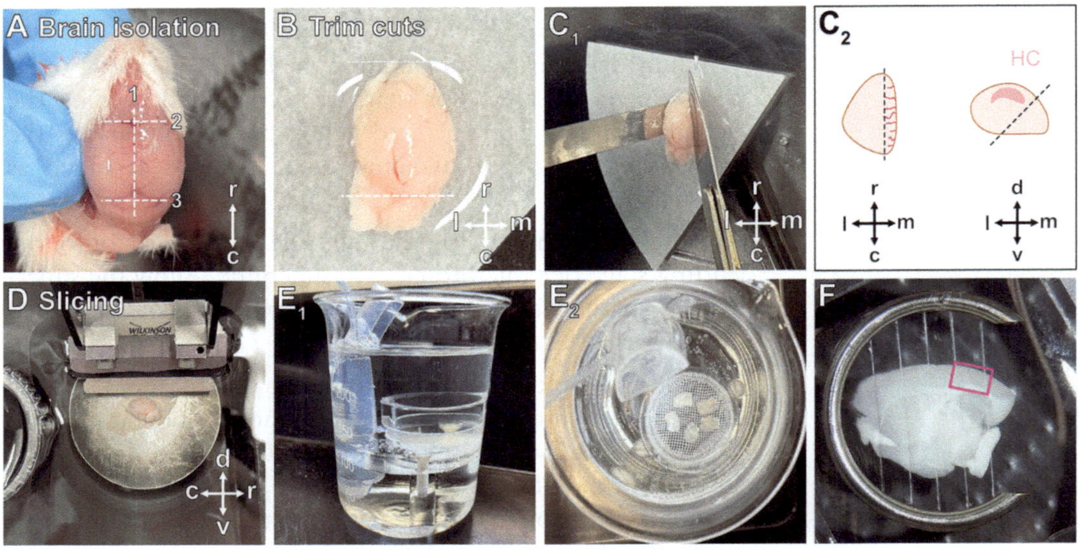

Fig. 1 Preparation of acutely isolated brain slices. **(a)** Step-by-step skull cuts for brain isolation are depicted in dashed lines. **(b)** Brain trim cuts are illustrated in dashed lines. **(c)** The parasagittal cut is achieved by cutting the hemisphere at a 45-degree angle (C_1). Graphical illustration of the hemisphere depicting the blood vessels used as a guide for the parasagittal cut and the hippocampus (HC) as a light pink area (C_2). **(d)** Hemisphere glued to the cutting stage of the vibratome. The buffer tray is filled with carbonated, ice-cold modified ACSF. **(e)** Beaker with the homemade slice-holding net from side view (E_1) and top view (E_2). Note the vertical plastic structure made from a 25 mL syringe that is used to carbonate the ACSF without disturbing the slices. **(f)** Acute slice fixed with a grid in the experimental chamber. The somatosensory cortex is highlighted with an orange rectangle

3.2 Determination of Ion Concentrations

Intracellular Na^+ concentration ($[Na^+]_i$) is measured in SBFI-AM bolus-loaded cells. This is achieved by bolus loading of a membrane-permeable form of the Na^+ fluorescent indicator that allows targeted bulk staining of large populations of living cells and simultaneous Na^- recordings in numerous individual cells.

Extracellular K^+ concentration ($[K^+]_e$) and extracellular Na^+ concentration ($[Na^+]_e$) are measured in acute tissue slices using double-barreled ion-sensitive microelectrodes.

3.2.1 Quantification of Intracellular Sodium in Astrocytes

1. Pull pipettes for dye bolus loading (tip diameter ~ 1 μm, resistance ~2–3 MΩ when filled with ACSF).
2. Backfill the loading pipette with approximately 3-4 μL of HEPES-based ACSF containing 150 μM SBFI-AM. Attach the pipette to the pressure application device. Adjust the holding pressure of the vacuum generator to the atmospheric pressure to prevent solution exchange from or into the pipette. Then, bring the pipette into the ACSF-perfused experimental chamber and position it under the objective.

3. Gently insert the pipette into the tissue to a depth of 10–20 μm. Apply positive pressure (~1-1.5 psi) for approximately 5 s to inject the dye into the tissue. Withdraw pipette and repeat dye injection into adjacent regions (*see* **Note 7**).
4. After bolus loading, keep the tissue in the bath for approximately 45 min with constant perfusion. This will allow the dye inside cells to de-esterify and excess extracellular dye to be washed out. Keep the tissue in the dark during this time to prevent dye bleaching.
5. Before starting the recording, evaluate the tissue and staining for quality. The presence of rounded and swollen neuronal bodies may indicate poor tissue quality, which can affect ion dynamics.
6. Set the imaging parameters for the experimental recording of SBFI-AM fluorescence, including excitation wavelength (400 nm; *see* **Note 8**), exposure time (100 ms), imaging frequency (0.5 Hz), and experimental length (~20 min).
7. Record SBFI-AM fluorescence for 3-5 min to establish a stable baseline. Then switch to the "chemical ischemia" solution for 2 min. If there is any movement in the z-axis due to cell/tissue swelling, carefully refocus during the recording. Wait for the Na^+ transient to fully recover baseline levels before ending the recording.
8. Collect SBFI-AM emission from regions of interest (ROIs) representing the cell bodies of SR101-positive astrocytes. Correct for potential movement artifacts by tracking the ROIs over time.
9. To perform dynamic background correction, draw a ROI in a region of the field of view where no fluorescent cells are present. Subtract the average intensity value measured in the background ROI from the values obtained in the cellular ROIs (*see* **Note 9**).
10. Normalize data by calculating the average value of the baseline period for each cellular ROI.
11. For non-ratiometric single wavelength imaging, it is necessary to perform bleaching correction. To do so, fit a monoexponential decay function to the traces by selecting points from the baseline and the recovery period in which the Na^+ transient is completely over.
12. Using established calibration protocols of SBFI-AM fluorescence [35–37], convert changes in fluorescence into changes in $[Na^+]_i$ (*see* **Note 10**).
13. Calculate the average time course for $[Na^+]_i$ (*see* Fig. 2a1).

3.2.2 Quantification of Extracellular Sodium and Potassium

1. Prepare double-barreled ion-sensitive microelectrodes from two borosilicate glass capillaries with filament glued and pulled out together. The tip diameter should be approximately 1 μm [12].
2. Silanize the ion-sensitive barrel with vaporized hexamethyldisilazane (Fluka, Buchs, Switzerland).
3. Fill the tip of the silanized ion-sensitive barrel with 1–2 μL of the ion-selective sensor solution. For K^+-sensitive microelectrodes, use a carrier based on valinomycin (Ionophore I, Cocktail B, Merck), and for Na^+-sensitive microelectrodes, use ETH 157 (Ionophore II, Cocktail A, Merck).
4. Fill the ion-sensitive barrel with 100 mM KCl for K^+-sensitive electrodes or 100 mM NaCl for Na^+-sensitive electrodes.
5. Fill the reference barrel with HEPES-based ACSF.
6. Insert chlorinated silver wires into both barrels, and seal them with dental wax.
7. Attach the microelectrode to the micromanipulator, and place it in the ACSF-perfused experimental chamber. Place the reference electrode in the pre-chamber. The reference barrel should have a resistance of 30–100 MΩ, and the ion-sensitive barrel should have a resistance of 10–20 GΩ (*see* **Note 11**).
8. Calibrate the ion-selective microelectrode before and after each experimental recording (*see* **Note 12**).
9. After calibration, insert the microelectrode in layer II/III of the somatosensory cortex at 40–60 μm depth (*see* **Note 13**).
10. Record a stable baseline of K^+ or Na^+ potential for 3–5 min before switching to the "chemical ischemia" protocol for 2 min. Allow the potential transient to completely recover baseline levels before finishing the recording.
11. Convert the voltage response of the ion-selective electrode to changes in $[K^+]_e$ or $[Na^+]_e$ using Eq. (3) derived from the Nernst equation:

$$\Delta[X^z]_e = [X^z]_B \cdot \left(10^{\frac{\Delta V_X}{s}} - 1\right), \tag{3}$$

where $\Delta[X^Z]_e$ is the change in extracellular ion concentration, $[X^Z]_B$ is the baseline ion concentration (2.5 mM for K^+ and 130 mM for Na^+), ΔV_X is the change in the potential of the ion, and s is the Nernst slope.

12. Calculate the average time course for $[Na^+]_e$ and $[K^+]_e$ (*see* Fig. 2a2–3).

3.3 Determination of Membrane Potential

The membrane potential (E_M) of astrocytes is measured using the cell-attached and whole-cell configurations of the patch-clamp technique.

3.3.1 Quantification of the Relative Ion Membrane Permeability

The E_M assessed using the whole-cell configuration is used to determine the relative membrane permeability for Na^+ versus K^+ (α). Since there is no exchange between the pipette solution and the cell, $[Na^+]_i$ and $[K^+]_i$ can be estimated later as values from the pipette solution. Recording only the baseline E_M is needed, as it is assumed that α remains constant.

1. Pull patch pipettes from borosilicate glass capillaries (input resistance (R_{in}) ~2.5–4.5 MΩ).
2. Backfill the pipette with approximately 5 μL of standard pipette solution. Attach patch pipette to the head stage of the micromanipulator. Apply positive pressure (~20 mbar) to prevent solution exchange.
3. To identify SR101-positive cells (i.e., astrocytes), switch to epi-fluorescence, and select a cell based on its morphology. Ensure that the cell body is not swollen or malformed.
4. Place the pipette into the ACSF-perfused experimental chamber under the objective and switch to voltage clamp mode. Correct offset potential (V_o) in "Patchmaster Next" by clicking "Auto." This will account for any potentials resulting from differences in concentration between the pipette solution and the ACSF. During the patching process, activate a test pulse of 50 ms for 5 pA to obtain the R_{in}.
5. Place the patch pipette on top of the chosen cell. Then, gradually lower the pipette. If a halo or/and an increase in R_{in} of around 0.3–0.5 MΩ are observed, release the positive pressure and directly apply negative pressure (<-30 mbar). R_{in} should increase instantly (*see* **Note 14**). If R_{in} exceeds 1 GΩ, a giga seal is formed. Set holding E_M to -85 mV immediately.
6. Rupture the membrane by applying a short, single suction with negative pressure (*see* **Note 15**). The whole-cell configuration is achieved when R_{in} is under 25 MΩ and decay currents are visible in the oscilloscope.
7. Switch to current clamp mode. Run a continuous current clamp measurement holding $I = 0$ pA. Record baseline values for at least 3–5 min.

3.3.2 Quantification of Time Course of Membrane Potential

The cell-attached configuration is used to measure the time course of changes in the astrocyte E_M during metabolic inhibition.

1. Repeat **steps 1–6** from Subheading 3.3.1. Instead of filling the pipette with pipette solution, use standard ACSF.

2. Switch to current clamp mode. Cell-attached configuration is achieved when a E_M of around −85 mV is stable for a couple of min.
3. Run a continuous current clamp measurement holding $I = 0$ pA. Record baseline values for 3–5 min. Switch to "chemical ischemia" solution for 2 min. Allow E_M to recover fully before finishing the recording.
4. Calculate the average time course for E_M (*see* Fig. 2a4).

3.4 Calculation of Intracellular Potassium Concentration

$[K^+]_i$ in astrocytes is determined based on an empirical–theoretical approach. Generally, baseline and peak values of $[K^+]_i$ can be estimated with the following calculations using values from the literature. However, to accurately determine the time course of astrocytic K^+ in situ during the "chemical ischemia" protocol, we will focus solely on experimentally obtained data.

1. We assume that α is constant during metabolic inhibition. To determine α, rearrange Eq. (4):

$$\alpha = \frac{[K^+]_e + e^{\frac{E_M F}{RT}} \cdot [K^+]_i}{e^{\frac{E_M F}{RT}} \cdot [Na^+]_i - [Na^+]_e}, \tag{4}$$

where R is the universal gas constant, T is the absolute temperature, F is the Faraday constant, $[X^Z]_{e/i}$ are the respective extra- and intracellular ion concentrations, and α is the relative membrane permeability for Na^+ versus K^+ (P_{Na+}/P_{K+}).

2. α can be estimated to 0.015, based on the literature [38]. However, we used experimental data for a more accurate determination. To this end, use the intracellular ion concentrations from the pipette solution (146 mM $[K^+]_i$ and 10 mM $[Na^+]_i$). For the extracellular ion concentrations, use data obtained from the ion-sensitive microelectrode measurements (2.59 mM $[K^+]_e$ and 157.25 mM $[Na^+]_e$). The baseline values for E_M (−90.85 mV) are obtained from the whole-cell measurements. This allows for the calculation of an $\alpha = 0.010$ for neocortical astrocytes.
3. To determine the time course of $[K^+]_i$ upon metabolic inhibition, rearrange Eq. (5) to calculate $[K^+]_i$:

$$[K^+]_i = \frac{[K^+]_e + \alpha[Na^+]_e}{e^{\frac{E_M F}{RT}}} - \alpha[Na^+]_i \tag{5}$$

4. To generate the confidence intervals, calculate the minimum and maximum confidence values for each time point using Eq. (6) (*see* Fig. 2):

$$\text{Confidence interval}^{\pm} = \text{mean} \pm z \cdot \frac{\text{SD}}{\sqrt{n}}, \tag{6}$$

where SD is the standard deviation of the mean, z is the degree of confidence, and n the sample number. Since we aimed for confidence intervals of 95%, we chose a z value of 1.96.

5. For analysis of key time points, load time courses in "OriginPro" software. Choose "data cursor" from the left toolbar. Pick baseline, peak, overshoot, and recovery values.

This analysis reveals a baseline astrocytic $[K^+]$ of 146 ± 4 mM in acute brain slices. Upon metabolic ischemic stress, $[K^+]_i$ drops to 103 ± 6 mM, followed by a transient K^+ overshoot that reaches 162 ± 6 mM, before returning to baseline values.

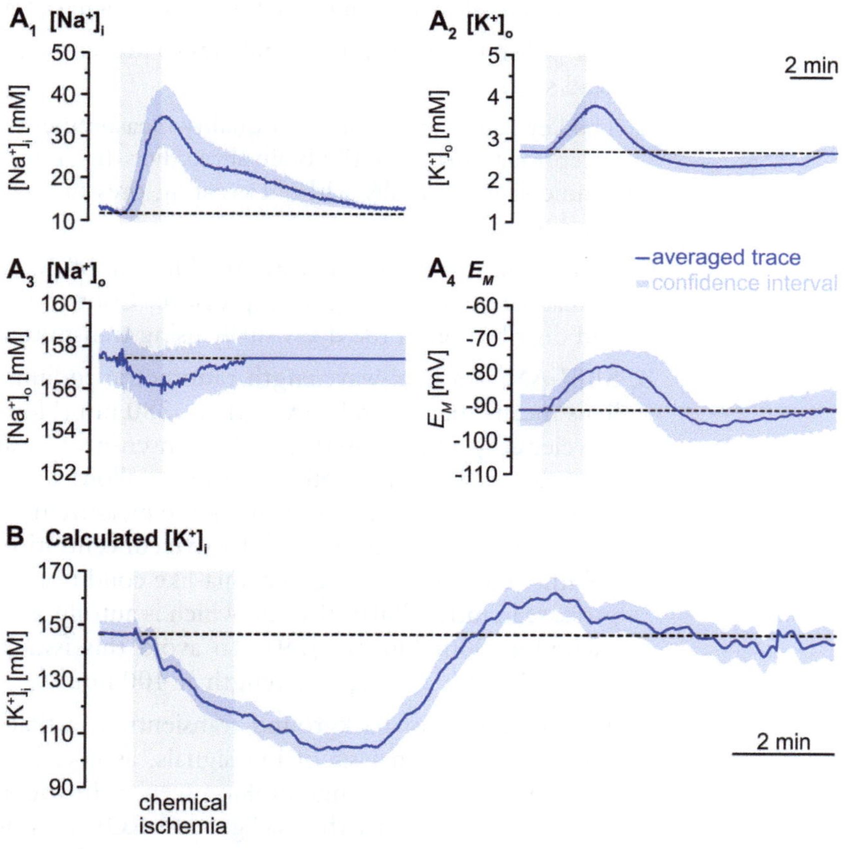

Fig. 2 Simulation of ischemia-induced changes in astroglial K^+. **(a)** Averaged traces of the experimentally measured $[Na^+]_i$ (A_1), $[K^+]_e$ (A_2), $[Na^+]_e$ (A_3), and E_M (A_4). **(b)** Calculated time course of astroglial K^+. Light blue areas indicate the 95% confidence intervals. Gray areas illustrate the 2-min period of metabolic ischemic stress. (Modified from [40])

4 Notes

1. All solutions are prepared in ultrapure water with a resistivity of 18.2 MΩ·cm, if not stated otherwise.
2. Solutions are prepared on the day of use or the day before. In the latter case, no glucose is added until the day of usage. Standard and modified ACSF should be carbonated for at least 30 min before usage to reach pH of 7.4.
3. SBFI-AM is prepared from a 1 mM stock solution dissolved in 20% pluronic/DMSO. The stock solution is stored at −20 °C and protected from the light. Caution: higher concentrations of pluronic may damage cell membranes.
4. We use a homemade CO_2 chamber. The chamber must be designed to have sufficient space for the animal, to avoid injury to the animals and to allow for controlled delivery of the gas.
5. Fire-polish the ends of the capillaries to prevent damage to the head stage.
6. The key factor to obtain high-quality measurements will be an exceptional quality of the brain slices. For this, it is essential to handle slices carefully without creating pressure, crushing, or bending the tissue.
7. If labelling is too low, increase the duration of dye application instead of increasing injection pressure. Control the injection and distribution of the dye visually using transmitted light.
8. SBFI-AM is a dual-wavelength ratiometric sodium indicator. This means that it can be excited at ~380 nm ("ion-sensitive" wavelength) and at ~340 nm ("ion-insensitive" wavelength). Alternate excitation at both wavelengths allows for the calculation of a fluorescence ratio, making the measurement independent from changes in intracellular dye concentration or in cell volume. However, during ischemia-like conditions, there is an increase in intracellular NADH, which is autofluorescent when excited at 300–380 nm [39]. To avoid this issue, we excite SBFI-AM with a single wavelength at 400 nm.
9. The correction of background transients or artifacts is a key step during the analysis of ion signals, as it can significantly affect the apparent change in fluorescence. In widefield imaging, it is essential that the background ROI does not register any real Na^+ signal but that it allows for the correction of refocusing-derived changes in fluorescence, out-of-focus fluorescence, or autofluorescence artifacts, among others.
10. Changes in SBFI-AM fluorescence can be converted into changes in $[Na^+]_i$ based on in situ calibrations. Perfuse bolus loaded cells with calibration-ACSF, containing different $[Na^+]$

(starting and ending in 0 mM Na^+) as well as ionophores (3 μm gramicidin, 10 μm monensin) and the NKA inhibitor ouabain (100 μM). Fluorescence changes from individual cells are normalized to its respective fluorescence levels at 0 mM Na^+. In the range between 10 and 50 mM $[Na^+]_i$, data can be fitted into a linear function, allowing for a direct conversion of fluorescence into $[Na^+]_i$.

11. If the resistance of the barrels is higher than indicated, a microforge can be used to polish the tip of the electrode before starting the calibration.
12. For the calibration of K^+-sensitive microelectrodes, perfuse the experimental chamber with HEPES-based calibration-ACSF with a total of 150 mM NaCl and KCl, modifying $[K^+]$ from 0 to 10 mM. Na^+-sensitive electrodes are calibrated in HEPES-based ACSF with a total of 160 mM NaCl and N-methyl-D-glucamine chloride (NMDG-Cl), with $[Na^+]$ ranging from 70 to 160 mM.
13. When inserting the ion-sensitive microelectrode in the tissue, it will register characteristic fluctuations. Wait until baseline has stabilized before starting the experiment. This time can vary depending on the quality of the microelectrode and the tissue, ranging from almost no time to up to 30 min.
14. If R_{in} does not go up instantly, gradually increase E_M to −85 mV in 10 mV steps. This will help to seal the cell.
15. If multiple suctions are required, wait a few seconds between attempts to allow the cell to rest before trying again.

Acknowledgments

The work was supported by grants from the Deutsche Forschungsgemeinschaft (DFG), Research Unit 2795 "Synapses under Stress" (Ro2327/13-2, 14-2), and from the German Federal Ministry of Education and Research (BMBF), Project SynGluCross (01GQ2204A). IAM was funded by a DFG-Walter Benjamin stipend (AL 2956/1-1). We thank Simone Durry for performing the extracellular recordings and Viola Bornemann for providing critical advice on Subheading 3.2.2.

References

1. Verkhratsky A, Nedergaard M (2018) Physiology of astroglia. Physiol Rev 98(1):239–389
2. Gage P, Quastel D (1965) Dual effect of potassium on transmitter release. Nature 206(4984):625–626
3. Erulkar S, Weight F (1977) Extracellular potassium and transmitter release at the giant synapse of squid. J Physiol 266(2):209–218
4. Malenka RC, Kocsis JD, Ransom BR, Waxman SG (1981) Modulation of parallel fiber excitability by postsynaptically mediated changes in

extracellular potassium. Science 214(4518): 339–341
5. Hertz L (1965) Possible role of neuroglia: a potassium-mediated neuronal—neuroglial—neuronal impulse transmission system. Nature 206(989):1091–1094
6. Wallraff A, Köhling R, Heinemann U, Theis M, Willecke K, Steinhäuser C (2006) The impact of astrocytic gap junctional coupling on potassium buffering in the hippocampus. J Neurosci 26(20):5438–5447
7. Hertz L, Song D, Xu J, Peng L, Gibbs ME (2015) Role of the astrocytic Na+, K+-ATPase in K+ homeostasis in brain: K+ uptake, signaling pathways and substrate utilization. Neurochem Res 40(12):2505–2516
8. Larsen BR, Stoica A, MacAulay N (2016) Managing brain extracellular K+ during neuronal activity: the physiological role of the Na+/K+-ATPase subunit isoforms. Front Physiol 7: 185841
9. Pannasch U, Vargova L, Reingruber J, Ezan P, Holcman D, Giaume C, Sykova E, Rouach N (2011) Astroglial networks scale synaptic activity and plasticity. Proc Natl Acad Sci USA 108(20):8467–8472
10. Chever O, Dossi E, Pannasch U, Derangeon M, Rouach N (2016) Astroglial networks promote neuronal coordination. Sci Signal 9(410):ra6
11. Van Putten MJAM, Fahlke C, Kafitz KW, Hofmeijer J, Rose CR (2021) Dysregulation of astrocyte ion homeostasis and its relevance for stroke-induced brain damage. Int J Mol Sci 22(11):5679
12. Haack N, Durry S, Kafitz KW, Chesler M, Rose R (2015) Double-barreled and concentric microelectrodes for measurement of extracellular ion signals in brain tissue. J Vis Exp 103
13. Kettenmann H, Sonnhof U, Schachner M (1983) Exclusive potassium dependence of the membrane potential in cultured mouse oligodendrocytes. J Neurosci 3(3):500–505
14. Ballanyi K, Grafe P, ten Bruggencate G (1987) Ion activities and potassium uptake mechanisms of glial cells in guinea-pig olfactory cortex slices. J Physiol 382:159–174
15. Coles JA, Schneider-Picard G (1989) Increase in glial intracellular K+ in drone retina caused by photostimulation but not mediated by an increase in extracellular K+. Glia 2(4):213–222
16. Minta A, Tsien RY (1989) Fluorescent indicators for cytosolic sodium. J Biol Chem 264(32):19449–19457
17. Rimmele TS, Chatton JY (2014) A novel optical intracellular imaging approach for potassium dynamics in astrocytes. PLoS One 9(10):e109243
18. Shen Y, Wu S-Y, Rancic V, Aggarwal A, Qian Y, Miyashita S-I, Ballanyi K, Campbell RE, Dong M (2019) Genetically encoded fluorescent indicators for imaging intracellular potassium ion concentration. Commun Biol 2(1):18
19. Wu SY, Wen Y, Serre NBC, Laursen CCH, Dietz AG, Taylor BR, Drobizhev M, Molina RS, Aggarwal A, Rancic V, Becker M, Ballanyi K, Podgorski K, Hirase H, Nedergaard M, Fendrych M, Lemieux MJ, Eberl DF, Kay AR, Campbell RE, Shen Y (2022) A sensitive and specific genetically-encoded potassium ion biosensor for in vivo applications across the tree of life. PLoS Biol 20(9):e3001772
20. Du Y, Wang W, Lutton AD, Kiyoshi CM, Ma B, Taylor AT, Olesik JW, McTigue DM, Askwith CC, Zhou M (2018) Dissipation of transmembrane potassium gradient is the main cause of cerebral ischemia-induced depolarization in astrocytes and neurons. Exp Neurol 303:1–11
21. Kimelberg HK, Bowman C, Biddlecome S, Bourke RS (1979) Cation transport and membrane potential properties of primary astroglial cultures from neonatal rat brains. Brain Res 177(3):533–550
22. Dietzel I, Heinemann U (1986) Dynamic variations of the brain cell microenvironment in relation to neuronal hyperactivity. Ann N Y Acad Sci 481:72–86
23. Halnes G, Østby I, Pettersen KH, Omholt SW, Einevoll GT (2013) Electrodiffusive model for astrocytic and neuronal ion concentration dynamics. PLoS Comput Biol 9(12):e1003386
24. Ullah G, Wei Y, Dahlem MA, Wechselberger M, Schiff SJ (2015) The role of cell volume in the dynamics of seizure, spreading depression, and anoxic depolarization. PLoS Comput Biol 11(8):e1004414
25. Breslin K, Wade JJ, Wong-Lin K, Harkin J, Flanagan B, Van Zalinge H, Hall S, Walker M, Verkhratsky A, McDaid L (2018) Potassium and sodium microdomains in thin astroglial processes: a computational model study. PLoS Comput Biol 14(5):e1006151
26. Kalia M, Meijer HGE, van Gils SA, van Putten M, Rose CR (2021) Ion dynamics at the energy-deprived tripartite synapse. PLoS Comput Biol 17(6):e1009019
27. Somjen GG (2004) Ions in the brain: normal function, seizures, and stroke. Oxford University Press

28. Hansen AJ (1985) Effect of anoxia on ion distribution in the brain. Physiol Rev 65(1): 101–148
29. Leis JA, Bekar LK, Walz W (2005) Potassium homeostasis in the ischemic brain. Glia 50(4): 407–416
30. Erecińska M, Silver IA (1994) Ions and energy in mammalian brain. Prog Neurobiol 43(1): 37–71
31. Meyer J, Gerkau NJ, Kafitz KW, Patting M, Jolmes F, Henneberger C, Rose CR (2022) Rapid fluorescence lifetime imaging reveals that TRPV4 channels promote dysregulation of neuronal Na+ in ischemia. J Neurosci 42(4):552–566
32. Pape N, Rose CR (2023) Activation of TRPV4 channels promotes the loss of cellular ATP in organotypic slices of the mouse neocortex exposed to chemical ischemia. J Physiol 601(14):2975–2990
33. Close B, Banister K, Baumans V, Bernoth EM, Bromage N, Bunyan J, Erhardt W, Flecknell P, Gregory N, Hackbarth H, Morton D, Warwick C (1996) Recommendations for euthanasia of experimental animals. DGXI of the European Commission. Lab Anim 30(4):293–316
34. Kafitz KW, Meier SD, Stephan J, Rose CR (2008) Developmental profile and properties of sulforhodamine 101 – labeled glial cells in acute brain slices of rat hippocampus. J Neurosci Methods 169(1):84–92
35. Gerkau NJ, Rakers C, Durry S, Petzold GC, Rose CR (2018) Reverse NCX attenuates cellular sodium loading in metabolically compromised cortex. Cereb Cortex 28(12): 4264–4280
36. Kelly T, Rose CR (2010) Ammonium influx pathways into astrocytes and neurones of hippocampal slices. J Neurochem 115(5): 1123–1136
37. Langer J, Rose CR (2009) Synaptically induced sodium signals in hippocampal astrocytes in situ. J Physiol 587(24):5859–5877
38. Stephan J, Haack N, Kafitz KW, Durry S, Koch D, Hochstrate P, Seifert G, Steinhauser C, Rose CR (2012) Kir4.1 channels mediate a depolarization of hippocampal astrocytes under hyperammonemic conditions in situ. Glia 60(6):965–978
39. Blacker TS, Mann ZF, Gale JE, Ziegler M, Bain AJ, Szabadkai G, Duchen MR (2014) Separating NADH and NADPH fluorescence in live cells and tissues using FLIM. Nat Commun 5: 3936
40. Eitelmann S, Stephan J, Everaerts K, Durry S, Pape N, Gerkau NJ, Rose CR (2022) Changes in astroglial K+ upon brief periods of energy deprivation in the mouse neocortex. Int J Mol Sci 23(9):4836

Chapter 4

Quantification of Astrocytic Sodium Signals Using Fluorescence Lifetime Imaging Microscopy (FLIM)

Jan Meyer, Karl W. Kafitz, and Christine R. Rose

Abstract

Intensity-based imaging with fluorescent indicators is a widely used method for monitoring ion transients, including Ca^{2+} or Na^+ signals, in astrocytes. However, using this technique, changes in ion concentrations cannot always be reliably separated from simultaneous changes in the concentration of the ion-sensitive fluorophores, e.g., due to changes in cell volume. An alternative approach is fluorescence lifetime imaging microscopy (FLIM), which is based on the time the fluorophore remains in the excited state rather than determining its emission intensity. Here, we describe the use of the chemical fluorescent indicator dyes ION Natrium Green 2 and CoroNa Green for FLIM of intracellular Na^+ in astrocytes. We also present different strategies for analyzing the FLIM data obtained and demonstrate a procedure for their calibration. Overall, Na^+-FLIM provides a reliable quantitative determination of changes in astrocyte Na^+ concentrations independent of changes in fluorophore concentrations.

Key words Fluorescence lifetime imaging microscopy, Astrocytes, Hippocampus, ING2, CoroNa Green

1 Introduction

The inwardly directed Na^+ gradient is a fundamental property that most animal cells have in common. The ubiquitously expressed Na^+/K^+-ATPase generates and maintains this gradient by constantly consuming ATP [1]. The Na^+ gradient is used by many secondary-active transport processes, such as the Na^+/Ca^{2+}-exchangers (NCX) [2] or the excitatory amino acid transporters (EAATs) [3]. These Na^+-dependent transporters are central for astrocyte function and are a central component for neuron–glia interaction in the brain [4, 5]. A thorough understanding of astrocyte Na^+ fluxes and Na^+ signals is therefore required.

The original version of the chapter has been revised. A correction to this chapter can be found at https://doi.org/10.1007/978-1-0716-4366-2_22

Barbara Di Benedetto (ed.), *Astrocytes: Methods and Protocols*, Methods in Molecular Biology, vol. 2896,
https://doi.org/10.1007/978-1-0716-4366-2_4,

Determination of astrocyte Na^+ has, however, always been limited by the limited toolbox available. Up to date, there are no suitable genetically encoded Na^+ indicators, and Na^+-sensitive fluorescent chemical indicators dyes, e.g., suffer from low quantum yields and increased bleaching compared to other dyes such as those used for Ca^{2+}. In the past, determination of changes in astrocyte Na^+ was mainly achieved by measuring changes in the emission intensity of these chemical indicators that occur upon binding of Na^+ [6]. However, this method has a drawback due to its dependence on the concentration of the fluorophore, which may change, e.g., due to bleaching and/or to changes in cell volume [7]. This issue can be resolved using fluorescence lifetime imaging microscopy (FLIM) instead of intensity-based imaging.

FLIM has become an invaluable tool for the analysis and quantification of intracellular ion dynamics [7–10]. It is based on the time the fluorophore stays in the excited state before releasing a photon and returning back to ground state, the so-called fluorescence lifetime (FL) [11]. For many ion-sensitive fluorescent dyes, this FL is dependent on the concentration of their respective ion. By relying on this intrinsic property of the dye, secondary effects that are based on the dye concentration can be canceled out. This enables reliable determination of ion concentrations even under conditions where the dye concentration is not stable (e.g., swelling during an ischemic stroke) [7]. As mentioned earlier, there are only few chemical indicators available for imaging of Na^+, and only few of them show measurable changes in their lifetime with changes in Na^+.

Two dyes that were mainly used so far for Na^+-FLIM are ION Natrium Green-2 and CoroNa Green. They exhibit a ~ 1- to 1.5-fold change in FL upon Na^+ binding and a K_d suitable for intracellular measurements [12]. The absorption maxima for ING2 and CoroNa Green are around ~488, and maximum emission is in the range from 500 to 600 nm [12, 13]. For two-photon excitation, both dyes have a relatively wide two-photon cross section and can be excited from 750 to 1000 nm [12]. The indicators can either be loaded into an astrocyte directly, e.g., using the patch-clamp technique or bulk-loaded into astrocytes using a membrane-permeable ester form of the dye [6]. Overall, FLIM of Na^+-sensitive fluorescent dyes is suitable for measuring dynamic changes in intracellular Na^+ concentrations and can help to elucidate the molecular and cellular pathways underlying astrocyte Na^+ signaling as well as their functional consequences.

2 Materials

2.1 Solutions and Dyes

1. Artificial cerebrospinal fluid ("ACSF") for preparation of brain slices ("preparation ACSF"): 130 mM NaCl, 2.5 mM KCl, 0.5 mM $CaCl_2$, 6 mM $MgCl_2$, 1.25 mM NaH_2PO_4, 26 mM $NaHCO_3$, and 20 mM glucose, bubbled with 95% O_2 and 5%

CO_2, resulting in a pH of 7.4 and an osmolarity of 312–320 mOsm (*see* **Note 1**).

2. "Normal ACSF" for experiments: 130 mM NaCl, 2.5 mM KCl, 2 mM $CaCl_2$, 1 mM $MgCl_2$, 1.25 mM NaH_2PO_4, 26 mM $NaHCO_3$, and 20 mM glucose, bubbled with 95% O_2 and 5% CO_2, titrated with NaOH to a pH of 7.4 and an osmolarity of 308–312 mOsm (*see* **Note 1**).
3. "HEPES-based ACSF" for dye dilution: 125 mM NaCl, 3 mM KCl, 25 mM HEPES, 2 mM $MgSO_4$, 2 mM $CaCl_2$, 1.25 mM NaH_2PO_4, and 10 mM glucose, resulting in a pH of 7.4 and an osmolarity of ~300 mOsm. Store in 1 mL reaction tubes at –20°C.
4. "Calibration ACSF." HEPES-based ACSF containing different concentrations of NaCl (e.g., 0, 10, 20, 30,... mM), with a total of 170 mM (NaCl + KCl). Add the NKA inhibitor ouabain at a concentration of 100 μM as well as the Na^+ ionophores gramicidin (3 μM) and monensin (10 μM) fresh on the day of the experiment.
5. 1.5 mM stock solution of sulforhodamine 101 (SR101) in ultrapure double-distilled water. Stored at –20 °C.
6. ION Natrium Green 2 AM dissolved in 20% Pluronic/DMSO or CoroNa Green AM dissolved only in DMSO, to a final concentration of 1 mM for bolus-loading of brain tissue slices (*see* **Note 2**). Keep the solution in the original sealed tube and store at –20 °C. Avoid unnecessary freeze and thaw cycles.
7. Dye stock solution diluted in HEPES-based ACSF to a final dye concentration of 150–200 μM (*see* **Note 3**). After mixing and micro-filtration with a 0.22 μm filter, the solution should be kept at 4 °C.
8. Solution to measure the internal response function (IRF) of the FLIM rig. First, dissolve potassium iodide in ultrapure double-distilled water until saturation. Second, dissolve Erythrosin B (approx. 5 μg) in the solution.

2.2 Experimental Devices

1. Microtome for slice preparation (e.g., Campden 7000smz-2, Campden Instruments, Loughborough, Leics., England).
2. Micropipette puller (e.g., Narishige PP-830, Narishige, Tokyo, Japan).
3. Micromanipulator (e.g., LN Junior, Luigs & Neumann, Ratingen, Germany).
4. Pressure application device for dye injection (e.g., PDES NPI Electronic GmbH, Tamm, Germany).
5. Multiphoton laser imaging system with a high-intensity pulsed laser source (*see* **Note 4**) (e.g., Nikon A1MP, Nikon Europe B. V., Amstelveen, the Netherlands with a 80 MHz MaiTai BroadBand, Spectra Physics, Milpitas, California, USA; also *see*

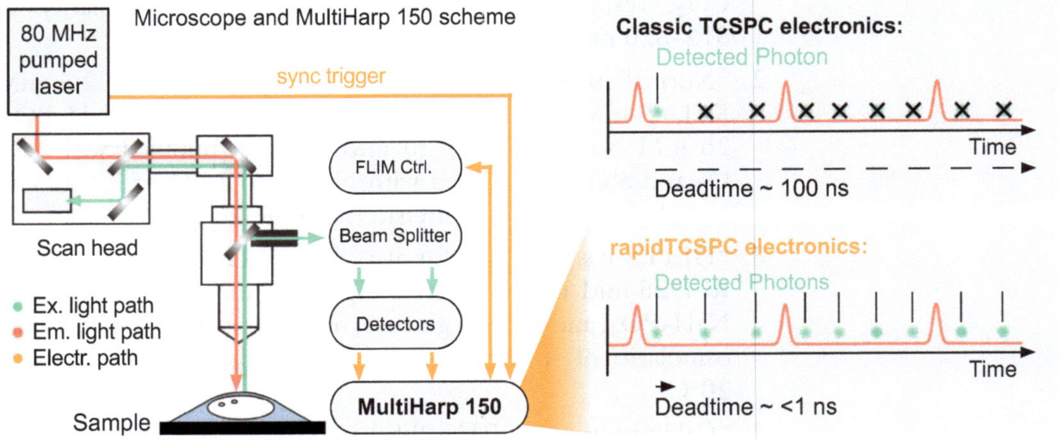

Fig. 1 Experimental FLIM setup and TCSPC principle. Left: Exemplary multiphoton setup with a pumped laser (also *see* Subheading 2.2), synchronized to a TCSPC device (MultiHarp 150). Right: Benefits of a low-dead time TCSPC device. Historically, long dead times have been a limitation in the registration of incoming photons. However, modern TCSPC solved this issue, enabling greatly increased photon counts. Taken and modified from [7].

Fig. 2). Alternatively, a confocal laser imaging system can be employed utilizing a pulsed laser excitation source (e.g., PicoQuant MicroTime 200).

6. Time-correlated single-photon counting (TCSPC) unit for FLIM data acquisition (e.g., MultiHarp 160, PicoQuant, Berlin, Germany; also *see* Fig. **1** and **Note 5**).
7. FLIM software (e.g., Symphotime64, PicoQuant, Berlin, Germany).

3 Methods

3.1 Acute Slice Preparation

Before starting experiments, ensure that all required permissions and approvals were acquired and that experiments are performed in accordance with all relevant institutional and national guidelines and regulations. The procedures described here are in strict accordance with the institutional guidelines of the Heinrich Heine University Düsseldorf, Germany, as well as the European Community Council Directive (86/609/EEC). All experiments were communicated to and approved by the Animal Welfare Office at the Animal Care and Use Facility of the Heinrich Heine University Düsseldorf, Germany (institutional act number: O52/05). In accordance with the German Animal Welfare Act (Tierschutzgesetz, Articles 4 and 7), no formal additional approval for the post-mortem removal of brain tissue was necessary.

1. To generate acute brain slices, mice older than postnatal day 14 were anaesthetized with CO_2 and decapitated, following the

recommendations of the European Commission's publication on Euthanasia of Experimental Animals (Luxembourg: Office for Official Publications of the European Communities, 1997; ISBN 92–827-9694-9).

2. Once decapitated, the brain has to be quickly removed and placed in a petri dish with ice-cold preparation ACSF. To keep pH at 7.4, bubble ACSF continuously with carbogen (95% O_2 and 5% CO_2).
3. Separate the hemispheres, perform another cut in parasagittal orientation, and glue the brain with the cut side downward onto to the cutting stage of a microtome.
4. Place the cutting stage into the cooled microtome (preferably a non-dissolving cooling element); submerge stage and tissue in ice-cold preparation ACSF.
5. Prepare hippocampal slices; thickness of slices should be around 250 μm.
6. Transfer slices to preparation ACSF containing 1 μM SR101 at 34 °C for 20 min to label astrocytes [14]. Subsequently, incubate at 34 °C for another 10 min in normal ACSF. Astrocytes should now be labelled brightly as shown in Fig. 1.
7. The slices can be kept at room temperature for 5–7 h, before use in an experiment (*see* **Note 6**).

3.2 Bolus-Loading of Ion Indicator Dyes into Tissue

1. This technique was introduced by Stosiek et al. in 2003 [15] for dye-loading of brain cells in vivo. Since then, it is the method of choice for many laboratories including ours. Example images resulting from using this staining technique with CoroNa Green AM in hippocampal brain slices are shown in Fig. 2.

Fig. 2 Sulforhodamin 101 and CoroNa Green AM counterstaining. Multiphoton emission images of the CA1 region of a mouse hippocampal tissue slice stained with SR101, which exclusively stains astrocytes (left), and CoroNa Green AM, which prominently stains neurons (middle). Right: Merge of the two channels. Scale bar is 20 μm

2. Prepare pipettes for dye injection from fire-polished borosilicate glass capillaries using a standard micropipette puller. Pipettes should have a tip diameter of approximately 1 μm and a resistance of 2–3 MΩ.
3. Carefully place a tissue slice into an experimental chamber, and fix it with a harp (e.g., Warner SHD-41/15).
4. Transfer the experimental chamber into your upright microscope, and keep the slice perfused with normal ACSF at rates of about 2 mL/min at all times.
5. Turn on the pressure application device to allow for pressure equilibration before attaching a pipette.
6. Load a pipette with HEPES-ACSF containing 125 μM ING2 or CoroNa Green, and attach the pipette to the pressure application device.
7. Carefully lower the pipette with the micromanipulator to your desired focal plane and area in the tissue slice, and apply ~1.5 psi for ~5 s to release the dye into the tissue. Withdraw the pipette and repeat the injection in adjacent areas.
8. Wait for 30–45 min to engage intracellular de-esterification, thereby trapping the dye inside of the cells. Additionally, excess extracellular dye will be cleared from the tissue.

3.3 Multiphoton FLIM Imaging of Na^+-Sensitive Fluorescent Dyes

1. The setup employed here for FLIM with multiphoton excitation is described in detail by Meyer and colleagues [7] . In principle, all components are tuned for collecting and recording a maximum of emitted photons. FLIM is always a trade-off between noise and speed/temporal resolution. The greater the number of photons that can be recorded, the lower the noise, and the higher the possible imaging frequency will be. The conditions required for proper FLIM measurements vary with the used dye. For Na^+ dyes, a minimum of approximately 5000 photons per region of interest (ROI) per frame, or 15–20 photons per pixel per frame, is typically needed for proper lifetime calculation.
2. Turn on the components for FLIM and multiphoton excitation (e.g., TCSPC devices, multiphoton laser).
3. Calibrate the laser brightness to approx. 2–5 mW under the objective, and adjust gain and offset of the detectors accordingly.
4. Make use of the appropriate emission filters and dichroic mirrors for the respective fluorescent dyes. ING2 and CoroNa Green emission can be collected between 530 and 580; SR101 emission can be collected >560 nm. It is important to note that when using SR101, it will be excited alongside the chosen Na^+ dye, and separation can only be achieved by splitting the emission (*see* **Note 7**).

5. Set the desired imaging parameters, including excitation wavelength (840 nm for ING2 and CoroNa Green), pixel dwell time (typically ~5 µs), and image size (typically 512×512). The imaging frequency can be adjusted by changing the image size or, to a limited extent, the pixel dwell time. Try to keep the sample illuminated as little as possible.
6. Measure the internal response function (IRF) using an appropriate solution (e.g., Erythrosin B in saturated K^+–iodide solution; also *see* **Note 8**).
7. Before proceeding, it is crucial to carefully evaluate the quality of your sample and staining. This is essential for conducting high-quality imaging experiments. Unhealthy tissue slices and/or low staining quality can result in decreased photon emission and increased noise during FLIM recordings.
8. Choose an area in the tissue slice which contains several clearly visible (SR-101-positive) astrocytes in the focal plane.
9. After starting the experiment, record the photon emission of chosen astrocytes in standard ACSF for a certain period. The duration of recording under "baseline conditions" may vary depending on the type of experiment, ranging from a few seconds (e.g., during high-speed line scans) to several minutes (e.g., during recording at a low imaging rate).
10. Manipulate the intracellular Na^+ concentration of astrocytes in your way of choice, and observe if this manipulation results in a detectable change in FL of the chosen Na^+ indicator.
11. After the astrocytic FL levels have returned to their pre-manipulation levels, continue to record the photon emission for a certain period of time.
12. If applicable, record a XYZ-stack of your sample to document the cellular morphology of the cells recorded from after the experiment. A deconvolution algorithm may be employed if necessary. FLIM or intensity images can be used for documentation purposes.

3.4 Manipulating Intracellular Na^+

1. Intracellular Na^+ can be manipulated in many ways (e.g., as illustrated in Fig. 4) and can be grouped into two main categories: global and local [4, 6, 16].
2. Global stimulation can, e.g., be achieved by application of inhibitory or excitatory transmitter agonists, such as GABA or glutamate via the bath perfusion system (*see* Fig. 4). Here, the application duration varies from a few seconds to several minutes, depending on the selected type of stimulation. Additionally, manipulations that mimic epileptiform discharges [17] or global energy deprivation in the brain [2, 7] result in global changes in intracellular Na^+ levels.

3. Local stimulation is usually accomplished by applying a substance focally (that is close to a chosen cell body or process) using a pressure application device and a glass pipette, as previously described for bolus injection. With this method, the cellular structure of interest is approached carefully with an agonist-filled pipette, and the chosen agonist is applied for a specific duration (e.g., 10–150 ms).

3.5 Calibration of Na^+-Sensitive Fluorescent Dyes In Situ

1. Wide field, intensity-based imaging in cultured astrocytes was introduced by Rose and Ransom [18]. Since then, it was adapted for astrocytes in tissue slice preparations also employing confocal as well as multiphoton imaging approaches [6]. FLIM of Na^+-sensitive indicator dyes in HEK cell culture and acute brain slice preparations of the hippocampus was first described by Meyer et al. [7, 9].
2. For full calibration, different calibration ACSFs should be prepared ranging from non-saturating (0 mM Na^+) to fully saturating (>100 mM Na^+) conditions.
3. The imaging frequency should be selected to ensure a high photon count for stable data fitting conditions also in non-saturated conditions.
4. For calibration, perfuse the slice with 0 mM Na^+ calibration ACSF without the ionophores for 30 min first, and start the measurement (*see* **Note 9**). Next, perfuse the slice with 0 mM Na^+ calibration ACSF containing ionophores until a stable plateau is reached, indicating complete equilibration of Na^+ between intra- and extracellular space.
5. Continue perfusion using different calibration ACSF containing different Na^+ concentrations, and measure the changes in FL accordingly. Before switching to the next solution, always wait until a stable plateau is reached in a certain condition (Fig. 3).

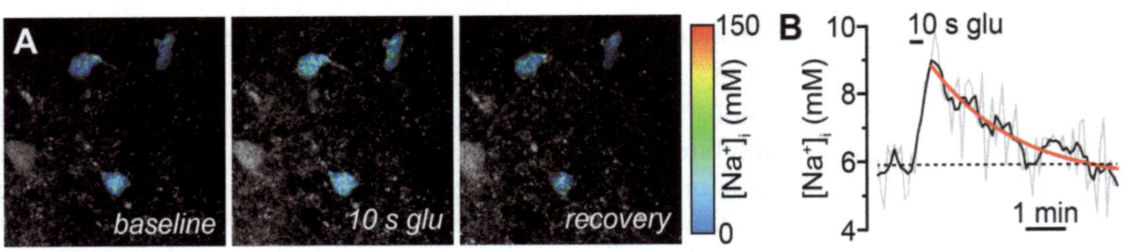

Fig. 3 Rapid FLIM measurement of intracellular Na^+ dynamics in hippocampal astrocytes in situ. (a): Fluorescence lifetime images of the hippocampal CA1 stratum radiatum showing astrocytes loaded with ING-2. Color code is shown on the right. Left image: Astrocyte Na^+ at baseline (control) conditions in standard ASCF. Center image: Change in astrocyte Na^+ after bath application of 1 mM glutamate for 10 s. Note the color change indicating a change in Na^+. Right image: Recovery of Na^+ at ~12 min after glutamate application. (b): Averaged glutamate-induced change in astrocyte Na^+ ($n = 33$ astrocytes). (Taken and modified from [7])

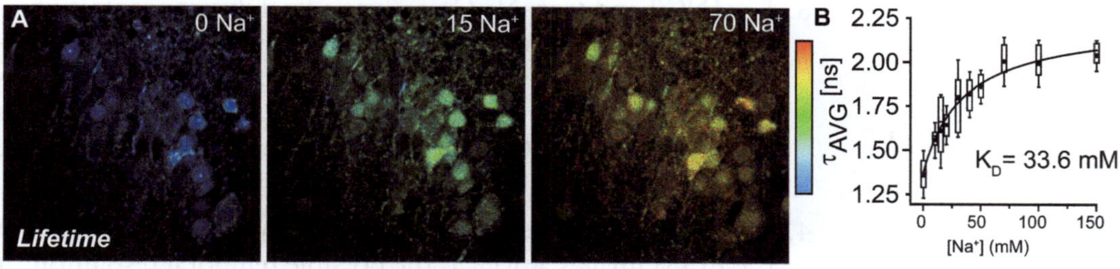

Fig. 4 Intracellular calibration of ING-2 in a hippocampal slice preparation. (a): Color-coded fluorescence lifetime images of ING-2-loaded cells taken at different intracellular Na+ concentrations with a temporal binning of 15–27 s. (b): Box plots illustrating ING-2ING-2 FL (τAVG) at different Na+ concentrations. Black line is a Michaelis–Menten fit (n = 21–135 cells). (Taken and modified from [7])

6. Plot the changes in FL against the Na^+ concentration, and fit the data with a Michaelis–Menten equation adjusted for an offset zero point (as described in [7]). Since the FL represents absolute values, the Na^+ concentration can be calculated directly from the calibration fit.

3.6 Data Analysis

1. The imaging sequence should be loaded by your analysis software of choice (e.g., Symphotime 64, from PicoQuant) along with the IRF, and photon counts should be checked on a pixel-by-pixel basis. In order to ensure stable and reliable fitting conditions, an appropriate binning should be selected to obtain a sufficient number of photons (*see* **Note 10**).
2. Select the cellular region of interests you would like to analyze.
3. Choose the optimal fitting parameters (e.g., number of exponents; mono- or bi-exponential for Na^+-sensitive dyes), and start calculation.
4. After fitting, select the desired parameters (e.g., FL, intensity, amplitudes), export the raw data to a suitable calculation software (e.g., Origin Pro, Mathworks), and use the calibration results to convert FL to Na^+ values (*see* **Note 11**).

4 Notes

1. In general, solutions for acute slice preparations and experiments should be freshly prepared on the day of their use. Standard and preparation ACSF have to be bubbled with carbogen for at least 30 min to ensure a correct pH.
2. Many chemical indicator dyes require Pluronic, which should be added fresh to the stock solution to allow sufficient passage across the cell membrane. For ING2, use a maximum of 20% Pluronic in DMSO. For CoroNa Green, loading with Pluronic is not recommended as it results in a loss of the dye over time.

3. For ING2 and CoroNa Green, use a final dye concentration of 150–250 μM. Ensure that you use freshly prepared HEPES-based ACSF, and avoid multiple freeze and thaw cycles as this can result in the formation of salt crystals. Remember to always filter the dye with a 0.22 μm net width before injection into the tissue.
4. The number of collected photons directly affects the quality of the FLIM images as well as the possible temporal resolution. It depends on the quantum efficacy of the dye used, the laser repetition rate, and the reading efficacy of the TCSPC electronics. A higher repetition rate leads to more excitation pulses and a greater chance of collecting a photon. The TCSPC electronics require time to register an incoming photon, known as dead time, during which no other incoming photon can be registered. Combining TCSPC devices with very low dead times (e.g., under 5 ns) and high repetition lasers (80 MHz and more) can significantly improve photon count rates and temporal resolution (*see* [7]).
5. The number of collected photons is a critical parameter. The system should be designed in a way that the absolute maximum of emitted photons can be collected. This requires TCSPC devices with very short dead times, detectors with high quantum efficacies, and a minimum number of mirrors and filters in the light path.
6. For imaging experiments such as FLIM, the quality of the slices is of utmost importance. Handle the slice during movement with greatest care, and restrain any movements and distortions of the tissue to the least amount necessary.
7. When using SR101 in combination with ion indicator dyes and FLIM, it is crucial to ensure correct separation of dye emissions for accurate FL calculation. The bleed-through of a secondary, Na^+-insensitive dye such as SR101, must be tested on separately. Typically, it should be below 1% to avoid affecting measured FL.
8. The internal response function (IRF) can be determined in different ways. One is probing a dye with a very short FL (e.g., quenched Erythrosine B), which enables detection of the IRF because your devices' response will be slower than the FL of this dye. Other possible ways to determine the IRF is measuring the reflection of gold particles or from crystalline samples (e.g., crystalized urea). Depending on the stability of your measurement devices, the IRF should be measured on a daily basis.
9. Ensure that the calibration process begins in saline devoid of Na^+. To achieve this, start by perfusing 0 mM Na^+ calibration ACSF without ionophores to wash our Na^+ from the tissue.

Only then, switch to 0 mM Na^+ calibration ACSF with ionophores to fully deplete intracellular Na^+.

10. As previously mentioned, photon numbers are crucial for achieving high-quality FLIM measurements. To ensure stable fitting conditions, it is common practice to bin multiple pixels into a single region of interest.
11. Choosing the correct parameters for fitting can be a tedious task. It is recommended to test fitting parameters in vitro and confirm them in the desired preparation.

Acknowledgements

The work presented here was supported by the Deutsche Forschungsgemeinschaft (DFG, German Research Foundation; Projekt# 461542557) and the Federal Ministry of Education and Research (BMBF), Germany (Project SynGluCross to CRR).

References

1. Sweadner KJ (1989) Isozymes of the Na+/K+-AtPase. Biochim et Biophys Acta (BBA)-Revi Biomembr 988(2):185–220
2. Gerkau NJ et al (2018) Reverse NCX attenuates cellular sodium loading in metabolically compromised cortex. Cereb Cortex 28(12): 4264–4280
3. Kirischuk S, Parpura V, Verkhratsky A (2012) Sodium dynamics: another key to astroglial excitability? Trends Neurosci 35(8):497–506
4. Felix L et al (2020) Sodium fluctuations in astroglia and their potential impact on astrocyte function. Front Physiol 11:871
5. Rose CR, Verkhratsky A (2024) Sodium homeostasis and signalling: the core and the hub of astrocyte function. Cell Calcium 117: 102817
6. Gerkau NJ, Kafitz KW, Rose CR (2019) Imaging of local and global sodium signals in astrocytes. Methods Mol Biol 1938:187–202
7. Meyer J et al (2022) Rapid fluorescence lifetime imaging reveals that TRPV4 channels promote dysregulation of neuronal Na(+) in ischemia. J Neurosci 42(4):552–566
8. Zheng K et al (2015) Time-resolved imaging reveals heterogeneous landscapes of nanomolar Ca^{2+} in neurons and astroglia. Neuron 88(2): 277–288
9. Meyer J et al (2019) Quantitative determination of cellular [Na^+] by fluorescence lifetime imaging with CoroNaGreen. J Gen Physiol 151(11):1319–1331
10. Untiet V et al (2017) Glutamate transporter-associated anion channels adjust intracellular chloride concentrations during glial maturation. Glia 65(2):388–400
11. Lakowicz JR (2013) Principles of fluorescence spectroscopy. Springer Science & Business Media
12. Naumann G, Lippmann K, Eilers J (2018) Photophysical properties of Na^+-indicator dyes suitable for quantitative two-photon fluorescence-lifetime measurements. J Microsc 272(2):136–144
13. Roder P, Hille C (2014) ANG-2 for quantitative Na^+ determination in living cells by time-resolved fluorescence microscopy. Photochem Photobiol Sci 13(12):1699–1710
14. Kafitz KW et al (2008) Developmental profile and properties of sulforhodamine 101-labeled glial cells in acute brain slices of rat hippocampus. J Neurosci Methods 169(1):84–92
15. Stosiek C et al (2003) In vivo two-photon calcium imaging of neuronal networks. Proc Natl Acad Sci USA 100(12):7319–7324
16. Rose CR, Karus C (2013) Two sides of the same coin: sodium homeostasis and signaling in astrocytes under physiological and pathophysiological conditions. Glia 61:1191
17. Karus C et al (2015) Astrocytes restrict discharge duration and neuronal sodium loads during recurrent network activity. Glia 63(6): 936–957
18. Rose CR, Ransom BR (1997) Gap junctions equalize intracellular Na+ concentration in astrocytes. Glia 20(4):299–307

Chapter 5

pH Sensor-Based Engulfment Assays to Measure Phagocytic Activity of Astrocytes

Gyuri Kim, Youkyeong Gloria Byun, and Won-Suk Chung

Abstract

Astrocytes play a crucial role in eliminating unnecessary synapses, neural debris, and pathogenic proteins such as amyloid β plaque through phagocytosis. Emerging evidence suggests that the phagocytic roles of astrocytes are critical for remodeling our neural circuits during development as well as maintaining brain homeostasis in adulthood. Moreover, abnormal astrocytic phagocytosis can have direct implications for the initiation and progression of various brain disorders. However, classical methods for identifying phagocytic events in the brains have utilized immunostaining-based and/or electron microscopy–based imaging analysis, which are very labor-intensive and often produce subjective results due to the inaccuracy in isolating the engulfment versus attachment to the plasma membrane of phagocytes. To overcome this problem, we have developed efficient in vitro and in vivo engulfment assays based on pH sensors, where only engulfed materials in astrocytes can be isolated by the changes in fluorescent signals, providing valuable tools for measuring phagocytic capacity and kinetics of astrocytes. In this chapter, we will explain and discuss two of our assays: (1) an in vitro engulfment assay using purified astrocytes and synaptosomes and (2) a mCherry-eGFP reporter system for monitoring in vivo synapse phagocytosis in the brains. Using these in vitro and in vivo assays, we have been able to accurately quantify the amount of synapses engulfed by multiple glial cells, including astrocytes, microglia, and oligodendrocyte precursor cells, revealing distinct roles and mechanisms of glial phagocytosis in eliminating specific synapse populations from defined neural circuits. These advanced assays provide accurate and efficient means of studying glial phagocytosis, contributing to a better understanding of neural circuit remodeling and the maintenance of brain homeostasis.

Key words Astrocytes, Microglia, Phagocytosis, Engulfment, pH sensors, Synaptosomes, Synapse elimination, Confocal microscopy

1 Introduction

Astrocytes, the major cell type in the mammalian central nervous system (CNS), express various phagocytic machineries, including MERTK and MEGF10 phagocytic receptors [1]. MERTK is one of the TYRO3/AXL/MER (TAM) receptor kinase family members and requires Protein S and GAS6 as bridging molecules to mediate binding between negatively charged phosphatidylserine and the TAM phagocytic receptor [2]. Similarly, MEGF10, an orthologue

Barbara Di Benedetto (ed.), *Astrocytes: Methods and Protocols*, Methods in Molecular Biology, vol. 2896,
https://doi.org/10.1007/978-1-0716-4366-2_5,

of Draper (*Drosophila*) and CED-1 (*Caenorhabditis elegans*), is well known for its phagocytic role in cellular debris clearance [3–6], although the identity of its ligands and bridging molecules remains unclear. In postnatal mouse brains, MERTK is expressed by both astrocytes and microglia, while MEGF10 is specifically expressed by astrocytes.

Previously, we have shown that astrocytes use these MERTK and MEGF10 phagocytic receptors to eliminate synapses in both developing and adult brains. In the developing retinogeniculate system, astrocytes in dorsal lateral geniculate (dLGN) use MERTK and MEGF10 to recognize and eliminate redundant synaptic connections from retinal ganglion cells (RGCs), mediating synapse remodeling and segregation between ipsilateral and contralateral eyes. In the adult hippocampus, astrocytes constantly eliminate unnecessary excitatory synapses through MEGF10, a process important for the homeostasis of hippocampal circuits and cognitive functions [7]. Recently, we also found that stress hormones can significantly enhance astrocytic phagocytosis by increasing *Mertk* transcription and that this stress-induced astrocytic MERTK and excessive developmental synapse pruning by astrocytes disrupt cortical neural circuitry, leading to an increased prevalence of mental disorder–related behaviors in mice that have experienced early life stress [8].

Importantly, throughout our studies, we have overcome technical challenges in accurately quantifying the phagocytic capacity of glial cells by developing pH sensor-based in vitro and in vivo engulfment assays. These assays allow the isolation of engulfed synaptic material in glial cells by measuring changes in the fluorescent signals of pH sensors. Incorporating pH sensors has been crucial in our studies due to two major problems in detecting synaptic phagocytosis: (1) the rapid protein degradation of engulfed synaptic material inside glial lysosomes and (2) the difficulty of accurately distinguishing engulfed synaptic materials from intact synapses in close proximity to glial fine processes. In this chapter, we will explain two of our assays, starting with an in vitro engulfment assay using purified astrocytes and synaptosomes, followed by a mCherry-eGFP reporter system for monitoring in vivo synapse phagocytosis in the brain.

To investigate in vitro astrocytic phagocytosis activity, we have purified astrocytes from wild-type rats using immunopanning methods [9, 10] and tested the phagocytic capacity of astrocytes (Fig. 1). Using pHrodo-conjugated synaptosomes in our assay has several advantages. pHrodo-red emits bright red fluorescence only when localized to cellular organelles with low pH, eliminating the need for quenching procedures to remove background signals. Additionally, this method allows real-time tracing of phagocytic events, enabling the monitoring of both engulfment and degradation kinetics by generating a graph showing changes in pHrodo-red

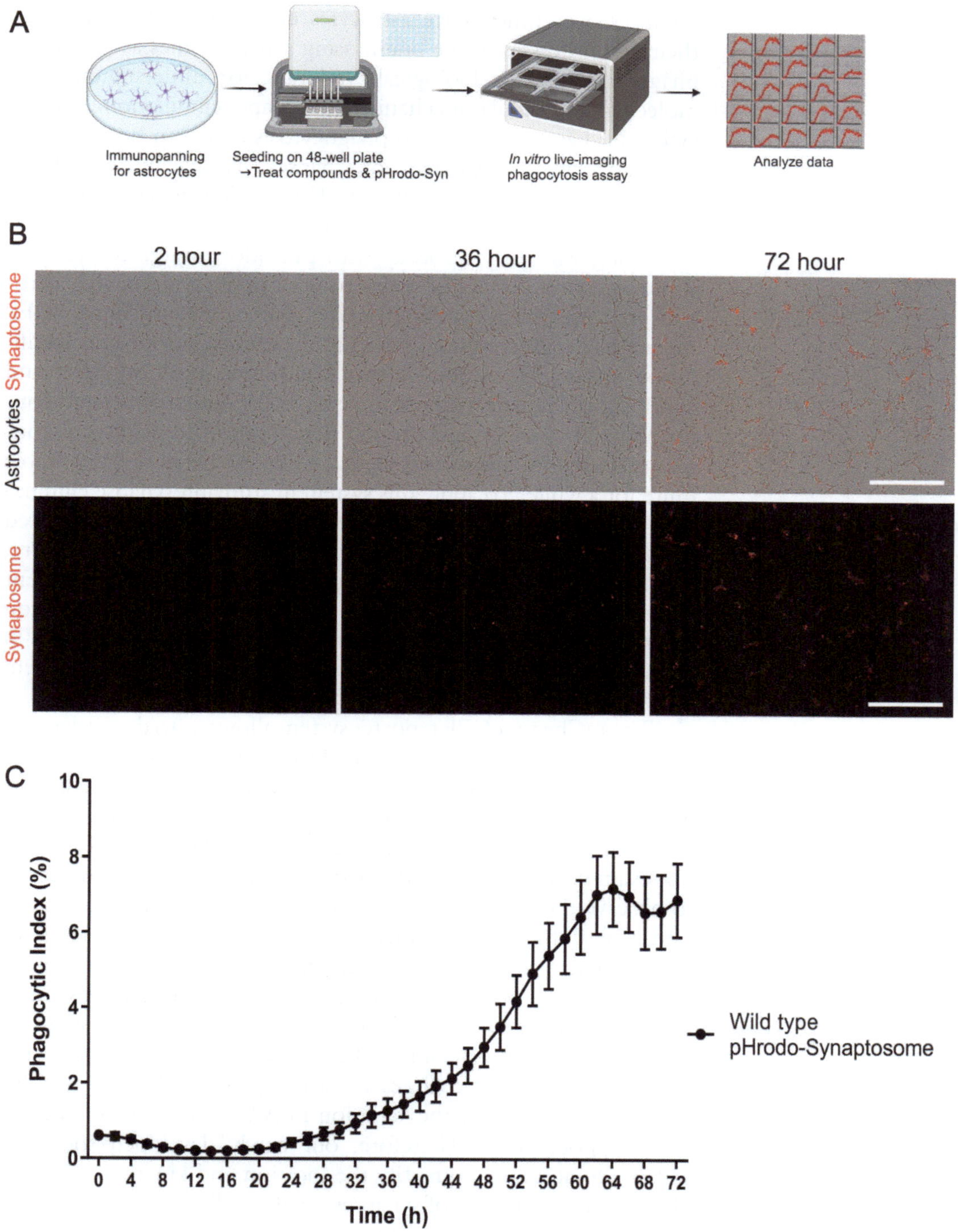

Fig. 1 In vitro astrocytic synaptosome phagocytosis assay. (**a**) Experimental scheme of the in vitro synaptosome phagocytosis assay (Created with BioRender). (**b**) Representative images of astrocytes (gray) with pHrodo-conjugated synaptosomes (red) after 2, 36, and 72 h of incubation. The upper panel shows bright field and red fluorescent images together, and the bottom panel shows red fluorescence images only from pHrodo-conjugated synaptosomes. (**c**) A representative graph showing the phagocytic activity of WT rat astrocytes. Data are shown as mean ± SEM. Scale bar = 400 μm

intensity over time. Since astrocytes eliminate synapses through their fine processes in the brain, using synaptosomes as bait in the phagocytic assay is biologically relevant for understanding the molecular and cellular mechanisms of synapse elimination by astrocytes [1]. Lastly, this in vitro phagocytosis assay can be applied to study astrocytic phagocytosis of synapses [8], myelin debris [11], and amyloid β oligomers/fibrils [12] following genetic manipulation of astrocytes or chemical treatment.

Along with in vitro assay with pHrodo-conjugated synaptosomes, we have developed a new tool for monitoring in vivo astrocytic synapse phagocytosis [7]. We utilized the mCherry-eGFP reporter system, commonly used in measuring autophagic acidification influx [13]. Under normal conditions, both mCherry and eGFP signals remain stable. However, eGFP undergoes rapid conformational changes and protein denaturation in acidic environments, losing its fluorescent signal, while mCherry remains stable quite for a while. To apply this system in studying synapse elimination by glial cells, we generated four different adeno-associated virus (AAV) harboring synapse phagocytosis reporters: ExPre, InhiPre, ExPost, and InhiPost (Fig. 2). Each reporter uses different promoters and synaptic proteins with mCherry-eGFP conjugation to specifically label excitatory or inhibitory pre-/post-synapses in the targeted neural circuits. This mCherry-eGFP reporter system can be further manipulated for various applications.

The mCherry-eGFP reporter system allows us to detect synapse phagocytosis using two different fluorescent proteins in live and fixed tissues. Since only the mCherry signal can be detected in glial lysosomes after being phagocytosed, we can specifically identify engulfed synapses inside brain phagocytes in defined neural circuits. For example, expressing ExPre AAV in the LGN neurons enables us to observe the engulfment of VGluT2+ excitatory pre-synapses by astrocytes in the mouse primary visual cortex (V1) layer 4 (Fig. 3). Using ImageJ and the Diana plugin, we can isolate the area of mCherry-alone puncta colocalized with astrocytes, representing the engulfed synaptic population (Fig. 4). It is also possible to calculate the phagocytic index after normalization to glial cell area to take account of the changes in reactive gliosis and/or total eGFP signal to take account the variation in AAV injection for labeling synaptic population. Therefore, our newly developed tools are suitable to accurately quantify the location and kinetics of in vivo phagocytic elimination of synapses by glial cells.

The in vivo synapse phagocytosis reporter system is versatile and adaptable to various conditions. By injecting our reporter AAVs into specific brain regions, we can investigate glia-mediated synapse elimination across diverse neural circuits, providing insights into region-specific synapse turnover and heterogeneity. In addition, this methodology enables us to investigate synapse phagocytosis across different temporal contexts, such as during development and aging or in various disease states.

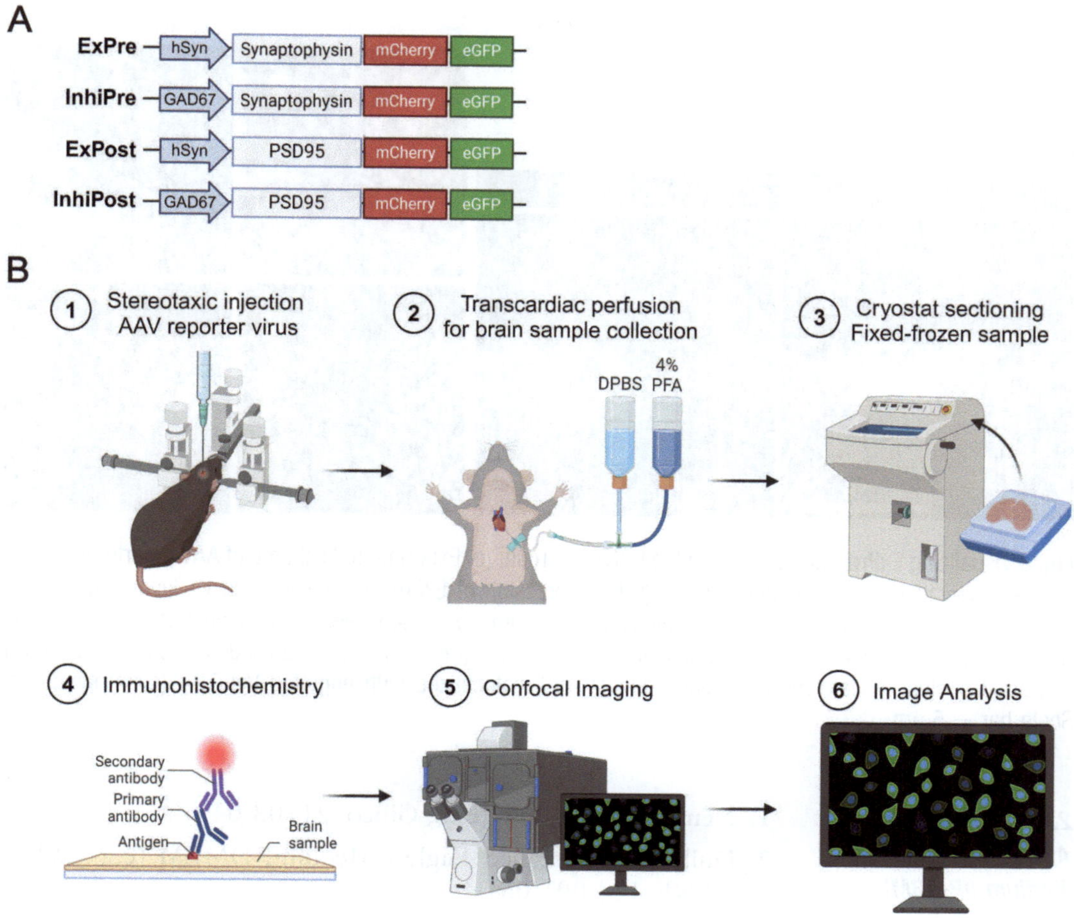

Fig. 2 In vivo synapse engulfment assay for astrocytic phagocytosis. (**a**) Schematic illustration of four glial synapse engulfment mCherry-eGFP reporters: ExPre, InhPre, ExPost, and InhPost. ExPre and InhPre can label pre-synapses in excitatory and inhibitory neurons. ExPost and InhPost can label post-synapses in excitatory and inhibitory neurons. (**b**) Schematic illustration of the in vivo synapse engulfment assay workflow using the mCherry-eGFP AAV reporter system. Figures created with BioRender

2 Materials

2.1 SATO (100×)

1. Apo-transferrin, human (e.g., Sigma, T1147).
2. Bovine serum albumin, BSA (e.g., Bovogen, BSA100).
3. Neurobasal medium (e.g., Gibco, 21103-049).
4. Progesterone stock: 2.5 mg progesterone (e.g., Sigma, P8783) in 100 μL of 100% ethanol.
5. Sodium selenite stock: 4 mg sodium selenite (e.g., Sigma, S5261) and 10 μL of 1 M NaOH in 10 mL of neurobasal medium.
6. Putrescine dihydrochloride (e.g., Sigma, P5780).
7. Sterile 150 mL vacuum filter unit with 0.22 μm polyether sulfone membrane (e.g., Corning, 431153).

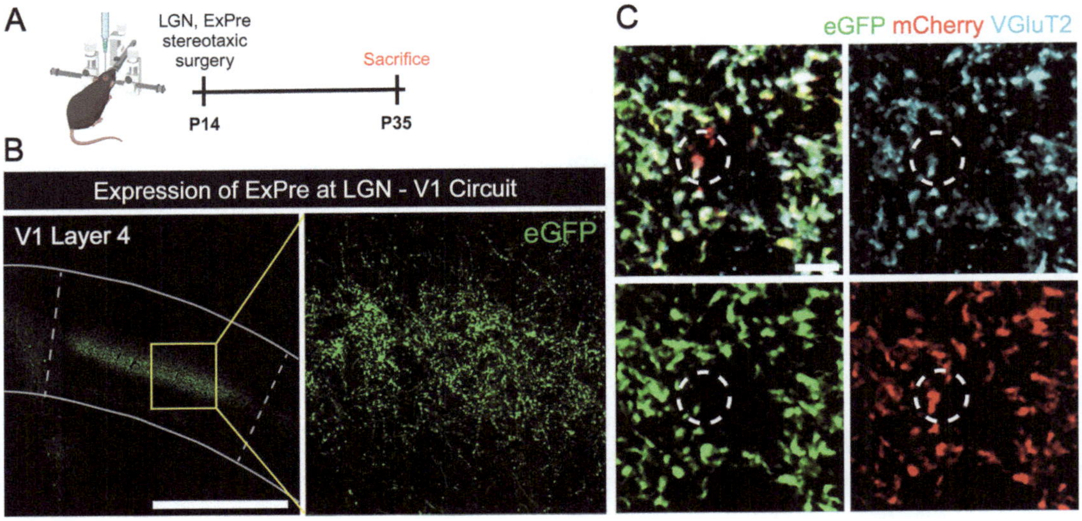

Fig. 3 Visualizing ExPre reporter virus in the LGN-V1 circuit. (**a**) Experimental scheme of AAV reporter injection. (**b**) Representative z-stack images showing the expression of ExPre reporter virus in V1 layer 4 (green for eGFP). Anatomical location of V1 with labeled ExPre (left) and expanded view (right). Scale bar = 1000 μm. (**c**) Representative confocal z-stack images of mCherry-eGFP (red and green) labeled synapses in V1 layer 4. Dotted circles indicate mCherry-alone puncta (red) colocalized with engulfed VgluT2+ (cyan) synapses. Scale bar = 5 μm

2.2 Immunopanning Astrocyte-Based Medium (IP-ABM)

1. Neurobasal medium (e.g., Gibco, 21103-049).
2. Dulbecco's Modified Eagle's Medium, DMEM (e.g., WELGENE, LM 001-05).
3. Penicillin/streptomycin, 10,000 U/mL (e.g., Gibco, 15140-122).
4. 100 mM sodium pyruvate (e.g., Gibco, 11360-070).
5. 200 mM L-glutamine (e.g., Gibco, 25030081).
6. N-acetyl-L-cysteine (NAC) stock (5 mg/mL, 1000×): 50 mg NAC (e.g., Sigma, A8199) in 10 mL of neurobasal medium. Make 130 μL aliquots and store at −20 °C.
7. Heparin-binding EGF-like growth factor (HBEGF), human, stock (10 μg/mL, 2000×): Make 20 μL aliquots and store at −80 °C. Mix HBEGF with IP-ABM before use (final concentration: 5 ng/mL).
8. Sterile 150 mL vacuum filter unit with 0.22 μm polyethersulfone membrane (e.g., Corning, 431153).

2.3 Stereotaxic Injection

1. Stereotaxic frame (KOPF or RWD).
2. Motorized Hamilton syringe pump (e.g., Harvard apparatus, 70-4507).

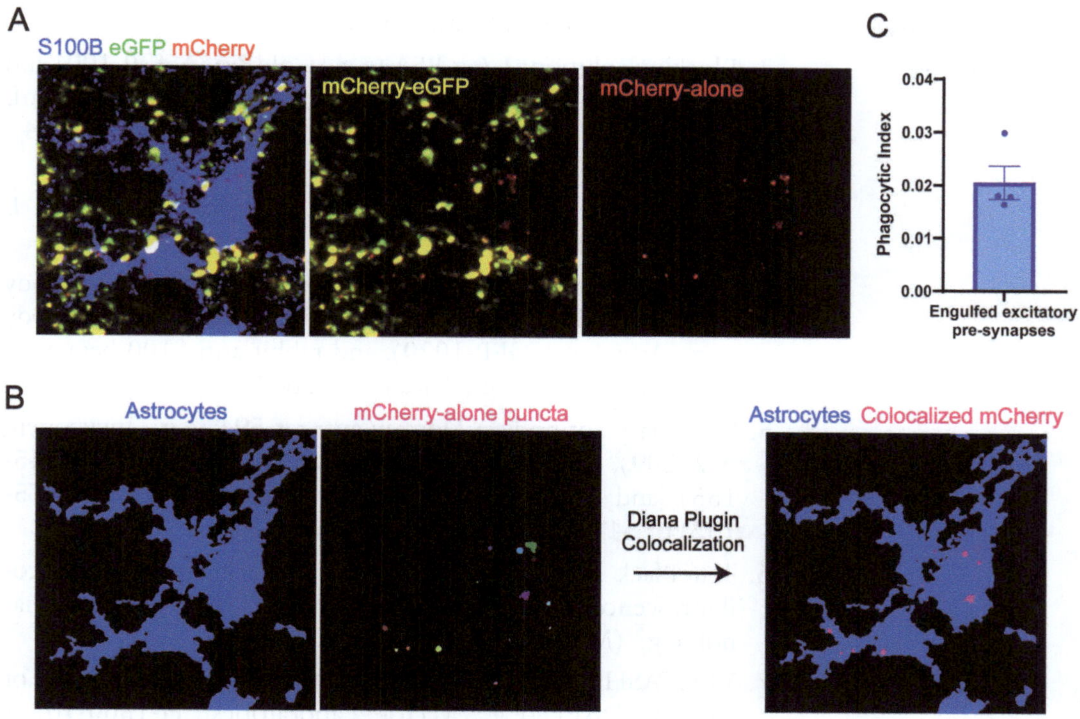

Fig. 4 Quantification of the mCherry-eGFP reporter system. (**a**) Representative single-plane confocal images of ExPre virus (yellow) stained with astrocytes (S100B, blue) (eGFP, green; mCherry, red). (**b**) Processed confocal images using ImageJ showing astrocytes (left), mCherry-alone puncta (middle), and colocalized mCherry-alone puncta in astrocytes (right). (**c**) A representative graph showing the phagocytic index of engulfed synapses by astrocytes from four wild-type mice ($n = 4$). Four images from each mouse were averaged. Data are shown as mean ± SEM

3. Anesthesia machine: Veterinary vaporizer (SurgiVet), isoflurane (Piramal).
4. Patch clamp glass capillary (e.g., World Precision Instrument, PG52165-4).
5. Polyethylene (PE)-20 tubing (e.g., Instech, BTPE-20).
6. Silk suture (e.g., Woori Medical, S714).

2.4 Transcardiac Perfusion and Cryostat Sectioning

1. Paraformaldehyde 4%/PBS (e.g., Biosolution, BP031a).
2. 1.2% Avertin: Make 100% Avertin with 0.8 g 2,2,2-tribromoethanol (e.g., Sigma, T48402) in 500 μL 2-methyl-2-butanol (Sigma, 240486). Dilute 500 μL 100% Avertin in 40 mL 1× DPBS and store at 4°C. Use within 2~3 weeks.
3. 30% sucrose solution: 30% sucrose (e.g., Sigma Aldrich, S1888) in 1× DPBS.
4. FSC 22 frozen section media (OCT compound) (e.g., Leica, 3801480).
5. Coating slide glass (MAS) (e.g., Matsunami, HMA-MAS-11).

2.5 Immunohistochemistry (IHC)

1. 24-well/48-well cell culture plate.
2. Blocking solution: 1.6 g BSA (e.g., GoldBio, A-420-100) and 1.2 mL 10% Triton ×-100/PBS (e.g., Sigma, ×100) in 40 mL DPBS. Filter with 0.22 μm syringe-driven filter and store at 4 °C.
3. Washing buffer: 1 mL Tween-20 (e.g., Sigma, P2287) in 1 L 1× PBS. Store at RT.
4. Primary antibodies: Rat anti-mCherry monoclonal antibody (16D7) (Invitrogen, M11217), chicken anti-GFP antibody (e.g., Aves Labs, GFP-1020), and rabbit anti-S100 beta antibody (EP1576Y) (e.g., Abcam, ab52642).
5. Secondary antibodies: Donkey anti-rat 594 (e.g., Invitrogen, A-21209), donkey anti-chicken 488 (e.g., Jackson, 703-545-155), and donkey anti-rabbit 405 (e.g., Jackson, 711-475-152) or 647 (Invitrogen, A-31573).
6. TrueBlack solution: Dilute 20× TrueBlack Lipofuscin Autofluorescence Quencher (Biotium, #23007) to 1× in 70% ethanol, e.g., (Merck, 100983).
7. VECTASHIELD hardset antifade mounting medium with or without DAPI (e.g., Vector Laboratories, H-1500-10 or H-1400-10).

2.6 Confocal Microscopy Imaging and Quantification

1. Confocal microscopy (e.g., Zeiss LSM 880).
2. ImageJ and Diana plugin (NIH).
3. GraphPad Prism 9 software.

3 Methods

All methods described here have been approved by the Korea Advanced Institute of Science and Technology Institutional Animal Care and Use Committee (IACUC), KA2023-053.

3.1 In Vitro Engulfment Essay: Long-Term Live-Imaging Assay

Reagent preparation for astrocyte culture and engulfment assay:

3.1.1 SATO (100×)

1. Weigh 800 mg transferrin, 800 mg bovine serum albumin (BSA), and 128 mg putrescine.
2. Bring a 200 mL glass beaker and add 80 mL of neurobasal medium into the beaker.
3. Transfer the transferrin into the beaker and dissolve it by stirring. Then, transfer BSA and putrescine one by one.

4. Add 20 μL of progesterone stock and 800 μL of sodium selenite stock into the beaker.
5. Mix well by stirring and filter through a 0.22 μm vacuum filter.
6. Make 1.3 mL aliquots and store them at −20 °C.

3.2 Immunopanning Astrocyte-Based Medium (IP-ABM)

1. 62.5 mL of neurobasal medium and 62.5 mL of DMEM into the 150 mL vacuum filter unit
2. Add 1.25 mL of penicillin/streptomycin, 1.25 mL of 100 mM sodium pyruvate, 1.25 mL of 200 mM L-glutamine, 1.25 mL of 100× SATO, and 125 μL of NAC (5 mg/mL, stock) into the filter top.
3. Filter the solution through the 0.22 μm vacuum filter.
4. Before using the medium, add heparin-binding epidermal growth factor, and incubate the tube in the 37 °C 5% CO_2 incubator.

3.3 In Vitro Engulfment Essay: Long-Term Live-Imaging Phagocytosis Assay with pHrodo-Conjugated Synaptosomes

1. Prepare a 48-well culture plate (or any type of plate/dish that prepared for live imaging) in which astrocytes are confluent (DIV 7~10 is ideal for stabilizing immunopanned astrocytes).
2. Prepare a new 1.5 mL tube.
3. Prepare cocktails by adding 300 μL of IP-ABM, 3~5 μL of pHrodo-conjugated synaptosomes, and additional factors into the tube (*see* **Note 1**).
4. Remove the maintaining media from each well.
5. Wash the wells with 500 μL of DPBS and repeat two times (*see* **Note 2**).
6. Remove the DPBS completely.
7. Mix the prepared cocktail well by gentle pipetting.
8. Add 300 μL of IP-ABM with pHrodo-conjugated synaptosomes and additional factors that can modulate astrocytic phagocytosis (*see* **Note 3**).
9. Incubate the plate in the 37 °C 5% CO_2 incubator for 40 min to allow pHrodo-conjugated synaptosomes to settle down and bind to phagocytic receptors on astrocytes.
10. Remove the media.
11. Wash each well three times with 500 μL of DPBS to remove unbound pHrodo-conjugated synaptosomes (*see* **Note 2**).
12. Add 500~600 μL of IP-ABM with additional factors to each well (*see* **Note 4**).
13. Place the plate into a live-imaging instrument (Incucyte).
14. Select the appropriate image channel, exposure time, objective, plate type, and vessel location (*see* **Note 5**).

15. Select wells and the number of images per well (*see* **Note 5**).
16. Set the time intervals (*see* **Note 6**).
17. Start the live imaging.
18. After finishing live imaging, start data analysis (*see* **Note 7**).

3.4 In Vivo Engulfment Assay: AAV mCherry-eGFP Reporter System

3.4.1 Plasmid Selection for mCherry-eGFP Reporter System

1. Choose one vector for the in vivo phagocytosis assay from among the four AAV mCherry-eGFP reporter vectors (*see* **Note 8**).
2. Generate AAV9 virus in HEK293T cells.
3. Aliquot the virus and store it at −80 °C.
4. Follow the experimental scheme outlined in Fig. 2.

3.4.2 Stereotaxic Injection: Detect Glial Phagocytosis Using the mCherry-eGFP System

1. Clean the surgical tools and reagents with 70% ethanol.
2. Anaesthetize mice with 1 mL isoflurane in a sealed plastic box.
3. Place the mouse to stereotaxic apparatus and fix its head with ear bars.
4. Shave the fur on the skull, and make a midline incision with small surgical scissors.
5. Clean the bregma and lambda areas.
6. Adjust the Z coordinates of bregma and lambda to equal.
7. Position the needle to target X and Y coordinates.
8. Aspirate the virus (AAV9-hSyn-Synaptophysin-mCherry-eGFP) until the desired volume is reached (*see* **Note 9**).
9. Slowly lower the microneedle to the desired Z coordinates.
10. Apply pressure to inject the virus.
11. Wait for 3 min and withdraw the microneedle slowly.
12. Clean the skull surface.
13. Suture the skin and keep the mouse on a heat blanket for recovery.
14. Monitor the mice carefully after surgery, and wait for viral expression for 3 weeks (*see* **Note 10**).

3.4.3 Transcardiac Perfusion for Collecting Virus-Injected Brain Samples

1. Prepare a stainless tray, 1× DPBS, and 4% paraformaldehyde (PFA) in a chemical fume hood.
2. Fix a three-way silicon hose in the perfusion pump, and connect a 24G needle to the hose.
3. Anesthetize mouse with avertin (0.2 g/mL) by intraperitoneal injection.
4. Wait for 3–5 min until the mouse is deeply anesthetized.

5. Fill each hose with DPBS and 4% PFA solution. Avoid bubbles in the silicon hose.
6. Place the mouse on the perfusion tray and pin its four paws.
7. Grip skin on the chest and make a small midline incision using scissors.
8. Carefully make lateral incisions to expose the diaphragm.
9. Make a full cut from small incision on the diaphragm and carefully expose the lung and the heart.
10. Insert the needle into the lateral ventricle, and make a small incision on the right atrium.
11. Infuse DPBS at a constant speed of 7 mL/min until the liver turns pale (*see* **Note 11**).
12. Switch to 4% PFA at the same speed for 3–4 min.
13. Remove the needle and decapitate the mouse head.
14. Make an incision along the midline and expose the skull.
15. Place fine scissors into the posterior skull and make a dual cut to the lateral side.
16. Carefully slide the fine scissors along the midline of the skull.
17. Remove the skull away from the brain.
18. Gently remove the brain and keep the brain in 4% PFA overnight at 4 °C.
19. Transfer the brain into a 30% sucrose solution until it sinks to the bottom.

3.4.4 Preparing Fixed-Frozen Brain Sections for Immunohistochemistry

1. Prepare dry ice, cryomold, and brain samples.
2. Prepare a 48-well culture plate and fill each well with 1 mL 1× DPBS.
3. Dissect brain tissue and place it in the cryomold.
4. Slowly fill OCT embedding compound, ensuring there are no bubbles and the brain tissue is completely embedded.
5. Freeze the tissue-embedded cryomolds on dry ice (*see* **Note 12**).
6. Place the cryomolds and equipment in a −20 °C cryostat chamber to equilibrate the temperature for 15–20 min.
7. Adjust the angle of brain sample stage so that the sample is not tilted.
8. Cut frozen sections at a thickness of 30–40 μm, and put four sections in each well. Ensure that collecting samples are in the region of interest (*see* **Note 13**).
9. Wrap brain section containing plates with aluminum foil to block the light (*see* **Note 14**).

3.4.5 Immunohistochemistry (IHC)

1. Prepare a 48-well plate, 4% BSA/0.3% Triton ×-100/DPBS blocking solution, and sectioned brain tissue (*see* **Note 15**).
2. Add 500 μL blocking solution and place two brain slices from one mouse.
3. Incubate the sections for 1 h in the blocking solution at RT.
4. Incubate the sections in the primary antibody diluted in blocking solution for 40 h at 4 °C shaker (*see* **Note 16**).
5. Wash the sections five times with PBS/0.1% Tween-20 for 20 min per wash.
6. Incubate the sections in the secondary antibody solution for 2 h at RT or O/N at 4°C shaker.
7. Wash the sections five times with PBS/0.1% Tween-20 for 20 min per wash.
8. Mount 4~6 brain sections on a coating slide glass.
9. Draw a barrier around the sections with a hydrophobic barrier pen.
10. Prepare a 20× TrueBlack stock solution. Dilute the TrueBlack 1:20 in 70% ethanol to make a 1× TrueBlack solution (*see* **Note 17**).
11. Cover each slide with 100~200 μL of 1× TrueBlack solution. Incubate for 1 min.
12. Wash the slides three times with 1× PBS.
13. Mount the slides in VECTASHIELD antifade mounting medium with or without DAPI.
14. Cover the slides with cover slips.
15. Keep the sections in a refrigerator until confocal imaging.

3.4.6 Confocal Microscopy Imaging

1. Turn on the confocal system and launch the Zen microscopy software.
2. Clean the 40× oil lens using 100% ethanol and lens cleaning tissue.
3. Position a slide on the sample stage, and adjust the sample holder to fit the slide without any displacement or misalignment.
4. Click the "Locate" tab on the Zen software to visualize the sample through eyepiece.
5. Lower the low-magnification object lens (10× or 20×) using the knob, and observe the sample for green or red fluorescence.
6. With low-magnification object lens, identify the region of interest within the brain. Adjust the X and Y positions using the joystick to target region.

7. Drop confocal oil onto the slide, and lower the 40× objective lens using the knob to the working distance (*see* **Note 18**).
8. Slowly adjust z-stack to acquire optimal focus on the sample.
9. Click the "Acquisition" tab, and set up the system for appropriate laser and detection parameters.
10. Click the "Smart Setup," and choose the dyes utilized in immunohistochemistry for each track. Apply the "Best signal" option in the proposals panel and complete the smart setup.
11. After the smart setup, manually confirm light paths in "channel mode." Confirm the excitation/emission wavelengths for each channel.
12. Click the "acquisition mode," and adjust each setting accordingly. We recommend the following settings: [frame size 1024 × 1024/scan speed 6–8/averaging number 4/dynamic range 8 bit] (*see* **Note 19**).
13. Click the "Channels tool," and adjust various parameters. Set the pinhole size to 1 AU and adjust laser power and gain (*see* **Note 20**).
14. Click "Live" mode for continuous scanning of the sample. Adjust X and Y positions and activate "range-indicator" button. This will display the live image in blue–red colors. (Pixels with a value of zero will be shown as blue, while saturated pixels will be shown as red.) The red color indicates out of range of detection, so adjust laser and gain intensity to minimize background and maximize fluorescent signal.
15. Select the "Z-stack" tool, and click the "Set First" button to set the upper position and "Set Last" button to set the bottom position of the z-stack (*see* **Note 21**).
16. Select all channels and click "Start Experiment" to acquire z-stack imaging.
17. Save the acquired images in the LSM 5 format.

3.5 Glial Engulfment Quantification

1. Install Image J and Diana (Distance analysis) plugin.
2. Open confocal images in Image J.
3. Pre-process images before further quantification (*see* **Note 22**).
4. Split red channel (mCherry), green channel (eGFP), and infrared channel (astrocytes) using Image J software.
5. Using the image calculator, subtract eGFP signals from mCherry signals. This will produce a new mCherry-alone puncta channel (*see* **Note 23**).
6. Segment objects from individual single-plane image stacks of mCherry-alone, eGFP, and astrocyte channels. Run the Diana

plugin [Plugins—Diana—Diana_Segment]. Set the adequate threshold, minimum object size, and maximum object size to segment objects within specific criteria.

7. Click [Go to analysis] or run Diana plugin [Plugins—Diana—Diana_Analyse]. Run colocalization assay for segmented mCherry-alone and astrocyte channels. This will produce colocalized objects (mCherry-alone signals in astrocytes).
8. Measure the area of mCherry-alone signal in astrocyte, eGFP signal, and astrocyte signal.
9. Normalize mCherry-alone in astrocyte with eGFP and astrocyte areas.
10. Phagocytic index represents the following equation: the area of mCherry-alone puncta in glia/(the area of eGFP puncta × the area of glial cells) (*see* **Note 24**).

4 Notes

1. If the washing steps take time, we recommend storing the tubes in the 37 °C incubator during the washing steps.
2. Astrocytes are vulnerable to air exposure. Therefore, wash the well as quickly as possible.
3. We use minimal volume of IP-ABM, so synaptosomes can rapidly settle down to the bottom of the plates and interact with astrocytes.
4. During the long-term live-imaging assay, the media can evaporate. Therefore, we recommend no less than 600 μL of IP-ABM with factors for each well of 24-well culture plates.
5. In the case of using in vitro phagocytosis assay with pHrodo-conjugated synaptosomes, bright field and red fluorescence image channels should be selected. We use a 10× objective to observe phagocytic events in a broad area. The settings for live imaging can be modified depending on the live-imaging instrument.
6. Set the time interval carefully to avoid cell damage caused by frequent laser/LED exposure in a short time.
7. Depending on the type of a live-imaging instrument, the data can be differently processed. In general, the fluorescent intensity or the area of fluorescent particles can be quantified at each time point. These numbers can be normalized by the number of astrocytes or the total area of astrocytes in the images.
8. There are four AAV9 synapse phagocytosis reporter vectors: ExPre, InhiPre, ExPost, and InhiPost (Fig. 2). For the hippocampal CA3-CA1 circuit, use ExPre for excitatory

pre-synapses, InhiPre for inhibitory pre-synapses, ExPost for excitatory post-synapses, and InhiPost for inhibitory post-synapses. However, in other brain regions, the hSyn promoter can label both excitatory and inhibitory neurons. Therefore, the CaMKII-promoter can be used for the specific expression in excitatory neurons and the mDlx or GAD67 promoter for inhibitory neurons.

9. Optimize the volume and titer of AAV virus for the desired brain regions. Too high AAV titer can lead to neuronal cell death.

10. Viral expression can be detected after 2 weeks of AAV virus injection. However, we suggest waiting until 3–4 weeks after injection due to the possibility of glial activation and to give sufficient time for mCherry-eGFP fusion protein maturation.

11. Adjust the flow rate of perfusion solutions based on the age of mice. For younger mice, a flow rate of 3–4 mL/min is recommended, or solutions can be manually infused using syringes.

12. If fixed-frozen samples are not immediately used, they can be stored at −80 °C for long-term storage.

13. We normally use floating samples for immunohistochemistry. Alternatively, samples can be directly attached slide to glass, when cutting samples with thinner thickness (10 μm thick).

14. Sectioned brain samples can be stored at 4 °C. To prevent bacterial contamination during long-term storage, we recommend adding sodium azide to DPBS as a preservative.

15. During the permeabilization step in immunohistochemistry, a concentration of 0.3% Triton ×-100 is recommended. However, depending on specific antibodies and staining conditions, Triton ×-100 concentration may be adjusted up to 0.5–1%.

16. Staining for eGFP is optional since its signal is bright enough to detect without staining. If desired, eGFP staining can be performed by diluting the primary antibody to 1:2000. Additionally, other cellular organelles such as lysosomes can be visualized using antibodies against Cathepsin D or LAMP2.

17. TrueBlack is a crucial reagent for quenching lipofuscin autofluorescence in mouse brain tissues. Lipofuscin accumulates in various cell types with age, and its autofluorescence can be detected in all channels of confocal microscopy. Therefore, during synapse imaging, lipofuscin quenching using TrueBlack reagent is essential. However, this step can be omitted for young mice under 1 month old.

18. When using a 40× oil lens, ensure that an adequate amount of confocal oil is applied.

19. To achieve clear imaging of engulfed mCherry puncta in astrocytes, consider using 2× zoom imaging. It is important to note that while cell area normalization is performed in the subsequent quantification, there should be a similar number of cells in each imaging field to reduce variation.
20. Too high gain may increase background signal during synapse imaging. Optimize laser power and laser gain settings accordingly.
21. It is recommended to capture ten stacks per image with an interval of 0.5 μm. Typically, four images are acquired per mouse. The averaged phagocytic index of four images is plotted as one data point.
22. If images are dark or dim, you may need to adjust intensity or remove background signals. Use ImageJ functions for image processing, such as intensity regulation, subtract background, and despeckle.
23. During quantification, any mCherry puncta overlaid with minimum eGFP signal should be excluded. Engulfed synapses are detected by the mCherry-alone signal, as the eGFP signal is not stable in acidic conditions, such as in lysosomes.
24. For the phagocytic index normalization, we normalize the mCherry-alone puncta area with the area of astrocytes to compensate for the density difference of astrocytes and the area of eGFP puncta to compensate for variance of AAV injection.

Acknowledgments

The authors thank Jinah Kim for images of synaptosomes and AAV ExPre virus expression. The authors thank all members of Chung's laboratory for helpful discussion. This work was supported by the Ministry of Science and ICT through the National Research Foundation in Republic of Korea (2020M3E5D9079912, 2021R1A2C3005704 and 2022M3E5E8081188), and Institute for Basic Science (IBS-R025-A1) (to W.-S.C.).

References

1. Chung W-S et al (2013) Astrocytes mediate synapse elimination through MEGF10 and MERTK pathways. Nature 504(7480): 394–400. https://doi.org/10.1038/nature12776
2. Lemke G, Rothlin CV (2008) Immunobiology of the TAM receptors. Nat Rev Immunol 8(5): 327–336. https://doi.org/10.1038/nri2303
3. Purice MD, Speese SD, Logan MA (2016) Delayed glial clearance of degenerating axons in aged Drosophila is due to reduced PI3K/Draper activity. Nat Commun 7:12871. https://doi.org/10.1038/ncomms12871
4. Ray A, Speese SD, Logan MA (2017) Glial draper rescues abeta toxicity in a drosophila model of alzheimer's disease. J Neurosci 37(49):11881–11893. https://doi.org/10.1523/JNEUROSCI.0862-17.2017
5. Tasdemir-Yilmaz OE, Freeman MR (2014) Astrocytes engage unique molecular programs

to engulf pruned neuronal debris from distinct subsets of neurons. Genes Dev 28(1):20–33. https://doi.org/10.1101/gad.229518.113

6. Wu HH et al (2009) Glial precursors clear sensory neuron corpses during development via Jedi-1, an engulfment receptor. Nat Neurosci 12(12):1534–1541. https://doi.org/10.1038/nn.2446
7. Lee JH et al (2021) Astrocytes phagocytose adult hippocampal synapses for circuit homeostasis. Nature 590(7847):612–617. https://doi.org/10.1038/s41586-020-03060-3
8. Byun YG et al (2023) Stress induces behavioral abnormalities by increasing expression of phagocytic receptor MERTK in astrocytes to promote synapse phagocytosis. Immunity 56(9):2105–2120 e2113. https://doi.org/10.1016/j.immuni.2023.07.005
9. Foo LC et al (2011) Development of a method for the purification and culture of rodent astrocytes. Neuron 71(5):799–811. https://doi.org/10.1016/j.neuron.2011.07.022
10. Byun YG, Chung WS (2018) A novel in vitro live-imaging assay of astrocyte-mediated phagocytosis using pH indicator-conjugated synaptosomes. J Vis Exp 132. https://doi.org/10.3791/56647
11. Brosius Lutz A et al (2017) Schwann cells use TAM receptor-mediated phagocytosis in addition to autophagy to clear myelin in a mouse model of nerve injury. Proc Natl Acad Sci USA 114(38):E8072–E8080. https://doi.org/10.1073/pnas.1710566114
12. Jung H et al (2022) Anti-inflammatory clearance of amyloid-beta by a chimeric Gas6 fusion protein. Nat Med 28(9):1802–1812. https://doi.org/10.1038/s41591-022-01926-9
13. Leeman DS et al (2018) Lysosome activation clears aggregates and enhances quiescent neural stem cell activation during aging. Science 359(6381):1277–1283. https://doi.org/10.1126/science.aag3048

Chapter 6

An Astrocyte Transplantation Method to Investigate Astrocyte Development and Diversity in the Central Nervous System (CNS)

Sabrina Chierzi and Keith K. Murai

Abstract

Astrocytes have rich structural, molecular, and physiological properties and show remarkable diversity in form and function that supports circuit-specific physiology within the central nervous system (CNS). However, the developmental mechanisms and cellular interactions that help to locally and globally diversify astrocytes with the CNS remain to be better understood. Understanding these processes will help to understand how astrocytes establish specialized microenvironments in the healthy brain and how they may differentially respond to CNS injury and disease. Here, we describe a cell transplantation method that enables the study of cell-autonomous and non-cell-autonomous determinants of astrocyte diversification by tracking the developmental trajectory of astrocytes transferred to the same or different regions of the mouse brain. With this method, it is possible to evaluate how astrocytes mature in "familiar" or "foreign" environmental contexts, for example, by transplanting cortical astrocytes into the cortex (a familiar context) or cerebellum (a foreign context). Also, with this method, the effect of developmental state on the ability of transplanted astrocytes to integrate into the brain environment can be studied, for example, by transplanting embryonically or postnatally derived cortical astrocytes into the cerebellum and monitoring their maturation within the recipient brain. We find that this transplantation method provides a flexible and robust approach to investigate how intrinsic cell properties and extrinsic cues from the extracellular environment shape astrocyte diversity.

Key words Astrocytes, Transplantation, Development, Heterogeneity, Cortex, Cerebellum

1 Introduction

Astrocytes are among the most abundant cells in the brain and play key roles in the central nervous system (CNS) in processes including ion homeostasis, synaptic development and function, and brain energetics. Numerous studies have demonstrated that astrocytes have complex and diverse molecular, structural, and functional properties which vary depending on their developmental origin, cellular positioning, and exposure to environmental signals [1–4]. This molecular and structural complexity and diversification of

Barbara Di Benedetto (ed.), *Astrocytes: Methods and Protocols*, Methods in Molecular Biology, vol. 2896,
https://doi.org/10.1007/978-1-0716-4366-2_6,

astrocytes enables them to create specialized microenvironments to support brain circuit-specific functions [5]. Despite our recognition of astrocytic diversity, the genesis of such diversity in the CNS requires further investigation. In particular, the contribution of cell-autonomous (intrinsic) and non-cell-autonomous (extrinsic) mechanisms driving astrocyte diversification remain to be better understood. To investigate these processes, an astrocyte transplantation approach can be helpful in examining how intrinsic and extrinsic mechanisms impact astrocyte integration, maturation, and stabilization within the mouse brain [6].

Here, we provide a straightforward, step-by-step astrocyte transplantation method that can be used to study astrocyte diversity. We specifically describe (1) how to prepare postnatal/embryonic astrocyte cultures from donor mice containing astrocyte-specific expression of the fluorescent protein Tomato (TOM) and (2) how to transplant TOM-positive astrocytes into the cortex or cerebellum of early postnatal day recipient mice to assess how transplanted astrocytes integrate and mature within the brain.

For this method, we prepare astrocytic cultures by mechanically triturating mouse cortices or cerebella into small fragments (explants) and collecting immature astrocytes after they have migrated away from explants and formed cell monolayers [7]. In our hands, this method is preferable to other methods for astrocyte isolation based on chemical dissociation, as cortical and cerebellar explants provide a high yield of healthy astrocytes even when prepared from pups as old as P8. We find that the yield of astrocytes from chemically dissociated cultures drops when donor mice are older than P2. With explant-derived cultures, it is possible to achieve cortical astrocytes from both embryonic day 17-18 (E17-18) and P4-8 mice, for example, to test the impact of developmental stage on the ability of transplanted astrocytes to integrate and mature in the postnatal cerebellum.

Overall, this transplantation approach has proven to be effective in assessing how intrinsic properties of astrocytes and cues from the host brain environment shape the development of transplanted astrocytes [6]. We observed that the properties of postnatally derived cortical and cerebellar astrocytes are largely guided by the tissue of origin (cortical or cerebellar) when transplanted into matched brain regions (Figs. 1 and 2a, b). Interestingly, cortical astrocytes transplanted to the cerebellum appear to have the capacity to develop cerebellar-like morphological features when derived from embryonic stage donor mice (Fig. 2c, d).

In addition to providing a useful method to study the cellular mechanisms contributing to astrocytic diversity, this transplantation method may also provide a suitable experimental framework to evaluate the role of astrocytic dysfunction in the context of neurodevelopmental disorders (NDDs), such as autism spectrum disorders [8], Down syndrome (DS) [9], Fragile X [10], Rett syndrome

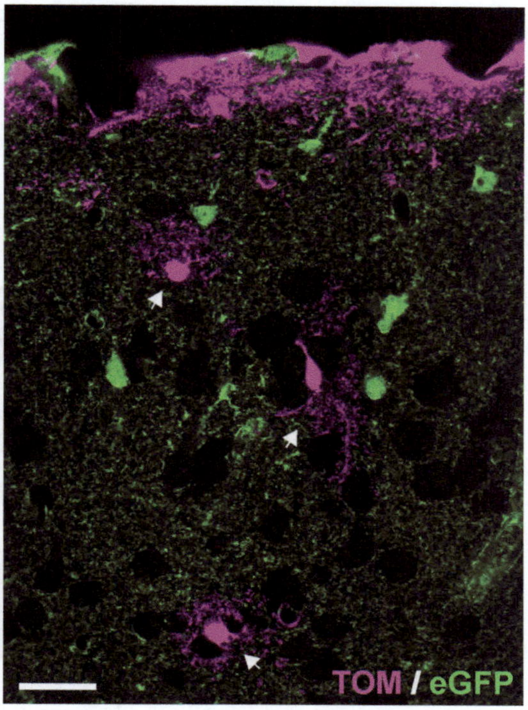

Fig. 1 Transplanted immature cortical astrocytes integrate into the developing cortex. Postnatal cortical astrocytes (magenta, arrows) transplanted into the developing cortex of Aldh1l1-eGFP mice display morphological features characteristic of mature protoplasmic astrocytes and form nonoverlapping territories with non-transplanted astrocytes (green) at 35–47 days post transplantation. See Chierzi et al. [6] for full details. *TOM*, Tomato. Scale bar is 20 μm

[11], and schizophrenia [12]. Astrocytes derived from NDD mouse models and transplanted into wild-type recipient mice could be a useful model to characterize the effect of NDD-affected astrocytes on developing brain circuits. Conversely, delivery of wild-type astrocytes into NDD mouse models could be used to counteract alterations to developing brain circuits.

2 Materials

2.1 Animals

All animal procedures used to develop this methodology were carried out according to policies set forth by the Canadian Council on Animal Care and the Montreal General Hospital Facility Animal Care Committee. Mice were maintained in standard housing conditions and with continuous access to food and water.

1. To aid in the identification of astrocytes following their transplantation to the brain, it is highly suggested to create astrocytic cultures from donor mice with a fluorescent reporter

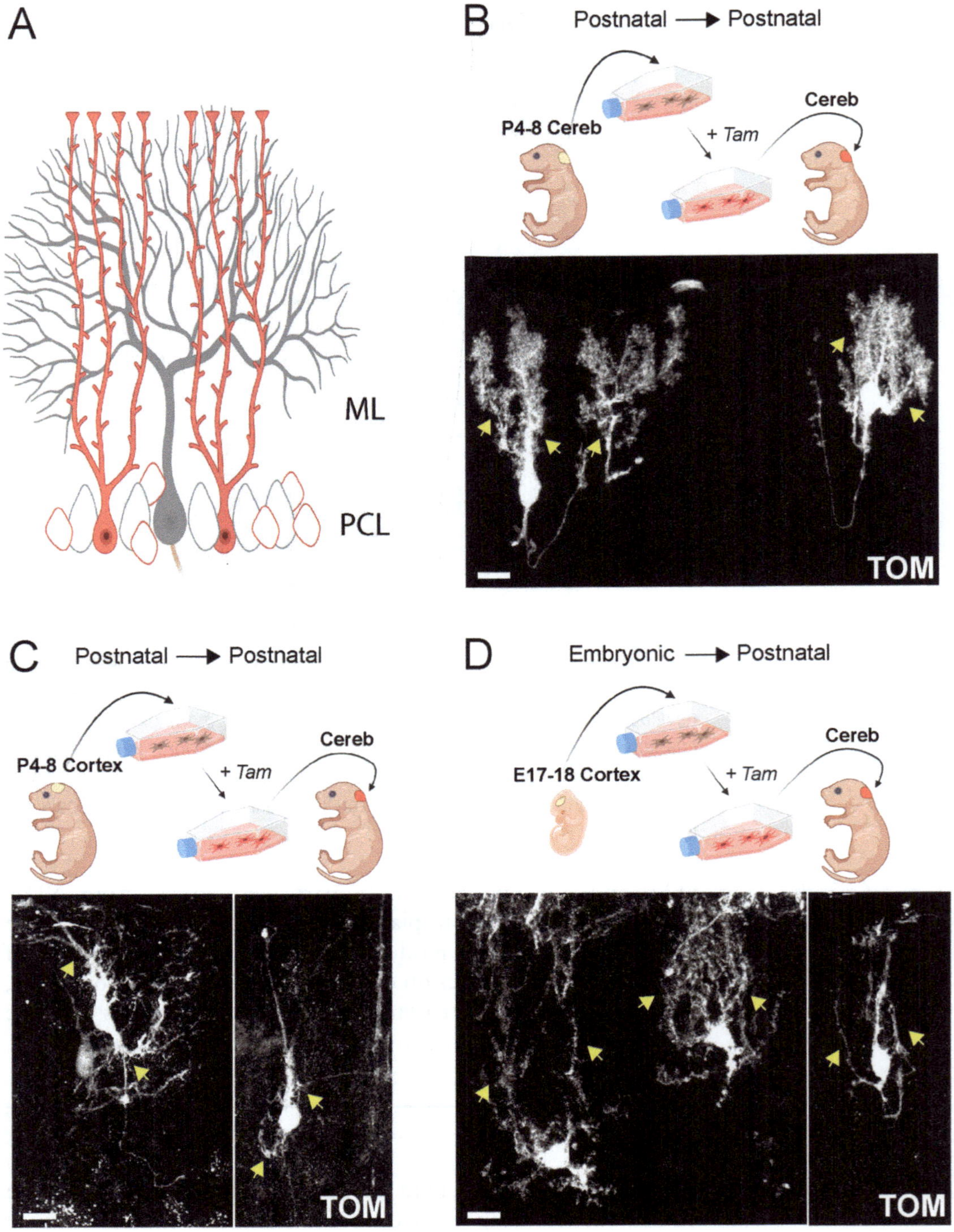

Fig. 2 Morphological phenotypes of cerebellar and cortical astrocytes transplanted in the developing cerebellum. (**a**) Bergmann glia (BG, red) are specialized cerebellar astrocytes which develop in tight contact with Purkinje cells (grey) and extend multiple polarized branches from the Purkinje cell layer (*PCL*) to the pial surface. (**b**) Postnatal cerebellar astrocytes prepared from P4–8 mice and transplanted in the cerebellar molecular layer display BG-like morphologies (arrows) 21 days following transplantation. (**c**) Postnatal cortical astrocytes transplanted to the cerebellar molecular layer develop altered branching patterns containing

(Fig. 3a). The example mouse line used here expresses tamoxifen-sensitive CreERT2 from the astrocyte-specific aldehyde dehydrogenase family 1, member L1 (Aldh1l1) promoter (Aldh1l1-CreERT2; Jackson Labs) [13] and Cre-inducible tdTomato (TOM) in the Rosa26 locus [14] (Ai9; Jackson Labs). In vitro tamoxifen induction allows permanent TOM labeling of astrocytes prior to their transplantation. Embryonic and postnatal astrocyte cultures are obtained from E17–18 embryos and early postnatal pups (aged postnatal day 4–8), respectively (*see* **Note 1**).

2. C57BL/6 mice or Aldh1l1-eGFP-Stop-DTA BAC (Aldh1l1-eGFP) [15] mice aged P3–P7 are used as transplant recipients (Jackson Labs). Aldh1l1-eGFP mice express eGFP in nearly all astrocytes in the CNS. Transplantation of TOM-labeled astrocytes into Aldh1l1-eGFP mice can be used to show how transplanted astrocytes interact with non-transplanted astrocytes and form nonoverlapping territories in the recipient brain and prove that TOM-labeled cells or debris are not internalized by neighboring eGFP-labeled astrocytes in the recipient brain (Fig. 1) [6].

2.2 Preparation of Astrocytic Cultures

All plasticware (i.e., Petri dishes, conical tubes, microcentrifuge tubes, pipettor tips, serological pipettes, and tissue culture flasks) must be sterile prior to use.

1. Dissection microscope (Zeiss Stemi 2000-C or any equivalent stereomicroscope with zoom range 0.65× – 5×).
2. Horizontal and vertical laminar flow hoods (*see* **Note 2**).
3. Standard cell culture CO_2 incubator.
4. Standard cell culture centrifuge.
5. Tissue culture microscope (Olympus CK2 or any equivalent inverted microscope equipped with 5×, 10×, and 20× lenses).
6. 70% ethanol (for disinfection of tools and surfaces).
7. Dissection tools: ToughCut surgical scissors (Fine Science Tools (FST) 14054-13), straight-tip forceps (FST 11210-20 or equivalent), fine-tip forceps (FST 11255-20 or equivalent), Bonn scissors (Moria 8143), and spatula (FST, 10091-12).
8. Surgical blades (#11) and blade holder.

Fig. 2 (continued) processes with unusual morphology (arrows) 41 days after transplantation. (**d**) Embryonic cortical astrocytes (i.e., prepared by E17–18 mice) transferred to the cerebellar molecular layer develop multiple polarized branches and small elaborated processes (arrows) and appear more similar to BGs (see BG diagram in **a**) 70–75 days after transplantation. *TOM* , Tomato. *Tam*, 4-hydroxytamoxifen. All diagrams are created with BioRender.com. Scale bars are 10 μm in (**b–d**)

9. Uncoated 6 cm and 10 cm dishes.
10. Standard 1.5 mL microcentrifuge tubes.
11. 1000 mL pipettor (Gilson).
12. 1000 mL pipettor tips.
13. Dissection medium: Hank's Buffered Salt Solution (HBSS) without Ca^{2+}/Mg^{2+}.
14. HBSS with Ca^{2+}/Mg^{2+}.
15. Phosphate buffered saline (PBS).
16. 40 μm cell strainers (Fisherbrand).
17. Standard 50 mL conical tubes.
18. 75 cm^2 tissue culture (T-75) flasks (Sarstedt).
19. Serological pipettes and pipette controller to transfer liquids from and to T-75 flasks.
20. Growth medium: F12-DMEM (Gibco) supplemented with 10% fetal bovine serum (Gibco), 5% horse serum (Gibco), 1% penicillin/streptomycin (Gibco), 20 ng/mL EGF (GenScript, Invitrogen), and 20 ng/mL FGF2 (Invitrogen). Store at 4°C for up to 1 month.
21. Trypsin/EDTA (0.25%) (Gibco).
22. 4-hydroxytamoxifen ready-made solution 5 mg/mL (Tam, Sigma) (*see* **Note 3**).

2.3 Cell Transplantation

1. Micropipette puller (Sutter Instrument P-97).
2. Biosafety cabinet equipped with ultraviolet (UV) lamp.
3. Stereo microscope (Leica M50, with 6.3×, 10×, 16×, 25×, 40× zoom steps) or equivalent surgery microscope.
4. Mouse stereotaxic frame (Stoelting 51730D).
5. Syringe pump (Pump 11 Elite Nanomite, Harvard Apparatus).
6. Heating pad.
7. Capillary glass tubing (Warner Instruments, Model No G150-4, ID = 0.86 mm, OD = 1.5 mm).
8. Thermoplastic rubber (TPR) tubing (Tygon, ACF00001, I.D. =1/32 IN).
9. Polyethylene (PET) tubing (Intramedic, 427406, ID = 0.38 mm, OD = 1.09 mm).
10. 5 μL Hamilton syringe (Model No 88000).
11. 70% ethanol.
12. 1 mL Luer Lock syringes.
13. G27 syringe needles.
14. Ice-cold ice pack.

15. Plastic wrap to cover the ice pack and heating pad and make them easier to disinfect.
16. Sterile mineral oil.
17. Sterile 24-well plates.
18. 20 μL pipettor (Gilson).
19. 20 μL pipettor tips.
20. Antiseptic 10% povidone iodine wipes.
21. Antiseptic 70% isopropyl alcohol wipes.

3 Methods

Astrocyte cultures are prepared from mouse pups or embryos using an adapted method [7]. Dissection is performed using a dissection microscope within a horizontal laminar flow hood. Surfaces and dissection tools must be disinfected with 70% ethanol prior to dissection. A timeline for the culture preparation procedure is shown in Fig. 3b.

3.1 Preparation of Astrocytic Cultures from Postnatal Mouse Cortex or Cerebellum

1. At least 20 min before starting the dissection, add 10 mL of growth medium to two T-75 tissue culture flasks, and place them in a cell culture incubator set to 37 °C, 5% CO_2.
2. At the start of dissection, fill two 6 cm dishes with chilled HBSS without Ca^{2+}/Mg^{2+}, and set aside.
3. Decapitate pups and place heads in a new 6 cm dish. Wipe skin on each head with 70% ethanol.
4. For each head, make a sagittal cut to open the skin and thin skull with scissors. Carefully dissect the brain, and transfer it to the 6 cm dish containing HBSS without Ca^{2+}/Mg^{2+} using a small spatula.
5. Carefully separate the cerebellum from the brain, transfer into one of the Petri dish containing dissection buffer, and set aside.
6. Split the remaining brain tissue in half along the midline with a surgical blade. For each half brain, separate the cortex from the diencephalon and brainstem, and detach the hippocampus from the cortex.
7. Using thin forceps, carefully peel off the meninges from the cortex/cerebellum.
8. Transfer dissected cortices and cerebella into separate 1.5 mL microcentrifuge tubes containing 1 mL dissection buffer, and allow tissue to settle at the bottom of the tube (*see* **Note 4**).
9. Under a vertical flow hood, replace supernatant in each 1.5 mL tube (HBSS without Ca^{2+}/Mg^{2+}) with 700–800 μL HBSS with Ca^{2+}/Mg^{2+}.

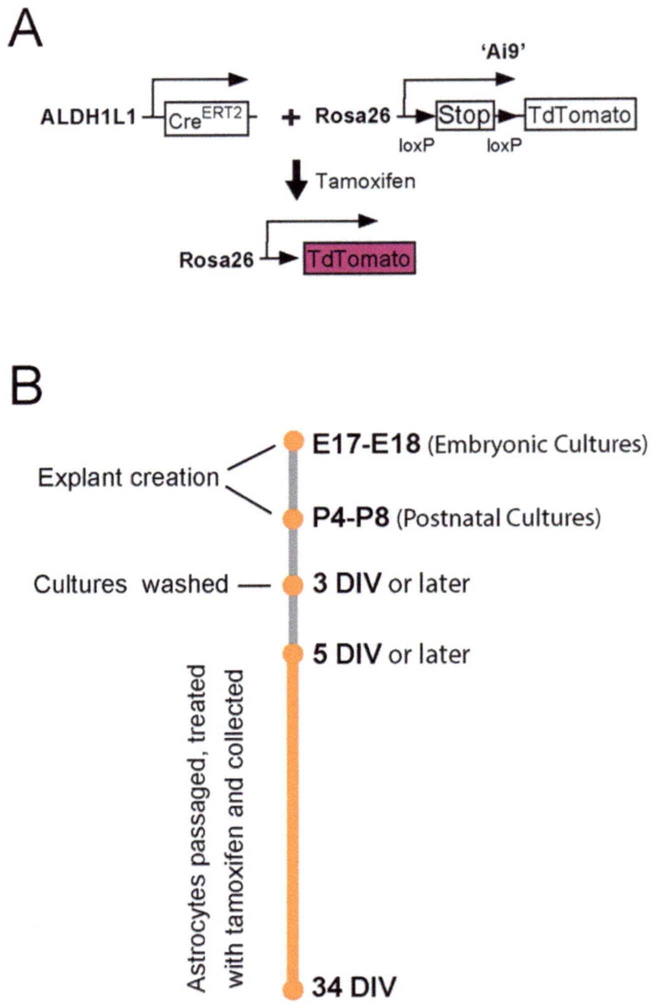

Fig. 3 Experimental design for preparing cultures of Tomato-expressing astrocytes from late embryonic and early postnatal mice. (**a**) Schematic showing the genetic background of the mice used as transplant donors. *Tamoxifen*, 4-hydroxytamoxifen. (**b**) Timeline of culture preparation

10. Using a 1000 μL tip and pipettor, gently triturate tissue into 1–2 mm-sized fragments by pipetting up and down 10–12 times. Perform the same procedure for cortical and cerebellar tissues.
11. Allow tissue fragments to settle to the bottom of the tube (*see* **Note 4**). After removing the supernatant, add 1 mL pre-warmed growth medium to the tube, and resuspend the tissue. With a 2 mL serological pipette, transfer the medium containing cortical and cerebellar fragments to separate T-75 flasks containing growth medium equilibrated to 37 °C, 5% CO_2. Swirl the flasks to disperse tissue fragments more uniformly.

12. Leave the flasks undisturbed in the incubator for at least 3 days to allow tissue fragments to attach to bottom of the flask.
13. Once most tissue fragments, now referred to as explants, have adhered to the flask bottom, aspirate the growth medium. Rinse the cultures three times with 10 mL room temperature PBS. Shake the flasks vigorously at every wash to induce detachment of loosely attached cells (i.e., microglia, oligodendrocytes, and neurons). Finally, replace the PBS with 10 mL of fresh, pre-warmed (37 °C) growth medium, and return the flasks to the incubator.
14. Maintain cultures at 37 °C, 5% CO_2 to allow astrocytes to migrate away from the explants.
15. When cultures reach 80–100% confluency (*see* **Note 5**), aspirate the growth medium from the flask. Wash the cells once with 10 mL room temperature PBS to remove residual serum, and briefly incubate the flask with 2 mL of Trypsin-EDTA solution at 37 °C and 5% CO_2. Monitor the cells under tissue culture microscope for detachment. Once most cells have detached (usually within 2–5 min), inactivate the Trypsin-EDTA by adding 8 mL of fresh, pre-warmed growth medium.
16. Transfer cell suspension to a sterile 50 mL conical tube through a 40 μm cell strainer to remove explant debris. Plate cells in new flasks T-75 flasks containing fresh growth medium equilibrated at 37 °C, 5% CO_2 (*see* **Note 6**), and maintain them in the incubator, until the day of transplantation.
17. At least 2 days before transplantation, add (z)4-hydroxytamoxifen to culture medium (final concentration: 2 μg/mL) to induce Aldh1l1-driven TOM expression.
18. On the day of transplantation, dissociate astrocytes with trypsin-EDTA, as described in Subheading 3.1, **step 15**, and collect cell suspension in a 50 mL conical tube. Rinse off growth medium twice, by spinning cells down and replacing supernatant with 10 mL HBSS with Ca^{2+} and Mg^{2+}.
19. Resuspend cells in fresh HBSS with Ca^{2+} and Mg^{2+} at a density of 40,000–100,000 cells/mL. Maintain the cell suspension on ice in a sterile 1.5 mL microcentrifuge tube throughout the transplantation procedure.

3.2 Preparation of Astrocytic Cultures from Embryonic Mouse Cortex

1. Equilibrate growth medium in T-75 flasks, and prepare materials for dissection as described in Subheading 3.1, **steps 1–2**.
2. Euthanize E17–18 pregnant mouse and make an incision into the abdominal cavity to extract uterine horns.
3. Place uterine horns in a 10 cm Petri dish containing cold dissection media, and cut the uterine horn wall to extract embryos.

4. Dissect brains out of embryos, and transfer them in a new 10 cm Petri dish with ice-cold dissection medium.
5. Proceed to preparation of cortical explants as described above in Subheading 3.1, **steps 5–18** (*see* **Note** 7).

3.3 Astrocyte Transplantation into the Mouse Cortex or Cerebellum During Early Postnatal Development

1. At least 1 day before transplantation, assemble the injection line composed of a glass needle and two pieces of tubing and needed for cellular delivery to the brain. Use a micropipette puller to create a glass needle from borosilicate glass capillaries, and adjust needle tip to get an internal diameter not larger than 100 μm. Cut TPR tubing (5 cm) and sterile PET tubing (25 cm) to desired length. First, insert one end of the PET tubing into the TPR tubing to create a leak-proof junction, and then, insert the blunt end of the pulled glass needle into the TPR tubing. Prepare a backup set of injection lines and pulled glass needles for the transplantation session, to minimize delays in case of a blocked line or broken needle during the transplantation procedure.
2. Prime the injection lines with sterile 70% ethanol. Use a sterile syringe with a 27 gauge needle to inject a small volume of ethanol into the PET tubing. Use the same syringe to inject 70% ethanol into loose glass needles. Sterilize external surface of injection lines and needles by keeping them under a UV lamp overnight.
3. On the day of transplantation, disinfect surfaces and instruments (including microscope, stereotaxic frame, injection pump, and plastic-covered heating pad) with 70% ethanol or 70% isopropyl alcohol.
4. Fit stereotaxic frame under the microscope. Wrap a freezer-cold ice pack in clean plastic film, disinfect it with 70% ethanol, and place it on the frame.
5. Install syringe pump by the stereotaxic frame. Set infuse rate at 1 μL/min and target volume at 1 μL.
6. Draw up 1 μL mineral oil into the 5 μL Hamilton syringe, and clamp the syringe into the syringe driver of the injection pump.
7. Secure the glass needle of the injection line onto the arm of the stereotaxic apparatus, and use a 1 mL syringe equipped with a G27 needle to inject mineral oil into the PET tubing. Fill the whole line with oil, making sure to displace any residual 70% ethanol. Connect the injection line to the Hamilton syringe, by inserting the Hamilton syringe needle into the PET tubing (*see* **Note 8**).
8. Verify that the injection line is working by making sure oil exits the tip of the glass needle when the fast infuse function of the injection pump is activated.

9. Pipette 2 μL cell suspension onto the internal side of a sterile 24-well plate lid. Place the lid on the gel pack positioning the cell suspension drop under the glass needle fastened to the stereotaxic arm. Lower stereotaxic arm, and draw cell suspension into the glass needle by using the fast withdraw function of the injection pump (*see* **Note 9**).
10. When ready to proceed with cryo-anesthesia, gently wrap the first pup in a piece of nitrile glove, and immerse it in crushed ice—the nitrile film will protect the pup skin from contact with ice. Monitor progression of anesthesia by assessing cessation of movements and reflexes (*see* **Note 10**).
11. Once the pup is anesthetized, disinfect its head with povidone iodine and 70% isopropyl alcohol wipes, and lay it in horizontal prone position on the ice pack. Check that the glass needle is not obstructed by running the fast infuse function for a few seconds and making sure cell suspension exits the needle tip. While gently holding the head with one hand, operate the stereotaxic arm to insert the glass needle through the skin and soft skull and into the forebrain at a depth of 0.7–1.2 mm from the skin (*see* **Note 11**).
12. Activate the infuse function of the injection pump to deliver 1 μL cell suspension (*see* **Note 12**). Once the injector stops running, keep the needle inserted into the cortex/cerebellum for one extra minute to allow any residual cell flow to diffuse into the brain.
13. After retracting the glass needle, disinfect head skin with isopropanol wipes, and warm up pups on a heating pad covered with plastic wrap until they regain normal movements and breathing (it usually takes a few minutes). The total injection procedure should take only a few minutes per animal. Following complete recovery from anesthesia, move the mice to a clean cage with fresh bedding material, and keep the cage in a warm environment (e.g., in a post-surgery rodent incubator or on a heating pad) until the end of the transplantation session (*see* **Note 13**).
14. Rub pups gently with cage bedding before returning them to the dam, to mask scents that may interfere with maternal care. Administer medications as indicated by your animal care agency/committees and monitor pups once daily for 3 consecutive days after the transplantation to check for abnormal motility/respiratory functions or signs of distress. In our experience, pups that wake up quickly (within a few minutes) from anesthesia have a high chance (more than 90%) of full recovery following transplantation (*see* **Note 14**).

The development of transplanted astrocytes can be followed by imaging Tomato-positive cells in the brains of transplant recipient mice collected at different times after cell injection. Cortical astrocytes transplanted into the cortex develop typical cortical morphologies within a few days and survive for more than 1 year in the host brain [6].

4 Notes

1. Use one to two pups or at least four embryos to get one 75 cm^2 flask of postnatal/embryonic astrocytes.
2. Dissection is carried out in a horizontal flow hood, whereas all steps including and following brain tissue trituration are carried out on a vertical flow hood. At any time, follow standard tissue culture procedures to avoid bacterial and fungal contamination.
3. Aliquot 4-hydroxytamoxifen stock and store aliquots at −20 ° C. To induce Cre-mediated recombination of the floxed-stop TOM allele in cultured cells, add 4 μL stock solution to 10 mL growth medium to get a final Tam concentration of 2 μg/mL. Recombination can also be induced prior to culture preparation by injecting Aldh1l1-TOM pups with nonmetabolized tamoxifen (Sigma), according to either of the following administration regimens: (1) 25 mg/kg tamoxifen for 2 consecutive days starting at P0–P1 or (2) 75 mg/kg tamoxifen for 2–3 consecutive days starting at P3.
4. Faster sedimentation of the tissue can be achieved by spinning the tubes for 30–60 s in a microcentrifuge.
5. Flasks prepared from the dissected brain tissue of two mouse pups can reach cell confluency as early as 3–4 days after PBS washes.
6. Choose a split ratio that aims for the cultures to reach 80–100% confluency on the day of transplantation.
7. The cerebellum has not yet developed in E17–18 mouse embryos.
8. A small air bubble may get trapped inside the line at this stage. This doesn't usually compromise the success of cell injection.
9. We recommend looking at needle under the microscope while lowering the stereotaxic arm, to minimize the risk of crushing the needle tip on the 24-well plate lid.
10. It usually takes 5–8 min for P3–P7 pups to get anesthetized through hypothermia.
11. For injections into the cerebellum, place pup on gel pack with head bent forward, and tilt the stereotaxic arm to make the

cerebellum more accessible to the glass needle. Insert needle at a depth of 1–1.5 mm from the skin.

12. Monitor cell delivery by observing the glass needle with the microscope. Focus on the junction between the cell suspension and the mineral oil, and check if this interface is moving downward during infusion, indicating successful injection of the cell suspension.
13. To prepare the glass needle between injections, eject any leftover cell suspension from the needle by using the fast infuse function, and proceed to load a fresh volume of cell suspension as described in Subheading 3.3, **step 8**.
14. The number of mice that can undergo transplantation in a single session depends on procedure efficiency and mouse age, with older mice taking more time for cryo-anesthesia. Performing transplantation on a large litter of mice or two average-sized litters is possible in one experimental session.

Acknowledgments

This work was supported by the Canadian Institutes of Health Research (PJT148569, 156247, and 180573 to K.K.M.), Natural Sciences and Engineering Research Council of Canada (408044-2011, 69404, and 2022-03395 to K.K.M.), and a Joint Canada-Israel Research Program Award provided by the International Development Research Centre (IDRC)/Israel Science Foundation/Canadian Institutes of Health Research/Azrieli Foundation. K.K.M is also supported by a Donald S. Wells Distinguished Scientist Award from the Montreal General Hospital Foundation.

References

1. Ben Haim L, Rowitch DH (2017) Functional diversity of astrocytes in neural circuit regulation. Nat Rev Neurosci 18(1):31–41. https://doi.org/10.1038/nrn.2016.159
2. Farmer WT, Murai K (2017) Resolving astrocyte heterogeneity in the CNS. Front Cell Neurosci 11:300. https://doi.org/10.3389/fncel.2017.00300
3. Schober AL, Wicki-Stordeur LE, Murai KK, Swayne LA (2022) Foundations and implications of astrocyte heterogeneity during brain development and disease. Trends Neurosci 45(9):692–703. https://doi.org/10.1016/j.tins.2022.06.009
4. Baldwin KT, Murai KK, Khakh BS (2023) Astrocyte morphology. Trends Cell Biol. https://doi.org/10.1016/j.tcb.2023.09.006
5. Khakh BS, Deneen B (2019) The emerging nature of astrocyte diversity. Annu Rev Neurosci 42:187–207. https://doi.org/10.1146/annurev-neuro-070918-050443
6. Chierzi S, Kacerovsky JB, Fok AHK, Lahaie S, Shibi Rosen A, Farmer WT et al (2023) Astrocytes transplanted during early postnatal development integrate, mature, and survive long term in mouse cortex. J Neurosci 43(9):1509–1529. https://doi.org/10.1523/JNEUROSCI.0544-22.2023
7. Chouchane M, Costa MR (2018) Culture and nucleofection of postnatal day 7 cortical and

cerebellar mouse astroglial cells. Bio Protoc 8(3):e2712. https://doi.org/10.21769/BioProtoc.2712
8. Petrelli F, Pucci L, Bezzi P (2016) Astrocytes and microglia and their potential link with autism spectrum disorders. Front Cell Neurosci 10:21. https://doi.org/10.3389/fncel.2016.00021
9. Ponroy Bally B, Murai KK (2021) Astrocytes in down syndrome across the lifespan. Front Cell Neurosci 15:702685. https://doi.org/10.3389/fncel.2021.702685
10. Jacobs S, Doering LC (2010) Astrocytes prevent abnormal neuronal development in the fragile x mouse. J Neurosci 30(12):4508–4514. https://doi.org/10.1523/JNEUROSCI.5027-09.2010
11. Jin XR, Chen XS, Xiao L (2017) MeCP2 deficiency in neuroglia: new progress in the pathogenesis of Rett syndrome. Front Mol Neurosci 10:316. https://doi.org/10.3389/fnmol.2017.00316
12. Dietz AG, Goldman SA, Nedergaard M (2020) Glial cells in schizophrenia: a unified hypothesis. Lancet Psychiatry 7(3):272–281. https://doi.org/10.1016/S2215-0366(19)30302-5
13. Srinivasan R, Lu TY, Chai H, Xu J, Huang BS, Golshani P et al (2016) New transgenic mouse lines for selectively targeting astrocytes and studying calcium signals in astrocyte processes in situ and in vivo. Neuron 92(6):1181–1195. https://doi.org/10.1016/j.neuron.2016.11.030
14. Madisen L, Zwingman TA, Sunkin SM, Oh SW, Zariwala HA, Gu H et al (2010) A robust and high-throughput Cre reporting and characterization system for the whole mouse brain. Nat Neurosci 13(1):133–140. https://doi.org/10.1038/nn.2467
15. Tsai HH, Li H, Fuentealba LC, Molofsky AV, Taveira-Marques R, Zhuang H et al (2012) Regional astrocyte allocation regulates CNS synaptogenesis and repair. Science 337(6092):358–362. https://doi.org/10.1126/science.1222381

Chapter 7

Primary Central Nervous System (CNS) Cultures with Mixed Neural Cell Types to Study Correlative Effects of High Glucocorticoids on Astrocytes, Oligodendrocytes, and Myelination Markers

José J. Miguel-Hidalgo and Yi Pang

Abstract

Dissociated glial and neuronal precursors from the spinal cord or cerebral cortex of late rat embryos are cultured on top of glass coverslips to ascertain the in vitro effects of high glucocorticoid levels on the process of myelination and on astrocyte markers during postnatal development and determine the dependence of those effects on glucocorticoid receptor activation. The study includes the immunohistochemical processing of culture-carrying coverslips with antibodies to astrocytic, neuronal, oligodendroglial, and myelin proteins to determine changes in those markers after glucocorticoid treatment. In addition, we describe oligodendrocyte–astrocyte cultures from neonatal embryos to determine the effects of high glucocorticoids on the morphology of oligodendrocytes in the absence of developing neurons.

Key words Mixed-cell culture, Oligodendrocytes, Astrocytes, Corticosteroid, Myelin

1 Introduction

Stress responses, and their major physiological hallmark, elevated glucocorticoid levels, have been demonstrated to cause morphological and molecular changes in neurons and glial cells. There is also evidence for the regulation of myelin formation or repair by glucocorticoids, while other research shows that the expression of some functionally and structurally relevant astrocytic proteins is influenced by stress or altered glucocorticoid concentrations [1]. During brain development, glucocorticoids further modulate myelin formation and glia differentiation. Here, we describe a method to obtain primary cultures from the spinal cord or cerebral cortex of rats that results in a mixture of astrocytes, neurons, and oligodendrocytes, where myelination of axons by oligodendrocytes and astrocyte differentiation occur following a time course similar

Barbara Di Benedetto (ed.), *Astrocytes: Methods and Protocols*, Methods in Molecular Biology, vol. 2896,
https://doi.org/10.1007/978-1-0716-4366-2_7,

to the in vivo myelination process in rodents [2]. The cultures are amenable to determine the effects of glucocorticoid levels and glucocorticoid receptor activation on the myelination process, the concomitant expression of major astrocytic markers such as glial fibrillary acidic protein (GFAP) or connexin 43, and the morphology of oligodendrocytes co-cultured with astrocytes.

2 Materials

2.1 Animals

1. Pregnant Sprague–Dawley or Wistar rats and 1-day-old postnatal offspring were obtained commercially and maintained and euthanized according to protocols approved by the Institutional Animal Care and Use Committee (IACUC) of the University of Mississippi Medical Center in accordance to the directives of the National Institutes of Health (NIH) Guidelines for the Care and Use of Laboratory Animals and the American Veterinary Medical Association (AVMA) Guidelines for the Euthanasia of Animals.

2.2 Dissecting Instruments and Cell Culture Vessels

1. Large scissors and forceps.
2. Small dissecting scissors, 3/4 in. blade.
3. Dissecting scissors: 1 1/2 in. blade.
4. Forceps, straight and curved.
5. Curved microdissecting forceps.
6. Straight microdissecting forceps.
7. Surgical scalpel and blades.
8. A dissecting microscope.
9. Glass coverslips (18 mm in diameter).
10. 12-well cell culture plates.
11. Petri dishes (35 and 90 mm in diameter).
12. Laminar flow hood.
13. CO_2 cell culture incubator.

2.3 Solutions and Cell Culture Media

1. Mixed N2 medium and neurobasal medium (NBM) with B27 supplement (1:1 ratio) is used for the first 2 weeks of culture. Later, insulin-free N2 is used, and its ratio to NBM-B27 is adjusted accordingly (see primary culture section). N2 medium is made in the laboratory, while NBM and B27 are obtained commercially. DMEM/F12.
2. HBSS (with Ca^{2+}/Mg^{2+}).
3. HBSS (without Ca^{2+}/Mg^{2+}).
4. Trypsin-EDTA solution.

5. NBM with B27 supplement.
6. N2 medium composition: Dilute stock solutions into DMEM/F12 (with glutamine, without $NaHCO_3$) to make N2 medium. To make 500 mL N2, add 10 mg N-acetyl-l-cysteine, 2.5 mL transferrin (10 mg/mL), 0.25 mL insulin (10 mg/mL), 0.5 mL sodium selenite (5.2 μg/mL), 0.18 mL hydrocortisone (50 μg/mL), 0.5 mL putrescine (16.1 mg/mL), 0.156 mL progesterone (20 μg/mL), 0.5 mL biotin (10 μg/mL), 5 mL penicillin-streptomycin (100×), 6.6 mL BSA (7.5%), 1.2 g $NaHCO_3$, and 0.9 g HEPES to 450 mL DMEM/F12. Gently stir with a pipet to mix, and add DMEM/F12 to bring the final volume to 500 mL. Sterilize the resulting medium using a 0.22 μm filter system. Store in 50 mL aliquots at 4 °C.
7. Insulin-free N2: N2 medium without insulin.
8. Soybean trypsin inhibitor solution (SBTI) (*see* **Note 4**): Add 25 mg soybean trypsin inhibitor, 2 mg DNase I, and 2 mL BSA (7.5%) to 48 mL DMEM/F12. Mix completely. Pass through a sterile 0.22 μm filter system. Store aliquots at −20 °C.
9. Plating medium: 100 mL horse serum, 40 mL 1× HBSS (containing Ca^{2+}/Mg^{2+}), and 60 mL DMEM/F12.
10. Poly-L-lysine (PLL) solution (for coating coverslips): Make a 0.1% PLL stock solution, sterilize with a 0.22 μm filter system, and store at 4 °C (~6 months). Dilute this stock solution 50× with ddH20 to make the working solution.
11. Antibody blocking solution: 10% normal goat serum and 0.2% Triton in PBS.
12. Antibody diluting solution: 10% normal goat serum and 0.1% Triton in PBS.

2.4 Drugs

1. Corticosterone.
2. Dexamethasone.
3. Mifepristone.

3 Methods

3.1 Primary Mixed-Cell Type Myelinating Cultures

1. The cultures are established on the top of PLL-coated glass coverslips housed within 12-well culture plates. Coverslips with an 18 mm diameter can be prepared according to our previous methods [2] or can be obtained from various commercial suppliers. Immediately before adding dissociated central nervous system (CNS) cells, individual coated coverslips are carefully placed at the bottom of 22 mm diameter wells of 12-well culture plates.

2. Embryos to prepare CNS cell cultures are obtained from timed-pregnant Sprague–Dawley or Wistar rats at gestation day 16 (embryonic day 16 or E16). Ten to 12 embryos can be obtained normally from a single pregnant rat, resulting in a sufficient number of cells to establish cultures in at least 36 coverslips housed in the corresponding wells of three 12-well culture plates.
3. Embryos are obtained from rats anesthetized with an overdose of isoflurane and immediately euthanized by pneumothorax. Immediately upon euthanasia, the uterus is removed and transferred to a Petri dish containing ice-cold HBSS (with Ca^{2+} and Mg^{2+}).
4. Under the hood, embryos are separated from their placentas, extracted, and transferred to another dish containing the same HBSS medium. Then, in each embryo, the cranium and/or the spine are separated from each other. The spinal cord is collected by cutting longitudinally the vertebrae along the spine using fine-tip scissors and carefully removing the split bone without searing the spinal cord. If cerebral cortex tissue is targeted, the cranium is split cutting along its dorsal aspect without damaging the underlying brain. The cortical mantle is then separated, and the frontal lobes (anterior to the optic chiasm) are dissected out. Transfer the blocks of cortex or spinal cord into a dish with fresh HBSS medium.
5. Mince either the spinal cords or the frontal cortex blocks with fine-tip scissors or a surgical blade attached to a scalpel, into smaller pieces, and transfer the fragments to 15 mL conical tubes with 1 mL of 0.1% Trypsin-EDTA solution while gently agitating. Following incubation at 37 °C for 15 min in a water bath, add 2 mL of the SBTI solution, and gently agitate the mixtures until tissue clamps separate.
6. Centrifuge the tube at 300 g for 5 min. Remove the supernatant, and add 2 mL of plating medium to resuspend the remaining tissue.
7. Triturate the tissue by slowly pipetting through a 1 mL fine-tip pipette about 15 times. Most tissue blocks will dissociate into a suspension of single cells. Continue pipetting if chunks of tissue are still conspicuous.
8. Filter the cell suspension through a cell strainer (40 μm) to exclude tissue debris, which will remain on the strainer. To increase yields of dissociated cells, wash the strainer with 1 mL of plating medium.
9. Mix 100 μL of cell suspension with 100 μL of trypan blue to assess the percentage of viable cells in a hemocytometer. A successful cell dissociation process should result in at least 80% of viable cells (trypan blue negative).

10. Adjust cell density to 4×10^5 cells/mL by diluting with the plate medium. Add about 250 μL of the suspension to each coverslip placed at the bottom of the culture plate well. Carefully spread the cell suspension on the top of coverslip without spilling over into the well (*see* **Note 1**).
11. Carefully bring the culture plates to the CO_2 incubator (avoid cells spilling out of the coverslips), and wait for 2 h to allow for strong attachment of the cells to the coverslip surface.
12. In the hood, add 750 μL pre-warmed N2/NBM-B27 medium (mixture of N2 and NBM-B27 at 1:1) with growth factors (NGF 50 ng/mL and NT3 10 ng/mL, used only for the first 10 days of culture to support neuronal development) to each culture coverslip, slowly delivering the medium at the wall of the culture well to avoid disturbing the recently seeded cells (*see* **Note 2**) .
13. After returning to the incubator, refresh the medium every 4 days by removing about 2/3 of the old medium with the same amount of fresh medium.
14. Cultures are maintained to the desired length of time depending on the goals of the research. For example, first, myelin sheaths, as determined by co-localization of MBP with axons (visualized by MBP and pNF double-immunolabeling), appear after 10 days in vitro, but markers of differentiating astrocytes and oligodendrocyte appear earlier. By 28 days in culture, well-developed MBP immunoreactive myelin sheaths (co-labeled with pNF) can be easily detected under the fluorescence microscope.

3.2 Treatment of Myelinating Cultures with Corticosteroids and Mifepristone

1. Starting at day 12 in vitro (DIV), coverslips are divided into six groups (each group with at least three coverslips). The medium is refreshed every 4 days and contains either of the following compounds: 5 μM corticosterone (group 1), 50 μM corticosterone (group 2), 50 μM dexamethasone (group 3), 10 μM mifepristone (group 4), 50 μM corticosterone plus 10 μM mifepristone (group 5), or vehicle (DMSO 1%; group 6, control). Before use, all these additional compounds are previously dissolved in 100% DMSO and further diluted to a 1% DMSO to be used in cultures at the mentioned concentrations. The control group of coverslips only receives refresh medium containing the 1% DMSO vehicle. This schedule of refresh medium application is partly based on procedures described in Mulholland et al. [3].
2. After 28 DIV, the coverslips, while still at the bottom of the culture plate wells, are rinsed with PBS (carefully removing and replacing solutions down the wall of the well) and fixed for 10 min with a 4% paraformaldehyde PBS solution at room temperature.

3. After washing with PBS, coverslips containing the fixed cells are carefully removed from the bottom of the wells (*see* **Note 3**) and placed preferably on the top of tube caps with flat endings and a diameter smaller than the coverslip itself (the caps are previously glued to the bottom of a box by one of the endings, while the free ending supports the coverslip as a platform) as described in Pang et al. [2].
4. The coverslips are immediately subjected to the various steps of fluorescence immunohistochemical staining, carefully applying and decanting solutions as per the specific procedure described below.

3.3 Immuno-fluorescent Labeling of Cell Cultures

1. Coverslips with fixed cells are washed with PBS three times (5 min each time) while in the wells of culture plates and then carefully removed from the bottom of the well using curved-tip tweezers. Each coverslip is placed on the top of one of the tube caps glued to the inside of a slide box as described elsewhere [2].
2. Block nonspecific binding with antibody blocking solution for 30 min at room temperature. In this and the following steps with solutions containing Triton-X, it is necessary to pay attention to the concentration of Triton-X that best allows for detection of myelin proteins (*see* **Note 4**).
3. Apply primary antibodies (Table 1) diluted in the preincubation (blocking) solution. If performing simultaneous labeling of two or more different antigens, use antibodies raised in different host species diluted in antibody preincubation solution. A total of 150 μL of antibody solution is sufficient to cover one coverslip. Incubate for 2 h at room temperature.
4. Wash three times with PBS.

Table 1
List of antibodies used to label specific glial and neuronal proteins in the cultures

Directed against	Provider	Catalog no.	Dilution	Host
Myelin basic protein (MBP)	Millipore	#AB9348	1:200	Chicken
Phosphorylated neurofilament (pNF)	BioLegend	#801602	1:1000	Mouse
Connexin 43 (Cx43)	BD Biosciences	#610062	1:300	Mouse
2′,3′-Cyclic nucleotide 3′-phosphohydrolase (CNPase)	Millipore	#MAB1580	1:1000	Mouse
Glial fibrillary acidic protein (GFAP)	Millipore	#AB5804	1:1000	Rabbit

5. Apply secondary antibodies. Fluorescently tagged secondary antibodies that specifically bind the primary antibodies (based on the animal species of the primary antibody hosts) are then diluted in blocking solution and applied to the coverslips, typically at a 1:200 dilution for 30–60 min.
6. Decant secondary antibody solution, and wash three times with PBS; drain excessive PBS by touching the edge of the coverslip against a stack of water-absorbent tissue.
7. Let the coverslips air dry.
8. As soon as the coverslips dry, apply a drop of mounting medium on a clean glass slide, and mount the coverslip on the slide being mindful that the fixed culture faces the drop of mounting medium on the slide. If visualizing cell nuclei is required, use a mounting medium containing DAPI.
9. Once mounted, the immunofluorescently labeled cells in the coverslip cultures can be best viewed at a laser confocal microscope. At the microscope, the ability of using different filter sets to detect fluorescent light of different wavelengths will allow to visualize different labels derived from secondary antibodies with different fluorophores. Since microscopes allow taking micrographs from light emitted at a specific wavelength (or channel), pictures of each channel provide an opportunity to determine/quantify the numbers or morphology of cells (or their parts) containing a specific molecular marker. In addition, the ability to merge channels imaged from a specific location allows to establish with higher confidence the identity of particular structures and quantify them (*see* Figs. 1 and 2 and **Note 5**).
10. The intensity of fluorescent signal registered in the eventual micrograph taken at the microscope depends on the nature of the label and the amount of label bound to cells. It also depends on the intensity of the laser beam directed to the tissue and the gain at the photomultiplier detector capturing the light reflected from the tissue. Thus, to ensure that the intensity and tissue extent of label in images from a particular experiment genuinely represent the actual presence of labeled structures, the laser intensity and detector gain for each wavelength channel must be uniform for all pictures taken from coverslips in the experiment.
11. Micrographs taken at the experiment then can be analyzed with software packages for image analysis, such as NIH-supported ImageJ. These applications allow to quantify intensity of labeling, number of labeled structures, and geometric parameters of labeled structures such as volume or area fraction occupied by them.

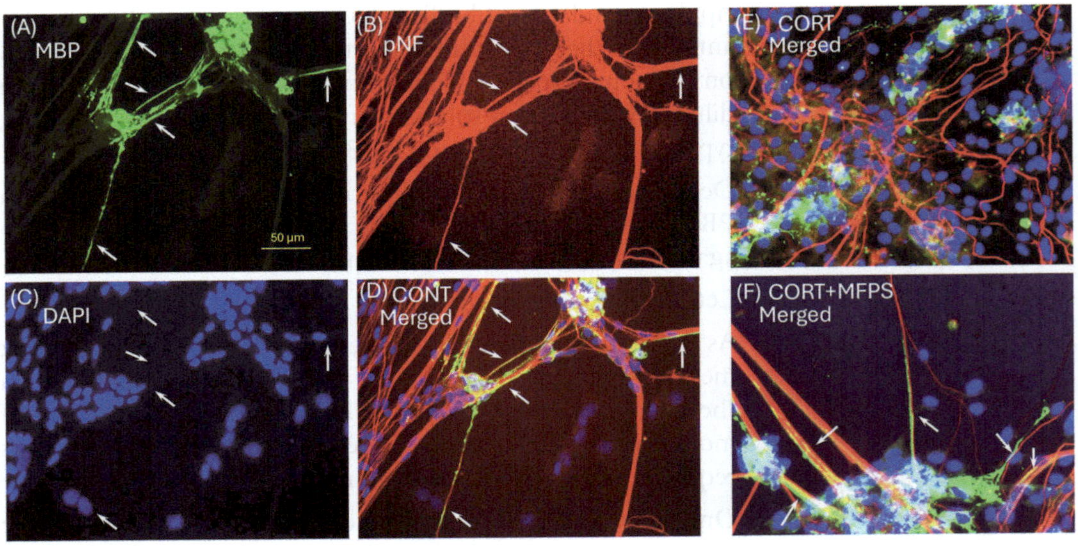

Fig. 1 Micrographs taken at a laser confocal microscope from a mixed culture from the cerebral cortex after 28 days in vitro. The cultures originated from a rat embryo collected at embryonic day 16. The immunofluorescent signals (green and red) correspond to antibodies to myelin basic protein (MBP, green) and axon marker phosphorylated neurofilament (pNF, red). Note the colocalization of MBP immunosignal along pNF-immunoreactive axon bundles (arrows) indicating myelinated axonal segments. DAPI is a fluorescent agent that binds to chromatin in the cell nucleus. The merged panels are overlays of MBP, pNF, and DAPI micrographs. The four panels to the left correspond to the same optic field from a control culture (CONT) viewed with different fluorescence filters to detect MBP (**a**), pNF (**b**), and DAPI (**c**), with those three channels merged in (**d**). The two panels to the right are merged pictures from cultures treated either with corticosterone (**e**, CORT) or corticosterone+mifepristone (**f**, CORT+MFPS). The 50 micrometer scale bar shown in A applies to all 6 micrographs

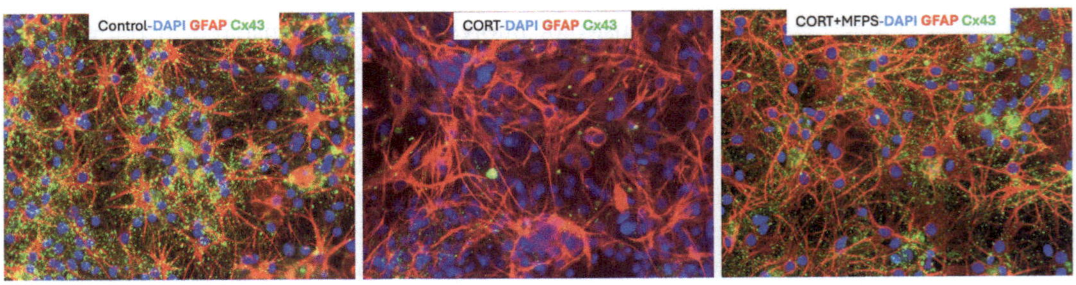

Fig. 2 Micrographs at a laser confocal microscope from three different primary mixed cultures from the cerebral cortex after 28 days in vitro. One of the cultures was treated only with vehicle (left panel), another with 50 μM corticosterone (CORT, middle panel), and the third one with 50 μM CORT and 10 μM mifepristone (CORT+MFSP, right panel). The culture originated from a rat embryo collected at embryonic day 16. The immunofluorescent signals correspond to antibodies to astrocyte marker GFAP (red) and Connexin 43 (Cx43, green). Note that CORT causes a drastically lower amount of Cx43 immunoreactive granules (middle panel) that is largely prevented by glucocorticoid receptor antagonist mifepristone (CORT+MFPS). DAPI (blue) is a fluorescent agent that binds to chromatin in the cell nucleus

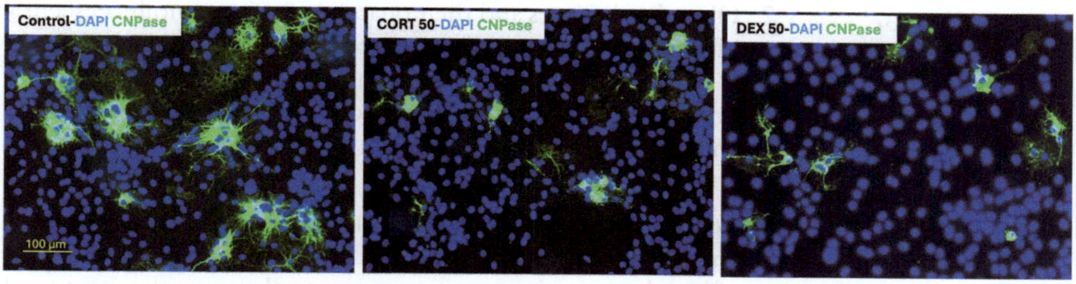

Fig. 3 Micrographs taken at a laser confocal microscope directed to three different oligodendrocyte–astrocyte cultures from the cerebral cortex. The cultures were originated from postnatal day 1 rats. Cultures were treated with vehicle only (left panel), 50 μM corticosterone (CORT 50, middle panel), and 50 μM of the corticosteroid analog dexamethasone (DEX 50). The immunofluorescent signal corresponds to antibodies to oligodendrocyte marker CNPase (green). Note that CORT and DEX cause significant shortening of cell processes from CNPase expressing oligodendrocytes as compared to the control cultures (left panel). DAPI (blue) is a fluorescent agent that binds to chromatin in the cell nucleus. The100 micrometer scale bar shown in the left panel applies to all 3 micrographs

3.4 Mixed Oligodendrocyte–Astrocyte Cultures to Study Oligodendrocytes' Response to High Corticosteroids

1. Mixed oligodendrocyte–astrocyte cultures allow to identify oligodendrocytes differentiated in cultures in the absence of neurons (Fig. 3). The cultures are originated from the cerebral cortex of postnatal day 1 rats using procedures similar to those described previously [4, 5], while immunofluorescence labeling is performed as described above. Specific details to prepare the cultures from the neonatal frontal cortex follow here.
2. The frontal cortices are separated from the brain and transferred to L15 medium. The meninges are separated and discarded, and the frontal cortical lobes are excised out, collected, minced, digested with trypsin-EDTA, and centrifuged as above to obtain a suspension of cells. Dissociated cells are cultured in DMEM with10% FBS in 75 mL flasks.
3. Cells are kept in culture until they reach confluency (about 10 days).
4. Then, culture flasks are shaken at 200 rpm at 37 °C overnight. The shaking results in mostly astrocyte and oligodendrocyte precursors remaining attached to the flask bottom, while other cell types go into suspension in the supernatant and can be removed.
5. The supernatant is then replaced with fresh medium, and the cells in the flask are cultured for at least 1 week. Cells in this culture are then dissociated with trypsin and plated on poly-L-lysine-coated glass coverslips to an approximate density of 1.5×10^5 cells per coverslip.
6. The mixture of cells on coverslips is maintained for 10 days and then treated once with either CORT5, CORT50, DEX50, or vehicle as described above and cultured for 4 days before fixing them with 4% PFA for immunofluorescence processing.

7. Oligodendrocytes and their processes are immunofluorescently identified using specific antibodies for MBP and CNPase. Under a laser scanning confocal microscope, micrographs of immunostained cells are obtained using the same laser intensity and fluorescence detector gain across samples as explained above.
8. To determine whether the extent and branching of processes and membrane expansions of cultured oligodendrocytes are affected by corticosterone or dexamethasone, the images of individual oligodendrocytes are analyzed using Sholl analysis as implemented in the Fiji version of the ImageJ image analysis software [6]. The analysis involves the overlaying of a series of concentric circles centered on the cell body, covering all visible fluorescently labeled cell processes. For the Sholl analysis, we have typically used circles with a radius difference between adjacent circles of 3 μm. The sampling consisted in all visible oligodendrocytes within at least 27 pictures (all under the 100× objective) per coverslip in 3 coverslips with cell culture.

4 Notes

1. When seeding the cell suspension on the coverslips at the bottom of the culture wells, particular attention must be paid to add the right volume of cell suspension (250 μL is sufficient), and do it while avoiding spilling the suspension out of the coverslip edge, because during 2 h, the cells will attach to any surface, and it is convenient that only the upper surface of the coverslip contains the attached culture cells when the additional culturing and immunohistochemistry steps are applied. The goal is to ensure a consistent starting cell density in all coverslips and preventing that cells attach to the lower surface of the coverslips or the sides of the well. This is less of a concern when the cells have already attached to the coverslip's upper surface.
2. Throughout the cell culture procedures, while the slides carrying seeded cells are at the bottom of the wells, it is of the utmost importance that washes and delivery of solutions are made with the transfer pipettes tips applied to the walls of the well without touching or disturbing the cultured cells, always striving that the flow of liquids slowly and smoothly happens at the wall itself and not on top of the cell culture. If solutions are directly delivered to or aspired from the middle of the slide, turbulence and liquid eddies may disturb or damage the cultured cells.
3. When lifting the coverslips containing paraformaldehyde-fixed cultures from the bottom of the culture wells, maximum

attention must be paid to only hold the slides by their edge with very fine-tip curved tweezers. Once transferred to the top of the plastic bottle cups, access to washing solutions or solutions with antibodies and/or fluorophores is straightforward, because after incubating with a solution, removal of it is easy by decantation. However, it is still important to add slowly solutions to the culture-containing side of the coverslips.

4. Binding of specific antibodies to MBP and other components of myelin can be impeded by the dense packing of the myelin membranes, so it is important for immunohistochemical demonstration of MBP to assess whether, in the particular preparation being processed, higher concentrations of Triton-X (up to 0.6% in PBS) and/or longer incubation times might be necessary, always being mindful that an excess of Triton-X may promote detachment.
5. As indicated, the extent of myelination in the mixed cultures was assessed at 28 DIV. At this stage, individual axons and some well-defined axon bundles that are labeled for both MAP and pNF are easily distinguishable, indicating the presence of myelinated axons. If MAP antibody is tagged using secondary antibodies carrying a fluorophore emitting green fluorescence and pNF is tagged to emit red fluorescence, then their coexistence in myelinated axons appears as a yellow fluorescence signal. Due to the simplicity, we chose to assess myelination by measuring the length of all yellow axon segments in a micrograph in relation either to the cell number or the total length of pNF (red) labeling within the micrograph. Later in culture period (beyond 28 DIV), many axons may become additionally myelinated and form a mesh in which length is more difficult to assess. In this case, the extent of co-labeled structures can be assessed with an image analysis system such as ImageJ [7] that can segment away the yellow portions of the micrographs, and the surface area of yellow labeled structures can be expressed in relation to the number of cells or the surface of pNF-labeled structures to calculate a myelination index.

Acknowledgments

The authors acknowledge partial funding support from NIMH grant R56MH113828 (JJMH), IRSP grants from the University of Mississippi Medical Center (JJMH), and the Imaging Core of NIGMS grant P30GM103328. We thank Kathleen Carter, Preston Hardin Deloach, and Leon Sanders for their excellent technical support in the studies reported.

References

1. Miguel-Hidalgo JJ, Hearn E, Moulana M, Saleem K, Clark A, Holmes M, Wadhwa K, Kelly I, Stockmeier CA, Rajkowska G (2023) Reduced length of nodes of Ranvier and altered proteoglycan immunoreactivity in prefrontal white matter in major depressive disorder and chronically stressed rats. Sci Rep 13(1):16419. https://doi.org/10.1038/s41598-023-43627-4
2. Pang Y, Simpson K, Miguel-Hidalgo JJ, Savich R (1791) Neuron/oligodendrocyte myelination coculture. Methods Mol Biol 2018:131–144. Epub 2018/07/15. https://doi.org/10.1007/978-1-4939-7862-5_10
3. Mulholland PJ, Stepanyan TD, Self RL, Hensley AK, Harris BR, Kowalski A, Littleton JM, Prendergast MA (2005) Corticosterone and dexamethasone potentiate cytotoxicity associated with oxygen-glucose deprivation in organotypic cerebellar slice cultures. Neuroscience 136(1): 259–267. https://doi.org/10.1016/j.neuroscience.2005.07.043
4. Pang Y, Cai Z, Rhodes PG (2000) Effects of lipopolysaccharide on oligodendrocyte progenitor cells are mediated by astrocytes and microglia. J Neurosci Res 62(4):510–520
5. Pang Y, Campbell L, Zheng B, Fan L, Cai Z, Rhodes P (2010) Lipopolysaccharide-activated microglia induce death of oligodendrocyte progenitor cells and impede their development. Neuroscience 166(2):464–475. Epub 2009/12/29. https://doi.org/10.1016/j.neuroscience.2009.12.040
6. Ferreira TA, Blackman AV, Oyrer J, Jayabal S, Chung AJ, Watt AJ, Sjostrom PJ, van Meyel DJ (2014) Neuronal morphometry directly from bitmap images. Nat Methods 11(10):982–984. https://doi.org/10.1038/nmeth.3125
7. Schroeder AB, Dobson ETA, Rueden CT, Tomancak P, Jug F, Eliceiri KW (2021) The ImageJ ecosystem: open-source software for image visualization, processing, and analysis. Protein Sci 30(1):234–249. https://doi.org/10.1002/pro.3993

Chapter 8

Postnatal Development of Perivascular Astrocytic Processes and Detection of Local mRNA and Translation

Katia Avila-Gutierrez, Heloïse Monnet, Philippe Mailly, Anne-Cécile Boulay, and Martine Cohen-Salmon

Abstract

Astrocytes' perivascular processes (PvAPs) entirely sheathe the brain's blood vessels and are equipped with an enriched molecular repertoire that sustains astrocytic regulatory functions at the vascular interface. We recently demonstrated that PvAP development in the mouse starts after birth and is essentially complete by postnatal day (P) 15. This time window also corresponds to a progressive molecular maturation of PvAPs with the acquisition of specific proteins. We proposed local translation of mRNAs in PvAPs as a molecular mechanism participating to their development and maturation. Here, we describe methods to quantify the development of PvAPs and detect local translation in developing PvAPs. We also show how to analyze quantitatively astrocytic-specific mRNAs and polysomal mRNAs in developing PvAPs. These experimental approaches allow to characterize important aspects of the dynamic postnatal development and molecular maturation of PvAPs.

Key words Astrocytes, Perivascular astrocytic processes, Local translation, Postnatal development, mRNA, Polysomal mRNA, Gliovascular unit

1 Introduction

Astrocytes are glial cells which greatly influence the development and functions of the cerebrovascular system. Astrocytes regulate the blood–brain barrier (BBB) integrity, endothelial transport between the brain and the periphery, the recruitment of immune cells by endothelial cells during inflammation, perivascular water ions and neurotransmitter homeostasis, neurovascular coupling, and metabolite transfer from blood to neurons [1, 2]. Most of these regulations are exerted at the level of perivascular processes (PvAPs) (also known as endfeet) that fully cover the brain's blood vessels [3, 4]. PvAPs in the mouse cortex and hippocampus are formed around postnatal (P) day 5. Astrocytic perivascular coverage increases rapidly between P5 and P10 and is almost complete by

Barbara Di Benedetto (ed.), *Astrocytes: Methods and Protocols*, Methods in Molecular Biology, vol. 2896,
https://doi.org/10.1007/978-1-0716-4366-2_8,

P15 [5, 6]. During this time window, PvAPs also accomplish their molecular maturation, with the progressive acquisition of specific proteins including the water channel aquaporin 4 (Aqp4) and the inward rectifying potassium channel Kir4.1, which control perivascular water and potassium homeostasis [7, 8], the glutamate transporter Glt1 which controls perivascular glutamate homeostasis, the gap junction proteins connexin (Cx)30 and Cx43 which regulate BBB integrity and immune system homeostasis [9, 10], and the MLC1/GlialCAM adhesion complex that controls PvAP development and cohesivity and influences vascular smooth muscle cell contractile development [11, 5, 12]. How these molecular properties are progressively acquired by PvAPs is unknown. We recently showed that mRNA distribution and local translation are occurring in astrocytes [13, 14] and proposed that local translation of mRNAs located in PvAPs might sustain their postnatal development and molecular maturation [15].

Here, we describe several experimental protocols to study PvAP postnatal development and detect and quantify local translation and total and polysomal astrocytic mRNAs in developing and mature PvAPs. These protocols allow to study the dynamic astrocytic postnatal mRNA perivascular localization and local translation which might contribute to the acquisition of these processes' molecular identity.

2 Materials

2.1 Mice

1. Tg(Aldh1l1-eGFP/Rpl10a) JD130Htz (MGI: 5496674) (Aldh1l:L10aeGFP) and Tg(Aldh1l1-eGFP) OFC789Gsat/Mmucd (Aldh1l1:eGFP) mice originate from Nathaniel Heintz (Rockefeller University, NY) and must be kept under pathogen-free conditions under the C57BL6 genetic background. These mouse strains have been generated by bacterial artificial chromosome transgenesis [16–18]. The Aldh1l1 promoter is used to induce expression of L10a-eGFP or eGFP in astrocytes and is also active in neurogenic areas [19].

2.2 Transcardial Perfusion and Tissue Sectioning

1. Fresh phosphate-buffered saline (PBS) 10× solution (e.g., Euromedex, Ref. 330-A) diluted 1× in Milli-Q distilled water.
2. Mix of ketamine (140 mg/kg) and xylazine (8 mg/kg) diluted in sterile NaCl 0.9% solution.
3. Perfusion pump system (flow rate: 5 mL/min).
4. Paraformaldehyde (PFA) 4% diluted in sterile PBS.
5. Cryoprotective solution: sucrose 30% diluted in PBS.
6. Freezing cryomicrotome (e.g., HM 450, Fisher Scientific, Ref. 12072999).

7. Solution for storage of the slices at −20 °C: 25% glycerol and 25% ethylene glycol in PBS.
8. 1.5 mL skirted tubes (e.g., Starlab, Ref. E1420-0316).
9. 50 mL tubes (e.g., Falcon, Ref. 352070).

2.3 Immunostaining

1. PBS 10× solution (Euromedex, Ref. 330-A) diluted in 1× in Milli-Q distilled water.
2. 24-well plaque (Costar, Ref. 3526).
3. Blocking/permeabilization solution: PBS 1×, normal goat serum 5% (Ozyme, Ref. 1000C), and Triton ×-100 (Sigma-Aldrich, Ref. 93443) 0.5%.
4. Primary and secondary solution: Blocking/permeabilization solution with primary or secondary antibodies.
5. Chicken polyclonal anti-GFP (1:500, AVES, Ref. GFP-1020).
6. Goat anti-chicken IgY (H + L) and Alexa Fluor 488 (1:1000, Jackson ImmunoResearch, Ref. 703-545-155).
7. Alexa Fluor 647-Isolectin GS-IB4 (1:100, Invitrogen, Ref. I32450).
8. Hoechst (1:1000, Thermo Fisher Scientific, Ref. H3570).
9. Mounting medium: Fluoromount-G™ (Thermo Fisher Scientific, Ref. 15586276).
10. 70% ethanol (EtOH).
11. Pre-cleaned microscopy glass slides (Thermo Fisher Scientific, Ref. 15998086) and coverslips (Thermo Fisher Scientific, Ref. 65.300.15).
12. Agitating platform (Orbi-Blotter, Benchmark).
13. Yokogawa W1 Spinning Disk confocal microscope (Zeiss) and a 40× oil objective.

2.4 Quantification of PvAPs Development on Brain Slices

eGFP detected by immunofluorescence in developing PvAPs is analyzed using a custom-written plug-in developed for Fiji software [20], using Bio-Formats [21], CLIJ [22], GDSC (https://github.com/aherbert/gdsc), and 3D Image Suite libraries [23] . The plug-in is available at https://github.com/orion-cirb/Vessel_IB4/ [15].

2.5 Fluorescent In Situ Hybridization (FISH) and Immunostaining

1. Tris-buffered saline 10× (TBS) (e.g., Euromedex, Ref. ET 220-B) diluted 1× in Milli-Q distilled water with 0.1% Tween® 20 (Euromedex, Ref. ET 2001-A).
2. 24-well cell culture plaque (e.g., Costar, Ref. 3526).
3. Superfrost® Plus-treated glass slides (e.g., VWR, Ref. 631-0108).
4. Dry oven at 60°C.

5. Milli-Q distilled water.
6. 100% pure EtOH.
7. ImmEdge® Hydrophobic Barrier Pen (e.g., Vector Laboratories, Ref. H-4000).
8. Food steamer.
9. RNAscope® Multiplex Fluorescent Reagent Kit v2 (e.g., Advanced Cell Diagnostics Inc., Ref. 323110).
10. HybEZ™ Hybridization System: HybEZ oven (e.g., PN 321710/321720), humidity control tray (e.g., PN 310012), and EZ-Batch Slide Holder (e.g., PN 321716).
11. FISH probes for some astrocyte-specific mRNAs (e.g., from Bio-Techne): Aqp4 (e.g., Ref. 417161); Gpr37l1 (e.g., Ref. 319301); Slc1a2 (e.g., Ref. 441341); Fabp7 (e.g., Ref. 414651).
12. Blocking/permeabilization solution: PBS 1×, normal goat serum 5% (e.g., Ozyme, Ref. 1000C), and Triton ×-100 (e.g., Sigma-Aldrich, Ref. 93443) 0.5%.
13. Alexa Fluor 647-Isolectin GS-IB4 (1:100, e.g., Invitrogen, Ref. I32450).
14. Hoechst (1:1000, e.g., Thermo Fisher Scientific, Ref. H3570).
15. Mounting medium: Fluoromount-G™ (e.g., Thermo Fisher Scientific, Ref. 15586276).
16. 70% EtOH pre-cleaned coverslips (e.g., Thermo Fisher Scientific, Ref. 65.300.15).

2.6 Astrocyte-Specific mRNA in Developing PvAPs Quantification

The astrocyte-specific mRNA in developing PvAPs FISH data were analyzed using a custom-written plug-in developed for Fiji software [20], using Bio-Formats [21], CLIJ [22], GDSC (https://github.com/aherbert/gdsc), and 3D Image Suite libraries [23]. The plug-in is available at https://github.com/orion-cirb/Astrocytes_InOut_Vessels [15].

2.7 In Situ Detection of Translation in Developing PvAPs by Puromycylation

2.7.1 On Brain Slices

1. Milli-Q distilled water.
2. Puromycin (PMY) (e.g., P7255, Sigma-Aldrich) 10 mg/mL diluted in Milli-Q distilled water.
3. PBS 1×.
4. PFA 4% diluted in PBS 1×.
5. 30% sucrose diluted in PBS 1×.

2.7.2 On Purified Brain Microvessels (MV) with Associated PvAPs

1. Low binding 1 mL pipette tips (e.g., Sorenson, Ref. 35260).
2. Low binding 1.5 mL tubes (e.g., Eppendorf, Ref. 022431021).

3. DMEM (e.g., Gibco, Ref. 31966–021).
4. PMY 20 μM (e.g., Sigma-Aldrich, Ref. P7255) diluted in DMEM.
5. Cell culture incubator at 37 °C, 5%CO_2.
6. Centrifuge for 1.5 mL plastic tubes at room temperature (RT).
7. PBS 1×.
8. PFA 4% diluted in PBS 1×.
9. 70% EtOH pre-cleaned microscopy slides (Thermo Fisher Scientific, Ref. 15998086) and coverslips (e.g., Thermo Fisher Scientific, Ref. 65.300.15).
10. Cell-Tak™ (e.g., Corning, Ref. 354240).

2.8 Digital qPCR Analysis of Astrocytic-Specific mRNAs and Polysomal mRNAs in Developing PvAPs

1. RNeasy Lipid Tissue Mini Kit (e.g., Qiagen, Ref. 74106).
2. Superscript™ III Reverse Transcriptase Kit (e.g., Thermo-Fisher, Ref. 12574026).
3. Mix dNTP (e.g., ThermoFisher, Ref. 10297-117).
4. Pdn6 random hexamers (e.g., Amersham Pharmacia Biotech, Ref. 27-2166-01).
5. RNAse out (e.g., Euromedex, Ref. 09-0312).
6. Milli-Q distilled water.
7. ddPCR Supermix for probes (e.g., Bio-Rad, Ref. 1863010).
8. Multiplate PCR plate 96 wells (e.g., Bio-Rad, Ref. MLP9601).
9. DG8 Cartridges for QX100/QX200 Droplet Generator (e.g., Bio-Rad, Ref. 1864008).
10. Droplet generator DG8 gaskets (e.g., Bio-Rad, Ref. 1863009).
11. ddPCR oil for probes (e.g., Bio-Rad, Ref. 1863005).
12. Pierceable foil heat seal (e.g., Bio-Rad, Ref. 1814040).
13. QX200 Droplet Generator (e.g., Bio-Rad, Ref. 1864002).
14. PCR thermocycler (e.g., Bio-Rad, Ref. 1861096).
15. QX200 Droplet Reader (e.g., Bio-Rad, Ref. 17005228).
16. QuantaSoft Software (e.g., Bio-Rad, Ref. 1864011).
17. TaqMan probes labeled with FAM (e.g., ThermoFisher) for some astrocytic mRNAs present in PvAPs (ref): Gpr37l1 (Mm00661872 m1); Slc1a2 (Mm01255814 m1); Ntsr2 (Mm00435426_m1); Mlc1 (Mm00453827 m1); Hepacam (Mm01151430 m1); Kcnj10 (Mm00445028 m1); Slc1a3 (Mm00600697 m1); Fabp7 (Mm00445225 m1); Aqp4 (Mm00802131 m1); Slc1a4 (Mm01223875 m1); and Ptprz1 (Mm00478484 m1).

3 Methods

3.1 Transcardiac Perfusion for Brain Fixation, Dissection, and Slice Preparation

1. Weigh the mouse to calculate the exact dose of anesthetics.
2. Administer the anesthetic mix by intraperitoneal injection at the lower left or right quadra of the abdomen.
3. Wait until the mouse gives no response to tailor toe pinches.
4. Place the mouse on its back, and tape the four paws on a surgical station under a chemical fume hood.
5. Expose the xiphoid, and make lateral incisions beneath the ribcage to expose the liver and the diaphragm.
6. Carefully make incisions on both sides of the ribcage, and expose the lungs and the heart by pulling the sternum over the head of the mouse.
7. Insert the perfusion needle connected to the peristaltic pump into the left ventricle, and cut off the right atrium.
8. Start the PBS (at RT) perfusion at a constant speed of ~5 mL/min (*see* **Note 1**).
9. Follow the perfusion with PFA 4% at RT (10 mL for P5 and P10 mice; 20 mL from P15).
10. Dissect out the brain, and transfer it into 50 mL tube previously filled with PFA 4%.
11. Leave the brain in PFA 4% at 4 °C for 24 h.
12. Wash the brain in PBS three times for 10 min at RT.
13. Transfer the brain into a 50 mL tube containing the 30% sucrose solution at 4 °C for 24 h (until it sinks).
14. Prepare 1.5 mL tubes filled with the brain slice storage solution.
15. Let the cryomicrotome reach −25 °C. Build a uniform platform with sucrose 30%.
16. Cut off the olfactory bulbs and the cerebellum, and place the brain on the sucrose platform in a vertical position, and embed it in 30% sucrose. Cut 40-μm-thick coronal slices.
17. Transfer brain slices progressively into 1.5 mL tubes (about 15 slices per tube).
18. Store the tubes at −20 °C.

3.2 Immunofluorescent Detection of Cytoplasmic eGFP or eGFP-Rpl10a (Ribosomes) in Postnatal Developing PvAPs

The main purpose of the following method is to analyze and quantify the development of PvAPs. To this end, we propose two different approaches: (i) to detect and quantify the presence of ribosomes in developing PvAPs from P5 using the Aldh1l1:L10a-eGFP transgenic mouse line which expresses the eGFP-tagged ribosomal protein Rpl10a specifically in astrocytes [18] and (ii) to detect and quantify the volume of developing PvAPs from P5 using the Aldh1l1:eGFP transgenic mice, in which the astrocyte's cytoplasm is filled with eGFP [24]. In both cases, eGFP signal is amplified and detected by immunofluorescence, and the brain vessel surface is stained with a fluorescent tagged-IB4. We perform these experiments on 40-μm-thick brain sections, and fluorescent signals are imaged using a spinning disk microscope. Analysis of the perivascular eGFP signal is done using a custom plug-in.

1. Distribute the brain sections in a 24-well plaque (one to four sections per well).
2. Wash brain sections in PBS three times for 10 min at RT under slow agitation.
3. Incubate the slices 1 h at RT in the blocking/permeabilization buffer under slow agitation.
4. Pipette out the buffer, replace by the blocking/permeabilization solution containing primary anti-eGFP antibodies and IB4, and incubate at 4 °C overnight (O/N) under slow agitation, protecting the plaque from light exposure (from this step to the end of the procedure).
5. Wash brain sections in PBS three times at RT for 10 min under slow agitation.
6. Incubate with Alexa 488-tagged secondary antibodies and Hoechst in the blocking/permeabilization buffer at RT for 2 h under slow agitation.
7. Wash brain sections in PBS three times at RT for 10 min.
8. Mount the brain sections using a small paintbrush on ethanol-cleaned microscopy glass slides. Let the sections dry. Add one drop of mounting medium, and gently mount the glass coverslip on top of the slices (*see* **Note 2**).
9. Let the slides dry at RT O/N.
10. Image the brain sections using a confocal microscope and a 40× immersion objective. We use a Spinning Disk CSU-W1 microscope and MetaMorph Premier 7.8 software. Perform imaging of a z-stack of 20–25 images (with a z-step of 0.3 μm).
11. Quantify the perivascular eGFP astrocytic signal using our custom plug-in Astrocytes InOut Vessels (https://github.com/orion-cirb/Astrocytes_InOut_Vessels) (*see* **Note 3**). This plug-in allows IB4-stained vessels to be detected by filtering slice-by-slice the channel with a 2D Laplacian of Gaussian filter ($\sigma = 14$) and segmenting it with the triangle method. The 3D segmented vessels thus obtained are filtered by volume. Cytoplasmic eGFP or L10a-eGFP signals are detected by applying first a 2D median filter (radius = 2) and then an automatic thresholding method using the Mean and Huang methods. Lastly, the volume of the contact between L10a-eGFP or eGFP signals and vessels is computed.

3.3 Detection of Astrocyte-Specific mRNAs in Developing PvAPs by FISH and Co-staining of the Vessel Surface

Detection of astrocyte-specific mRNAs is performed on brain sections by FISH. We then quantify FISH dots in PvAPs surrounding the IB4-labeled blood vessels.

1. Distribute the brain sections in a 24-well plaque (one to four section per well).
2. Wash the brain sections in PBS three times at RT for 10 min.
3. Remove PBS and add three to five drops of RNAscope® hydrogen peroxide solution for 10 min at RT. This step blocks endogenous peroxidase activity and results in the formation of small bubbles.
4. Wash the sections in TBS-T three times at RT for 5 min.
5. Fill a small glass petri dish with TBS-T and immerse the brain slices. Gently agitate the liquid to remove remaining bubbles that the tissue might retain.
6. Mount the brain slices on the Superfrost® Plus glass slides with a small paintbrush (maximum four sections per slide). Leave some space between each slice.
7. Carefully and without touching the brain slices, drain the excess of liquid with an absorbent paper.
8. Dry the slices at RT for 1 h in the dark.
9. Immerse slides in a water bath for 3 s to wash out the TBS-T.
10. Leave the slides to dry at RT for 1 h in the dark.
11. Incubate slides for 1 h at 60 °C in a dry oven.
12. Dry the slices at RT in the dark O/N.
13. Rehydrate the brain sections by immersing the slides no more than 3 s in a water bath.
14. Remove excess liquid with an absorbent paper.
15. Carefully and without touching the brain slices, draw a hydrophobic barrier around every section using the ImmEdge® Hydrophobic Barrier Pen, leaving a small space between the tissue and the barrier to allow a proper liquid contention. Leave the hydrophobic barrier dry for at least 5 min before proceeding to the next steps.
16. Add a drop of 100% pure EtOH on each slice for less than 3 s, and remove it using an absorbent paper.
17. Turn on the steamer and allow it to reach 100 °C.
18. Add a drop of preheated diluted RNAscope® Target Retrieval Reagent (TRR) on each brain section, and incubate the slides in the steamer for 15 min. Protect the slices from condensed water placing them in a closed chamber (for instance, an empty pipette tip plastic box). This step allows mRNA demasking (*see* **Note 4**).

19. Wash slides three times in cold (4 °C) Milli-Q distilled water.
20. Remove excess liquid with an absorbent paper.
21. Add a drop of 100% pure EtOH and leave it for 3 min at RT.
22. Remove EtOH with an absorbent paper and let the slides at RT for 5 min.
23. Turn on the HybEZ™ Hybridization oven and allow it to reach 40 °C. The incubation chamber should be moistened with paper dampened in water to prevent the slices from drying out.
24. Place the slides on the tray in the incubation chamber, and cover them with two to five drops of protease+ reagent (depending on the size of the brain section). Incubate the slides for 30 min at 40 °C in the HybEZ™ Hybridization oven. Incubation times may vary depending on the analyzed tissue and targeted mRNA.
25. Wash the slides three times in Milli-Q distilled water. Let them dry at RT for 5 min.
26. Perform the FISH using the RNAscope® Multiplex Fluorescent Reagent Kit v2 protocol according to the manufacturer's instructions.
27. Following the FISH procedure, incubate the slides in blocking/permeabilizing solution (PBS, normal goat serum 5%, and Triton ×-100 0.5%) for 1 h at RT. Incubate the slides in blocking/permeabilizing solution containing Alexa Fluor 647-Isolectin GS-IB4 and Hoechst for 2 h at RT or 4 °C ON.
28. Wash brain sections in PBS three times for 10 min under slow agitation in the dark.
29. Let the slides to dry for at least 5 min at RT.
30. Add a drop of mounting medium on each brain section. Slowly mount a coverslip on top. Be careful not to create any bubbles that could interfere with the image acquisition.
31. Let the slides dry at RT O/N.
32. Image the brain sections using a confocal microscope and 40× immersion objective. We use a Spinning Disk CSU-W1 microscope and MetaMorph Premier 7.8 software. Image a z-stack of 20–25 images (with a z-step of 0.3 μm).
33. Quantify the perivascular FISH signal using our custom plug-in (https://github.com/orion-cirb/Vessel_IB4/) developed for Fiji software [20], using Bio-Formats [21], CLIJ [22], GDSC (https://github.com/aherbert/gdsc), and 3D Image Suite libraries [23]. Before starting the plug-in, all the images to be analyzed must be on one same folder. In this plug-in, the vessel channel is first filtered slice-by-slice with a two-dimensional (2D) Laplacian of Gaussian filter ($\sigma = 20$)

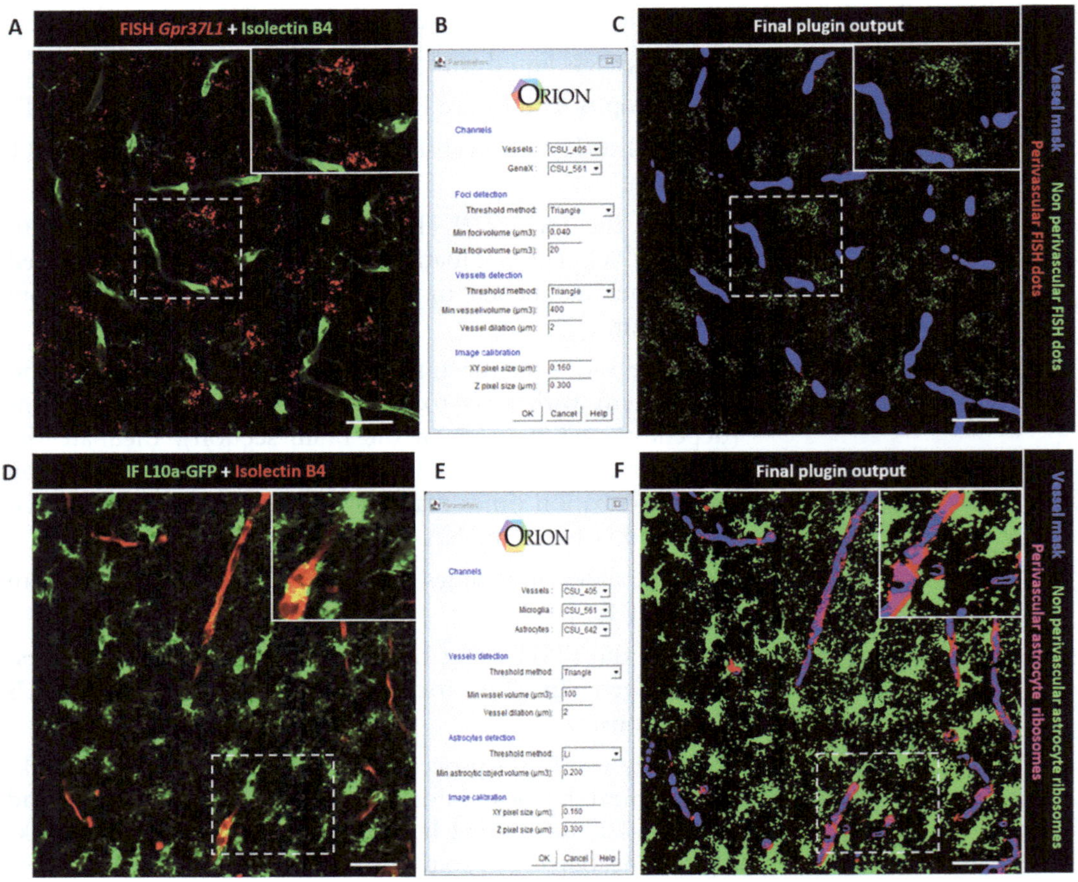

Fig. 1 Image processing to quantify perivascular astrocyte mRNAs and ribosomes. (**a**). A representative confocal microscopy image of *Gpr37l1* mRNAs detected by FISH (in red) in the cortex of a wild-type (WT) mouse on P5. Blood vessels are labeled with Isolectin B4 (IB4) (in green). (**b**). IB4 and FISH parameters are entered manually in the dialog box. (**c**). Plug-in-based interpretation of the results. The mask created by the plug-in for the vessels appears in blue. mRNAs FISH dots at more than 2 μm from the IB4 signal appear in green. mRNAs FISH dots at less than 2 μm from the IB4 signal appear in red and are designated as perivascular. (**d**). Representative confocal microscopy image of the immunofluorescent (IF) detection of astrocytic ribosomes tagged with GFP in the Aldhl1:L10a-eGFP (in green) mouse cortex at P15. Blood vessels are stained with IB4 (in red). (**e**). Parameters for vessels (IB4), microglia (optional), and astrocytes (eGFP) are entered manually in the InOut vessels dialog box. (**f**). Plug-in-based interpretation of the results for astrocytic eGFP ribosomes. The mask obtained using the plug-in for the vessels appears in blue. The astrocytic eGFP ribosome signal located far from the vessels is represented in green. The astrocytic eGFP ribosome signal localized at less than 2 μm from the IB4 signal appears in pink. Scale bars = 50 μm

and segmented with the triangle method. The binary mask thus obtained is labeled in three dimensions, and three-dimensional (3D) vessels were filtered by volume to avoid false positives (check that the output vessel image corresponds to the IB4 signal) (Fig. 1). In a second step, RNA FISH dots are detected by successively applying a 3D difference of Gaussian filter ($\sigma 1 = 1, \sigma 2 = 2$). The thresholding methods used for the RNA signal is triangle. Dots are then labeled in 3D, filtered by volume (again, check that the output FISH image corresponds to the FISH dots), and classified as perivascular or not as a function of their distance to the closest vessel: dots with a distance below 2 μm are considered to be perivascular. For each category, the number of dots, the dots' volume, and their integrated intensity are computed. The plug-ins open a dialog box (Fig. 1b, e) that enables to define (i) a specific channel to every component of the image (for instance, Vessels 650, GeneX (mRNA) 561, Astrocytes 488), (ii) the threshold method, and (iii) the detection parameters for every component (for instance, minimum and maximum volume for mRNA dots or vessel dilation). Of note, these parameters can vary depending on the confocal microscope and objective. When all the parameters are set, the plug-in can start. A new folder named "results" is automatically created containing a .TIF image of the analysis (Fig. 1c, f) as well as an Excel sheet computing all information.

3.4 In Situ Detection of Translation in Developing PvAPs by Puromycylation on Brain Sections and Purified Brain MVs

PMY is an antibiotic that mimics tRNA-Tyr incorporated into the translating ribosome, inducing translation termination. Hence, immunodetection of PMY allows the detection of nascent protein chains [25]. We present here the method to perform this labeling on brain sections or on purified brain MVs with attached PvAPs [26].

1. For the study on brain section, anesthetize the mouse as described in Subheading 3.1.
2. Inject intraperitoneally 50 μL of PMY solution or water only for the control condition.
3. Wait for 25 min and proceed to brain fixation and brain sectioning as described in Subheading 3.1.
4. For the study on MVs with associated PvAPs, proceed as previously described [26], and resuspend them using 1 mL low binding pipette tips in DMEM containing PMY or DMEM alone for the control condition in a low binding 1.5 mL Eppendorf tube.
5. Incubate the MVs 30 min in a 5% CO_2 incubator at 37 °C.
6. Pellet the MVs at 2000 g for 10 min.
7. Spread 1 μL of Cell-Tak using a pipette tip on a small area of a glass microscope slide (Cell-Tak allows MVs to stick to the glass slide). Carefully draw a hydrophobic barrier around Cell-Tak using the ImmEdge® Hydrophobic Barrier Pen. Leave the Cell-Tak and the hydrophobic barrier dry for at least 5 min.

8. Resuspend MVs in 100 μL PBS by pipetting up and down carefully with a low binding tip, and plate them on the glass microscope slide area coated with Cell-Tak.
9. Wait until MVs are stuck to the slides.
10. Carefully drain out the liquid with an absorbent paper. This will allow all MVs to adhere to the slides.
11. Fix the MVs with PFA 4% for 15 min at RT.
12. The following steps are identical for purified MVs and brain sections.
13. Rinse three times with PBS at RT.
14. Incubate in blocking/permeabilizing buffer (PBS, normal goat serum 5%, and Triton ×-100 0.5%) for 1 h at RT.
15. Incubate in blocking/permeabilization solution containing the anti-PMY primary antibodies and Alexa Fluor 555 conjugated IB4 at 4 °C O/N.
16. Wash in PBS three times for 10 min under slow agitation in the dark.
17. For brain slices only: mount the slices on slides as previously described, and let the slides dry for at least 5 min at RT.
18. Add a drop of mounting medium on each brain section or MVs. Slowly mount the coverslip on top. Be careful not to create any bubbles that could interfere with the image acquisition.
19. Let the slides dry at RT O/N.
20. Image using a confocal microscope and 63× immersion objective. We use a Spinning Disk CSU-W1 microscope and MetaMorph Premier 7.8 software. Image a z-stack of 20–25 images (with a z-step of 0.3 μm).

3.5 Quantitative PCR Analysis of Astrocytic-Specific mRNAs and Polysomal mRNAs in Developing PvAPs

PvAPs stay attached to mechanically purified MVs, and their content in mRNAs selectively or specifically expressed in astrocytes can therefore be assessed. Here, we described the method to detect and quantify total astrocytic mRNAs as well as polysomal mRNAs by quantitative RT-PCR (qPCR). This procedure can be developed at all postnatal stages allowing to compare the level of astrocytic mRNAs between stages.

1. Brain MVs (with attached PvAPs) are mechanically purified in RNase-free conditions as previously described in [13]. Polysomal mRNA extraction is performed on MVs purified from Aldhl1:L10a-eGFP mice and following the TRAP protocol previously described [13, 15].
2. mRNAs are extracted from MV homogenates or TRAP extracts using the RNeasy Lipid Tissue Mini Kit (*see* **Note 5**).

3. Reverse transcription is performed on 100 ng of total mRNAs or the whole TRAP extract for polysomal mRNAs as followed: for one sample, in PCR tubes, mix mRNAs with 1 μL of pdN6 random hexamers and 1 μL of dNTP mix (10 mM), and add Milli-Q distilled water to a final volume of 13 μL. Incubate in a PCR cycler, at 65 °C 5 min and then 4 °C. Keep the tube at 4 ° C, add 4 μL of first-strand Superscript III mix, 1 μL of 100 mM DTT (provided in the reverse transcriptase Superscript III kit), 1 μL RNase out, and 1 μL reverse Superscript III (200 U), and incubate in the PCR cycler at 25 °C for 5 min, 50 °C for 1 h, and 70 °C for 15 min. Stock cDNAs at −80 °C.
4. cDNA level is measured here using a Droplet Digital PCR System developed by Bio-Rad. Briefly, a qPCR mix with cDNA and TaqMan primers are distributed into approximately 10,000–20,000 droplets. cDNAs are then PCR-amplified in a thermal cycler, and the number of positive and negative droplets is read.
5. For each reaction, mix 1 μL cDNAs, 11 μL ddPCR Supermix, and 9 μL with Milli-Q distilled water (*see* **Note 6**).
6. Generate the lipid droplets in the Bio-Rad cartridge with 20 μL of the reaction mix and 50 μL of oil. Avoid making bubbles.
7. Carefully transfer the 40 μL emulsion in a qPCR 96-well plate.
8. Seal the plate, and immediately perform the PCR following the droplet PCR Bio-Rad procedure. Note that after ddPCR, droplets are stable for a week and can be kept at 4 °C before reading.
9. Droplet reading is performed using the QuantaSoft Bio-Rad Software. The number of positive droplets/μL for each tested gene is normalized against the number of positive droplets/μL RNA 18S (in the case of polysomal mRNAs extracted with TRAP) or Gapdh (in the case of total mRNA).

4 Notes

1. During perfusion, the liver gradually turns pale as PBS flushes out the blood, and the whole mouse body must become stiff.
2. For immunofluorescence, be careful not to create bubbles between the glass slide and the coverslip that could interfere with the image acquisition.
3. During immunofluorescence analysis, carefully verify the correspondence between the plug-in readout and the initial confocal acquisition.
4. For FISH experiments, the time of incubation with TRR may vary depending on the targeted mRNA.

5. For TRAP extractions of PvAPs mRNAs, it is necessary to pool tissues from more than one mouse as the level of mRNA in PvAP is low compared to the astrocyte soma as previously described.
6. For ddPCR, it is not necessary to perform triplicates of the same reaction. It is important to perform all mixes at room temperature. Do not use treated RNAse-free water as DEPC treatment is incompatible with this procedure. To compare cDNAs samples, it is very important to prepare them at the same time. Indeed, they might be variation of RT efficacy which would then artificially modify the results. For the quantification, it is important to calibrate the level of cDNAs and primers in the qPCR mix to have positive and negative droplets.

Acknowledgments

We are grateful to the donors who support the charities and charitable foundations cited below. This work was funded by grants from the *Fondation Vaincre Alzheimer* to Katia Avila (FR-21011), the *Fondation pour la recherche sur la sclérose en plaques* (ARSEP) and the *Fondation pour la Recherche Médicale* (AJE20171039094) to Anne-Cécile Boulay, and the *Association Européenne contre les Leucodystrophies* (ELA) (2012-014C2B and 2022-007C4) and the *Fondation pour la Recherche Médicale* (EQU202303016292) to Martine Cohen-Salmon.

References

1. Cohen-Salmon M, Slaoui L, Mazare N, Gilbert A, Oudart M, Alvear-Perez R, Elorza-Vidal X, Chever O, Boulay AC (2021) Astrocytes in the regulation of cerebrovascular functions. Glia 69:817–841
2. Diaz-Castro B, Robel S, Mishra A (2023) Astrocyte endfeet in brain function and pathology: open questions. Annu Rev Neurosci 46: 101–121
3. Mathiisen TM, Lehre KP, Danbolt NC, Ottersen OP (2010) The perivascular astroglial sheath provides a complete covering of the brain microvessels: an electron microscopic 3D reconstruction. Glia 58:1094–1103
4. Hosli L, Zuend M, Bredell G, Zanker HS, Porto de Oliveira CE, Saab AS, Weber B (2022) Direct vascular contact is a hallmark of cerebral astrocytes. Cell Rep 39:110599
5. Gilbert A, Elorza-Vidal X, Rancillac A, Chagnot A, Yetim M, Hingot V, Deffieux T, Boulay AC, Alvear-Perez R, Cisternino S, Martin S, Taib S, Gelot A, Mignon V, Favier M, Brunet I, Decleves X, Tanter M, Estevez R, Vivien D, Saubamea B, Cohen-Salmon M (2021) Megalencephalic leukoencephalopathy with subcortical cysts is a developmental disorder of the gliovascular unit. elife 10:e71379
6. Freitas-Andrade M, Comin CH, Van Dyken P, Ouellette J, Raman-Nair J, Blakeley N, Liu QY, Leclerc S, Pan Y, Liu Z, Carrier M, Thakur K, Savard A, Rurak GM, Tremblay ME, Salmaso N et al (2023) Astroglial Hmgb1 regulates postnatal astrocyte morphogenesis and cerebrovascular maturation. Nat Commun 14: 4965
7. Lunde LK, Camassa LM, Hoddevik EH, Khan FH, Ottersen OP, Boldt HB, Amiry-Moghaddam M (2015) Postnatal development of the molecular complex underlying astrocyte polarization. Brain Struct Funct 220:2087–2101
8. Iliff JJ, Wang M, Liao Y, Plogg BA, Peng W, Gundersen GA, Benveniste H, Vates GE, Deane R, Goldman SA, Nagelhus EA, Nedergaard M (2012) A paravascular pathway

facilitates CSF flow through the brain parenchyma and the clearance of interstitial solutes, including amyloid beta. Sci Transl Med 4: 147ra111

9. Ezan P, Andre P, Cisternino S, Saubamea B, Boulay AC, Doutremer S, Thomas MA, Quenech'du N, Giaume C, Cohen-Salmon M (2012) Deletion of astroglial connexins weakens the blood-brain barrier. J Cereb Blood Flow Metab 32:1457–1467
10. Boulay AC, Mazeraud A, Cisternino S, Saubamea B, Mailly P, Jourdren L, Blugeon C, Mignon V, Smirnova M, Cavallo A, Ezan P, Ave P, Dingli F, Loew D, Vieira P, Chretien F, Cohen-Salmon M (2015) Immune quiescence of the brain is set by astroglial connexin 43. J Neurosci 35:4427–4439
11. Gilbert A, Vidal XE, Estevez R, Cohen-Salmon M, Boulay AC (2019) Postnatal development of the astrocyte perivascular MLC1/GlialCAM complex defines a temporal window for the gliovascular unit maturation. Brain Struct Funct 224:1267–1278
12. Slaoui L, Gilbert A, Rancillac A, Delaunay-Piednoir B, Chagnot A, Gerard Q, Letort G, Mailly P, Robil N, Gelot A, Lefebvre M, Favier M, Dias K, Jourdren L, Federici L, Auvity S, Cisternino S, Vivien D, Cohen-Salmon M, Boulay AC (2023) In mice and humans, brain microvascular contractility matures postnatally. Brain Struct Funct 228: 475–492
13. Boulay AC, Saubaméa B, Adam N, Chasseigneaux S, Mazaré N, Gilbert A, Bahin M, Bastianelli L, Blugeon C, Perrin S, Pouch J, Ducos B, Le Crom S, Génovésio A, Chretien F, Declèves X, Laplanche JL, Cohen-Salmon M (2017) Translation in astrocyte distal processes sets molecular heterogeneity at the gliovascular interface. Cell Discovery 3: 17005
14. Mazare N, Oudart M, Moulard J, Cheung G, Tortuyaux R, Mailly P, Mazaud D, Bemelmans AP, Boulay AC, Blugeon C, Jourdren L, Le Crom S, Rouach N, Cohen-Salmon M (2020) Local translation in perisynaptic astrocytic processes is specific and changes after fear conditioning. Cell Rep 32:108076
15. Avila-Gutierrez K, Slaoui L, Alvear-Perez R, Kozlowski E, Oudart M, Augustin E, Claveau C, Mailly P, Monnet H, Mignon V, Saubamea B, Boulay AC, Cohen-Salmon M (2024) Dynamic local mRNA localization and translation occurs during the postnatal molecular maturation of perivascular astrocytic processes. Glia 72:777
16. Doyle JP, Dougherty JD, Heiman M, Schmidt EF, Stevens TR, Ma G, Bupp S, Shrestha P, Shah RD, Doughty ML, Gong S, Greengard P, Heintz N (2008) Application of a translational profiling approach for the comparative analysis of CNS cell types. Cell 135: 749–762
17. Gong S, Zheng C, Doughty ML, Losos K, Didkovsky N, Schambra UB, Nowak NJ, Joyner A, Leblanc G, Hatten ME, Heintz N (2003) A gene expression atlas of the central nervous system based on bacterial artificial chromosomes. Nature 425:917–925
18. Heiman M, Schaefer A, Gong S, Peterson JD, Day M, Ramsey KE, Suarez-Farinas M, Schwarz C, Stephan DA, Surmeier DJ, Greengard P, Heintz N (2008) A translational profiling approach for the molecular characterization of CNS cell types. Cell 135:738–748
19. Foo LC, Dougherty JD (2013) Aldh1L1 is expressed by postnatal neural stem cells in vivo. Glia 61:1533–1541
20. Schindelin J, Arganda-Carreras I, Frise E, Kaynig V, Longair M, Pietzsch T, Preibisch S, Rueden C, Saalfeld S, Schmid B, Tinevez JY, White DJ, Hartenstein V, Eliceiri K, Tomancak P, Cardona A (2012) Fiji: an open-source platform for biological-image analysis. Nat Methods 9:676–682
21. Linkert M, Rueden CT, Allan C, Burel JM, Moore W, Patterson A, Loranger B, Moore J, Neves C, Macdonald D, Tarkowska A, Sticco C, Hill E, Rossner M, Eliceiri KW, Swedlow JR (2010) Metadata matters: access to image data in the real world. J Cell Biol 189: 777–782
22. Haase R, Royer LA, Steinbach P, Schmidt D, Dibrov A, Schmidt U, Weigert M, Maghelli N, Tomancak P, Jug F, Myers EW (2020) CLIJ: GPU-accelerated image processing for everyone. Nat Methods 17:5–6
23. Ollion J, Cochennec J, Loll F, Escude C, Boudier T (2013) TANGO: a generic tool for high-throughput 3D image analysis for studying nuclear organization. Bioinformatics 29: 1840–1841
24. Srinivasan R, Lu TY, Chai H, Xu J, Huang BS, Golshani P, Coppola G, Khakh BS (2016) New transgenic mouse lines for selectively targeting astrocytes and studying calcium signals in astrocyte processes in situ and in vivo. Neuron 92:1181–1195
25. Pestka S, Brot N (1971) Studies on the formation of transfer ribonucleic acid-ribosome complexes. IV. Effect of antibiotics on steps of bacterial protein synthesis: some new ribosomal inhibitors of translocation. J Biol Chem 246:7715–7722
26. Boulay AC, Saubamea B, Decleves X, Cohen-Salmon M (2015) Purification of mouse brain vessels. J Vis Exp 105:532208

Chapter 9

Imaging Techniques for the Analysis of the Blood–Brain Barrier and Gliovascular Unit Dysfunction

Samantha Golf, Carmen Muñoz-Ballester, Benjamin Heithoff, Kijana George, and Stefanie Robel

Abstract

Proper blood–brain barrier (BBB) function is essential to maintain homeostasis, keeping the central nervous system healthy. Disruption of the BBB has been described in nearly all neurological disease and injury contexts. Given the various physical and metabolic roles of the BBB and gliovascular unit (GVU), dysfunction should be assessed across multiple levels to pinpoint the exact nature of the impairment. Here, we describe imaging techniques for evaluating various aspects of BBB and GVU integrity and function, tailored for use in mouse models, and accompanied by protocols for quantitative image analysis. These imaging techniques complement classical dye tracer experiments and reveal underlying molecular causes of BBB dysfunction.

Key words Blood–brain barrier, BBB, Impairment, Tight junctions, Astrocyte endfeet, Endothelial cells, Basement membrane

1 Introduction

The vessels which deliver blood throughout the central nervous system exhibit specialized barrier properties. This barrier restricts and controls the entry of ions, molecules, and cells into the brain from the periphery to protect the brain from potentially harmful substances or toxic concentrations (e.g., concentrations of the excitatory neurotransmitter and amino acid glutamate are 5–10× higher in the blood compared with the brain). Yet, the barrier must be permissible to the entrance of oxygen, nutrients, hormones, and other substances important to support and regulate brain function. The exclusionary properties of this blood–brain barrier (BBB) were first described in early rodent dye tracer experiments [1, 2]. Since initial identification of the homeostatic role of the BBB, disruption

Samantha Golf and Carmen Muñoz-Ballester have equally contributed to this work.

Barbara Di Benedetto (ed.), *Astrocytes: Methods and Protocols*, Methods in Molecular Biology, vol. 2896, https://doi.org/10.1007/978-1-0716-4366-2_9,

of BBB integrity and function has been identified as a hallmark of various neurological diseases and injuries, reviewed in [3]. While tracer studies are able to determine whether peripheral factors nonspecifically enter the brain, they cannot distinguish whether leakiness is due to disruption of BBB properties or dysfunction in scar/border formation by glial cells. Further, the restriction of peripheral factors from the CNS is dependent upon the combined function of endothelial cells, astrocytes, and pericytes, as well as the specialized extracellular matrix secreted by these cells. The selection of protocols described here provides a toolset for assessing multiple aspects of BBB and gliovascular unit (GVU) dysfunction. These methods encompass techniques for the quantification of the distribution of (a) BBB proteins expressed within the endothelium (tight junction protein zona occludens-1 (ZO-1) or glucose transporter type 1 (GLUT1), which is part of the metabolic barrier); (b) proteins within astrocytic endfeet, including aquaporin 4 (AQP4) and β-dystroglycan; (c) quantification of basement membrane protein expression, including laminin and collagen IV; and (d) quantification of BBB permeability using the glycoprotein fibrinogen that we have successfully optimized in the last years [4, 5]. Additionally provided are methods for the qualitative assessment of tight junctions, astrocytic endfeet, and basement membrane by electron microscopy (EM), as well as astrocytic endfoot coverage of vasculature using astrocyte reporter mice.

2 Materials

2.1 Study of Proteins/Structures Related to BBB and GVU Function

1. 1× phosphate buffered saline (PBS), pH 7.3.
2. 4% paraformaldehyde (PFA), diluted in 1× PBS pH 7.3.
3. Surgical tools: large surgical scissors, small dissection scissors, curved forceps, spatula (with microspoon), serrated blunt forceps, and vaculet blood collection set (*see* **Note 1**).
4. 15 mL Falcon tubes.
5. Vibratome.
6. 24-well plates.
7. 10× PO_4 buffer, pH 7.2–7.4: 65 g $NaH_2PO_4 \times H_2O$, 15 g NaOH, 2 N HCl, and distilled H_2O to total volume of 500 mL.
8. Glycerol-based storing solution: 150 mL glycerol, 150 mL ethylene glycol, 50 mL 10× PO_4 buffer pH 7.2–7.4, 150 mL distilled H_2O.
9. Rocking platform at room temperature.
10. Blocking buffer: 10% normal goat serum (NGS), 0.5% Triton ×-100, diluted in PBS pH 7.3.
11. Antibodies (*see* Tables 1 and 2).

Table 1
List of primary antibodies required for the outlined methods for analysis of BBB and GVU dysfunction

Primary antibodies						
Name	**Manufacturer**	**Catalog #**	**RRID**	**Species raised in**	**Concentration**	**Function/relevance**
Fibrinogen	Agilent	A008002	AB_578481	Rabbit	1:500	Plasma protein that can only penetrate the brain parenchyma when the blood–brain barrier is disrupted. It causes brain inflammation and damage (reviewed in [15]).
ZO-1	Abcam	ab96587	AB-18.0006	Rabbit	1:100	Tight junction protein involved in the maintenance of the physical barrier between blood and brain.
Aquaporin-4 (AQ4)	Sigma-Aldrich	A5971	AB_258270	Rabbit	1:400	Water channel protein expressed in astrocytes' endfeet in association with the vessel. It can be used to assess the integrity of astrocyte endfeet.
B-dystroglycan	Leica Biosystems	NCL-b-DG	AB_442043	Mouse	1:200	Glycoprotein with a structural role involved in linking the basement membrane components and the astrocyte endfeet. It can be used to assess the integrity of astrocyte endfeet.
Pan-laminin	Bio-Rad	AHP2491	AB_2133777	Rabbit	1:1000	Secreted basement membrane protein participating in cell attachment. This specific antibody recognizes the isoform secreted from both endothelial cells and astrocytes.
Collagen IV	Bio-Rad	2150-1470	AB_2082660	Rabbit	1:1000	Major protein of the basement membrane.
GLUT1	Novus Biologicals	NB300-666	AB_10000485	Rabbit	1:1000	Main glucose transporter in endothelial cells. It is an essential part of the metabolic barrier in the brain.
CD-31	BD Biosciences	550274	AB_393571	Rat	1:100	Membrane protein involved in adhesion expressed at the surface of endothelial cells.

Table 2
List of secondary antibodies required for the outlined methods for analysis of BBB and GVU dysfunction

Secondary antibodies					
Name	**Manufacturer**	**Catalog #**	**RRID**	**Species raised in**	**Concentration**
Rabbit AF 488	Jackson ImmunoResearch	111-546-144	AB_2338057	Goat	1:1000
Mouse AF 488	Jackson ImmunoResearch	115-546-003	AB_2338859	Goat	1:1000
Rat AF 488	Jackson ImmunoResearch	112-546-003	AB_2338364	Goat	1:1000
Rabbit Cy3	Jackson ImmunoResearch	111-166-144	AB_2338011	Goat	1:1000
Mouse Cy3	Jackson ImmunoResearch	115-166-003	AB_2338699	Goat	1:1000
Rat Cy3	Jackson ImmunoResearch	112-166-003	AB_2338253	Goat	1:1000
Rabbit AF647	Jackson ImmunoResearch	111-606-144	AB_2338083	Goat	1:1000
Mouse AF647	Jackson ImmunoResearch	115-606-003	AB_2338921	Goat	1:1000
Rat Alexa-647	Jackson ImmunoResearch	112-606-003	AB_2338406	Goat	1:1000

12. DAPI 1:1000.
13. Rocking platform at 4 °C.
14. Paintbrushes.
15. Positive charged glass microscope slides.
16. Mounting media (DAPI-free).
17. Coverslips.
18. Clear nail polish.
19. Epifluorescence microscope.
20. Confocal microscope.
21. Image analysis software.
22. Aldh1l1-tdTomato mice: Tg(Aldh1l1-tdTomato)TH6Gsat.

2.2 Evaluation of ZO-1 Expression

1. Pepsin solution: 100 mg Pepsin, 100 mL of 10 mM HCl.
2. Thermomixer.
3. PBS-T: 150 μL Tween-20, 1 L of 1× PBS pH 7.3.
4. 3% H_2O_2, diluted in millipore H_2O.
5. Antibodies (*see* Tables 1 and 2).

2.3 Electron Microscopy to Assess BBB and GVU Components

1. 1× PBS, pH 7.4.
2. Fixation solution: 2.5% glutaraldehyde, 4% PFA in 0.1 M sodium cacodylate (Caco) buffer pH 7.4.

3. Vibratome.
4. Dissection microscope.
5. Tissue puncher.
6. 1% osmium diluted in 0.1 M Caco buffer, pH 7.4.
7. 0.2 M Caco buffer, pH 7.4.
8. Rotator.
9. Acetone (100%, 95%, 80%, 50% in H_2O).
10. EPON 812 with accelerator.
11. Embedding solution: 1 part 100% acetone to 1 part EPON 812 with accelerator.
12. Incubator at 60–70 °C.
13. Ultramicrotome.
14. EM mesh grid.
15. Transmission electron microscope.

3 Methods

3.1 Assessment of Proteins/Structures Related to BBB and GVU Function

To determine which aspects of BBB or GVU function may be affected (tight junctions, endothelial transporter expression, extracellular matrix and basement membrane, pericytes, or proteins expressed in astrocyte endfeet), immunohistochemistry against proteins involved in these structures and proteins that can permeate a disrupted BBB (e.g., fibrinogen) can be performed (Fig. 1, Table 1). Importantly, while the absence of tight junctional proteins indicates a loss or impairment of tight junctions and thus a broken BBB, astrocyte endfeet proteins such as AQP4 or β-dystroglycan can be absent, while endfeet are still present on the vasculature. Their absence may affect the BBB in yet to be determined ways but is not a direct readout for BBB dysfunction. Whether endfeet proteins are absent while endfeet are still in place along the vasculature can be assessed using astrocyte reporter mice, such as Aldh1l1-tdTomato, expressing a fluorescent protein in astrocytes (Fig. 1c). Alternatively, astrocyte endfeet coverage can be assessed with electron microscopy (EM) (Fig. 2).

1. Perfuse the mouse transcardially as described in [6] with PBS (*see* **Note 2**) for several minutes (min) until the liver is clear of blood. Follow with 4% paraformaldehyde (PFA) perfusion for 7 min or until the tail and neck are stiff.
2. Extract the brain from the skull and immerse it in 15 mL Falcon tube with 10 mL of 4% PFA overnight at 4 °C.

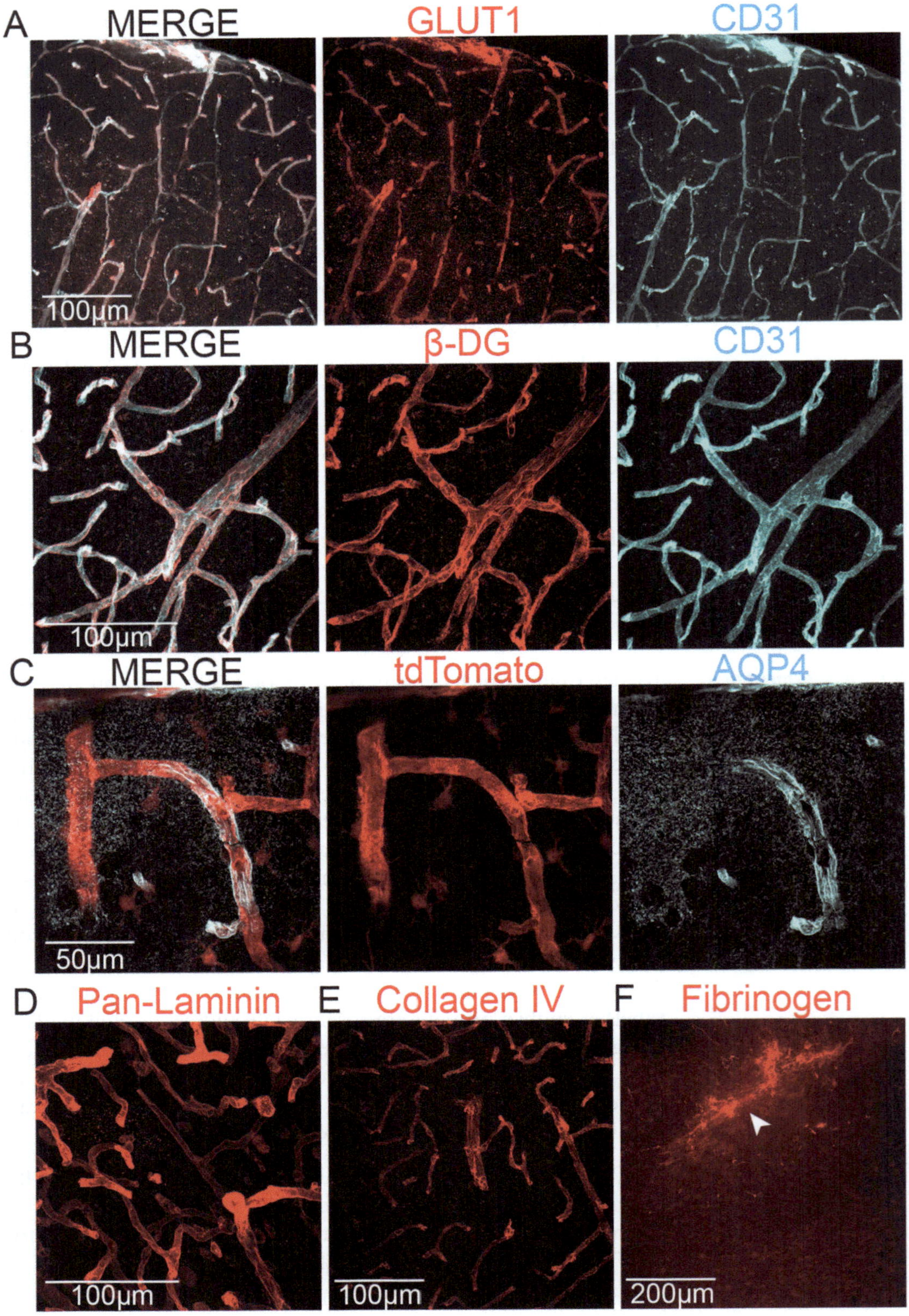

Fig. 1 Immunohistochemistry for proteins related to BBB and GVU function. Representative images from mouse cortex of (**a**) glucose transporter type 1 (GLUT1, red) and cluster of differentiation 31 (CD31, cyan), (**b**) β-dystroglycan (β-DG, red) and CD31 (cyan), (**c**) astrocytic endfeet in reporter mice (Aldh1l1-tdTomato, red) and aquaporin 4 (AQP4, cyan), (**d**) pan-laminin, (**e**) collagen IV, and (**f**) fibrinogen

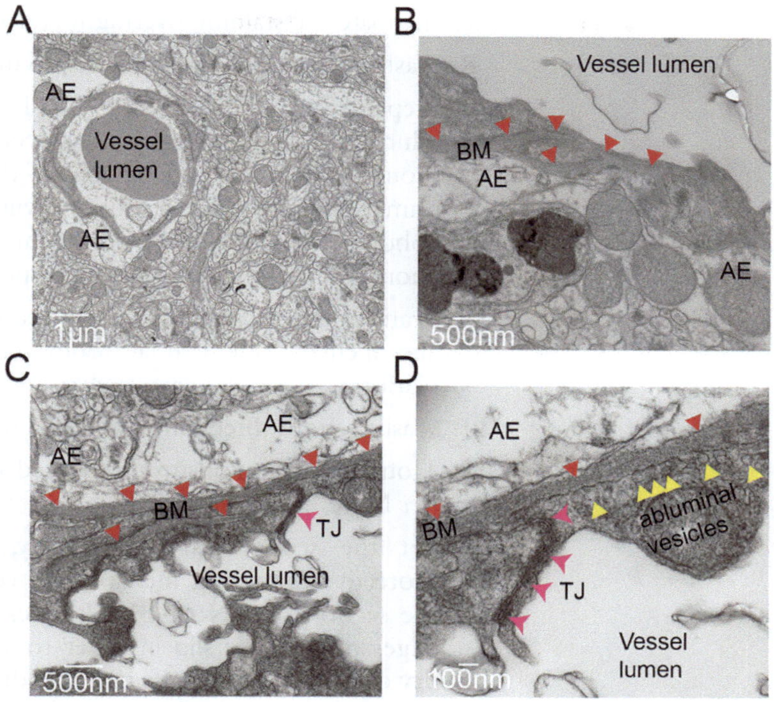

Fig. 2 Electron microscopy for the qualitative assessment of BBB and GVU components. Representative images of (**a**) astrocytic endfeet (AE) in contact with a vessel, taken at low magnification; (**b**) basement membrane (BM, red arrows) sandwiched between astrocytic endfeet (AE) and a vessel, captured with mid-range magnification; (**c**) astrocytic endfeet (AE) and two distinct layers of basement membrane (BM, red arrows) wrapping a vessel with tight junctions (TJ, pink arrows) present at the interface between two adjacent endothelial cells comprising the vessel, also taken with mid-range magnification; and (**d**) higher magnification image of a portion of the field of view in **c**, enabling visualization of abluminal vesicles primed for transcytosis (yellow arrows). Basement membrane (BM, red arrows) and tight junctions (TJ, pink arrows) can also be seen clearly at this higher magnification

3. The following day, wash the brain with filtered PBS, and store the washed brain in filtered PBS (*see* **Note 3**).
4. Section the brains using a vibratome in 50 µm slices. Place the slices in PBS for short-term storage at 4 °C. For long-term storage, *see* **Note 4**.
5. Prepare 24-well plate with 300 µL of primary antibodies diluted in blocking buffer (*see* **Note 5**). This volume is sufficient for eight to ten slices. Look at Table 1 for antibodies.
6. Transfer the slices of one animal to a well with the primary antibody dilution using a paint brush. Leave one section in blocking buffer without primary antibody. This will be the negative control of the staining.

7. Incubate the staining overnight at 4°C on a rocking platform.
8. Wash sections three times in PBS, 5 min per wash (*see* **Note 6**).
9. Prepare 24-well plate with DAPI and secondary antibody diluted in blocking buffer. The secondary antibody selected should be directed against IgG of the host of the associated primary antibody. Every antibody will be conjugated to a fluorophore. Antibodies against different species should not have a fluorophore in the same emission spectrum.
10. Transfer the slices to wells with secondary antibody dilution, and incubate for 1 h at room temperature on a rocking platform.
11. Wash sections three times in PBS, 5 min per wash (*see* **Note 6**).
12. Mount brains on positively charged slides using a paint brush (*see* **Note 7**).
13. Let the sections dry completely, and add fluorescence-protective mounting media and a coverslip that fully covers the sections (*see* **Note 8**). Use clear nail polish to coat the edges of the coverslip in order to seal it in place. Let slides dry overnight. If they are not fully dry, the coverslip can move when using oil objectives.
14. Acquire images using a confocal microscope, using the negative control of the staining to determine the laser power and photomultiplier gain. Z-stacks encompassing the thickness of the section are recommended. Comparable neuroanatomical regions should be sampled across groups. The imaging strategy may vary depending on the type of experiment, but for having an accurate estimation of expression across the entire cortex, we most often collect a total of five images across three to five sections. Laser power and photomultiplier gain must remain constant within the whole experiment. Only stainings within the same batch can be directly compared.
15. Images can be processed in image analysis software, such as FIJI. Z-projections are recommended for better visualization and analysis. Different types of analysis can be performed. Some of the most common ones are quantifying fluorescence intensity or distribution of the staining along vasculature. For the latter, we recommend using cluster of differentiation 31 (CD31) staining as a marker for the structure of the vessels.

3.2 Evaluation of ZO-1 Expression

Endothelial cells lining the vasculature within the central nervous system, unlike those in the periphery, are largely not fenestrated and exhibit high levels of tight junctions distributed along vessels [7]. Tight junctions are essential for reducing paracellular transport between the blood and neural tissue [8]. These tight junctions are comprised of several different proteins including ZO-1, with

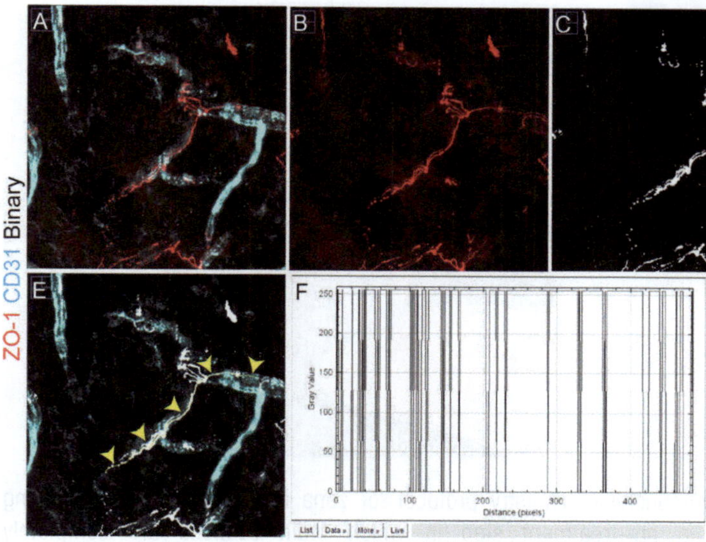

G

Distance_(pixels)	Gray_Value
0	0
1	255
2	255
3	0
4	0
5	0
6	0
7	255
8	0
9	0
10	0

Fig. 3 Step-by-step procedure for quantitative analysis of zona occludens 1 (ZO-1) coverage over a vessel. The panels are in the sequential order needed for the analysis. (**a**) Initial staining with ZO-1 (red) and CD31 (cyan) in cortex. (**b**) Image of ZO-1 (red) after splitting the channels. (**c**) Same image after binarization. Positive pixels (255) are shown in white. (**d**) Overlap of binarized ZO-1 and CD31. (**e**) Yellow freehand line following the profile of ZO-1 over the vessel that will be used to determine the percentage of coverage of ZO-1. The yellow arrows point to the yellow line. (**f**) Plot profile of the drawn line. It represents the value of each pixel (either 0 or 255) along the distance of the drawn line. Positive pixels get a value of 255 and negative pixels a value of 0. (**g**) Example of the values taken for each pixel. These values will be entered into the equation in 3.2, **step 22**.

dissociation of ZO-1 from the tight junction complex known to produce increased BBB permeability, reviewed under [9]. Due to the expression pattern of ZO-1 along vessels (Fig. 3b), in which it winds around the vessel like a zipper, quantification of ZO-1 staining requires a specifically tailored methodology, as described below (Fig. 3).

1. Perfuse anesthetized mouse transcardially.
2. Extract the brain from the skull, and immerse it in 15 mL Falcon tube with 10 mL of 4% PFA overnight at 4 °C.
3. The following day, wash the brain with filtered PBS, and store the washed brain in filtered PBS (*see* **Note 3**).
4. Section the brains using a vibratome in 50 μm slices. Place the slices in PBS for short-term storage at 4 °C. For long-term storage, *see* **Note 4**.
5. Prepare in 1.5 mL tubes with pepsin solution (up to 1.4 mL). Prepare one tube per two to three brain sections.
6. Add sections to the tubes using a paint brush, and incubate them for 20 min at 37 °C (no shaking). Make sure that the sections are submerged in the solution (*see* **Note 9** and Fig. 4).

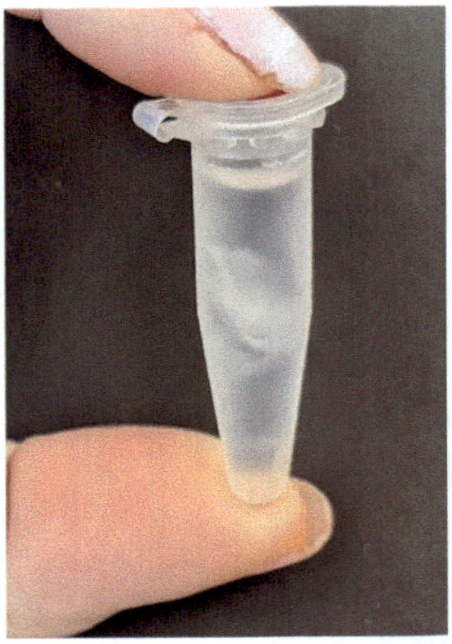

Fig. 4 Immunohistochemistry protocol for zona occludens 1 (ZO-1) staining requires a pre-treatment step in which brain sections are completely submerged in pepsin solution in 1.5 mL Eppendorf tubes and subsequently heated at 37 °C for 20 min

7. Transfer the slices to a 24-well plate, and wash the sections in PBS-T two times for 5 min each, rocking at room temperature (*see* **Note 6**).
8. Incubate slices in 3% H_2O_2 for 10 min at room temperature (with no shaking).
9. Wash sections three times in PBS, 5 min per wash (*see* **Note 6**).
10. Incubate anti-ZO-1 primary antibody diluted 1:100 and anti-CD31 1:100 in blocking buffer for at least 48 h at 4 °C in a rocking platform.
11. Wash sections three times in PBS, 5 min per wash (*see* **Note 6**).
12. Incubate in DAPI and fluorescent secondary antibodies against rabbit and rat IgG antibodies for 1 h at room temperature.
13. Wash sections in PBS three times, 5 min per wash (*see* **Note 6**).
14. Mount brains on positively charged slides using a paint brush. Let sections dry and add fluorescence-protective mounting media, and let them dry overnight.
15. Use a confocal microscope at high magnification, for example 60×, making sure that comparable neuroanatomical regions are sampled across groups. As in the previous section, the imaging strategy may vary depending on the type of experiment, but we recommend sampling multiple regions of interest (ROIs) across several sections. For having an accurate estimation of expression across the entire cortex, we most often collect a total of five images across three to five sections.

16. Open acquired images in image analysis software (for example, FIJI), and make a maximum intensity z-projection of all channels.
17. Split the different channels.
18. In the ZO-1 channel, threshold the staining and binarize the image. Normally, the automated threshold and binarization of the image analysis software accurately distinguish the signal from background, but the experimenter may need to modify these parameters to ensure that the pixels highlighted correspond to the real staining. For example, this determination can be made based on negative control staining.
19. Create a new composite image containing the CD31 staining channel, as well as the binarized ZO-1 staining as a separate channel. For this, both images will need to have the same depth (8-bit). For example, in the FIJI software, select the image to be converted and click Image > Type > 8-bit.
20. Use the freehand line tool to trace the ZO-1 signal along the vessel, using the CD31 as a guide and including areas where ZO-1 signal is discontinuous.
21. Export the gray value of each pixel along the drawn line (pixel values should either be 0 or 255) (*see* **Note 10**). To do this in FIJI, select Analyze > Plot Profile > List.
22. To calculate the percent of vessel length covered by ZO-1, count the number of pixels with a gray value of 255 (these are the ZO-1-positive pixels), and divide it by the total number of pixels comprising the line, following the equation below:

$$\frac{\text{Number of 255 value pixels}}{\text{Number of total pixels measured}} \times 100 = \%\text{vessel length covered by ZO} - 1$$

The higher the percentage obtained, the higher the coverage of the vasculature by the tight junction protein ZO-1.

3.3 Electron Microscopy to Assess BBB and GVU Components

Electron microscopy provides the highest resolution assessment of some BBB properties (tight junctions and transcytosis), basement membrane structure, and gliovascular unit. It is useful to assess whether astrocyte endfeet cover the vasculature when endfeet proteins localization is impaired. It can be used to quantify transcytosis and assess localization of vesicles within the endothelium (luminal, abluminal). It also allows for a closer look at tight junctions and can help determine structural abnormalities that are impossible to see using IHC combined with light microscopy.

1. Transcardially perfuse the anesthetized mouse as described in Subheading 3.1 first with PBS for 5 min followed by fixation solution.
2. Extract the brain from the skull and section into 100–500 μM slices.

3. Under a dissection microscope, capture tissue punches from areas of interest.
4. Place tissue punches in fixation solution overnight at 4 °C.
5. Wash samples two times for 15 min in 0.1 M Caco buffer, pH 7.4.
6. Place samples in 1% osmium diluted in 0.1 M Caco buffer, pH 7.4 for 50 min at room temperature on a rotator in the dark.
7. Wash samples four times in 0.2 M Caco buffer, pH 7.4.
8. Dehydrate samples in 50%, 80%, and 95% acetone in water (5 min at each concentration).
9. Wash in 100% acetone for 15 min for four times.
10. Equilibrate samples for embedding in a solution containing equal parts of 100% acetone and EPON 812 with accelerator on a rotator overnight at room temperature.
11. Embed samples in 100% EPON 812 with the accelerator at 60–70 °C overnight.
12. Cut ultrathin sections.
13. Image on a transmission electron microscope (we used a Tecnai T12 120 kV to obtain the images in Fig. 2).

4 Notes

1. Once tools have come in contact with PFA, they can no longer be used for in vivo procedures or with unfixed tissue and cells.
2. PBS should be filtered through a sterile 0.2 μm filter.
3. Brains can remain stored at 4 °C for months until sliced.
4. For long-term storage, slices can be kept in a glycerol-based storing solution. Before starting the staining, wash the slices in PBS twice for 10 min at room temperature in 24-well plates.
5. It is important that antibodies raised in the same host are not used at the same time in the staining or it will not be possible to distinguish them. For example, if rabbit anti-GLUT1 is used, any other antibody raised in rabbit should not be used.
6. For washing steps, transfer the sections to a different well with the washing solution using a paint brush.
7. Prior to mounting all sections, check the staining on one section on the slide in a fluorescence microscope. If it worked, mount all sections. If the staining did not work, identify potential mistakes in following the protocol, and repeat antibody staining.

8. Make sure that the thickness of the coverslip is compatible with the objective that will be used. Most objectives require coverslips of number 1.5, which have a thickness between 0.16 and 0.19 mm.
9. After treatment, slices can be fragile. Handle with care.
10. If the values offered by the analysis software differ from 0 or 255, the software is probably interpolating results. Disable this option in your software. Also, make sure that the results are provided in pixels to have discrete values in the distance parameter.

Acknowledgments

The work was supported by the National Institute of Neurological Disorders and Stroke (grant R01NS121145).

References

1. Ehrlich P (1885) Das Sauerstoff-Bedürfniss des Organismus: eine farbenanalytische Studie. August Hirschwald
2. Goldmann EE (1909) Die äussere und innere Sekretion des gesunden und kranken Organismus im Lichte der "vitalen Färbung": Teil 1. H. Laupp
3. Profaci CP, Munji RN, Pulido RS, Daneman R (2020) The blood–brain barrier in health and disease: important unanswered questions. J Exp Med 217:e20190062. https://doi.org/10.1084/jem.20190062
4. Heithoff BP, George KK, Phares AN et al (2021) Astrocytes are necessary for blood-brain barrier maintenance in the adult mouse brain. Glia 69:436–472. https://doi.org/10.1002/glia.23908
5. George KK, Heithoff BP, Shandra O, Robel S (2022) Mild traumatic brain injury/concussion initiates an atypical astrocyte response caused by blood-brain barrier dysfunction. J Neurotrauma 39:211–226. https://doi.org/10.1089/neu.2021.0204
6. Marinković P, Godinho L, Misgeld T (2015, 2015) Generation of tissue sections for screening Thy1 mouse lines. Cold Spring Harb Protoc. https://doi.org/10.1101/pdb.prot087684
7. Dotiwala AK, McCausland C, Samra NS (2024) Anatomy, head and neck: blood brain barrier. In: StatPearls. StatPearls Publishing, Treasure Island (FL)
8. Luissint A-C, Artus C, Glacial F et al (2012) Tight junctions at the blood brain barrier: physiological architecture and disease-associated dysregulation. Fluids Barriers CNS 9:23. https://doi.org/10.1186/2045-8118-9-23
9. Hawkins BT, Davis TP (2005) The blood-brain barrier/neurovascular unit in health and disease. Pharmacol Rev 57:173–185. https://doi.org/10.1124/pr.57.2.4

Chapter 10

Assessment of Blood–Brain Barrier Integrity Using Dyes and Tracers

Carmen Muñoz-Ballester, Samantha Golf, Benjamin Heithoff, Kijana George, and Stefanie Robel

Abstract

Under physiological conditions, the blood–brain barrier (BBB) is required for regulating the passage of substances between the bloodstream and the brain. To date, disruption of the BBB has been associated with nearly all neurological diseases. Dye and tracer experiments have been used to assess BBB integrity for decades, yet quantitative assessment of these experiments has only recently been implemented. Further, interpretation of these results requires the consideration of tracer-specific properties. Here, we describe a protocol for evaluating BBB leakage using the small molecule cadaverine in a quantitative way. We also cover advantages/disadvantages of using some reagents and highlight limitations in data interpretations based on assays used to assess BBB permeability.

Key words Blood–brain barrier, BBB, Dye, Tracer, Cadaverine, Impairment, Leakage, Permeability

1 Introduction

Why is the blood–brain barrier (BBB) important?

The vasculature of the central nervous system is unique because of its barrier properties that prevent unregulated passage of ions, molecules, and cells. These barrier properties include low transcytosis levels and a lack of paracellular passage [1]. Tight junctions are expressed in endothelial cells to ensure the physical exclusion of the blood content from the brain. Disruption of these endothelial tight junctions compromises the integrity of the blood–brain barrier (BBB), resulting in increased permeability. This permeability is associated with almost all brain disorders (reviewed in [2]).

BBB permeability is often assessed using intravenous injection of fluorescent tracers or dyes of different molecular weights. These tracers are normally excluded from entering the brain due to the

The authors Carmen Muñoz-Ballester and Samantha Golf have equally contributed to this work.

Barbara Di Benedetto (ed.), *Astrocytes: Methods and Protocols*, Methods in Molecular Biology, vol. 2896,
https://doi.org/10.1007/978-1-0716-4366-2_10,

function of the BBB. This is an excellent assay to assess BBB permeability as long as the experimenter is clear on the limitations of the respective tracers and interpretations that can be drawn from the results of the experiments.

Traditionally, Evans blue was used to assess BBB permeability. Evans blue is an azo dye that appears blue in brightfield microscopy but also fluoresces with emission at 680 nanometer (nm; far red) and can thus be analyzed using fluorescent microscopy. Evans blue binds to albumin (66.5 kDa) and is thus often thought to reflect BBB permeability to molecules of the size of albumin. Yet, contrary to common belief, there is a significant amount of free Evans blue (~960 Da) present at the typically used concentration of 2% (w/v) and 4 mL/kg bodyweight [3], which needs to be considered when interpreting the extent and nature of BBB damage. In its free form, Evans blue can be taken up by cells (often neurons), which makes assessment of small BBB damage easier. Bound to albumin, Evans blue may be entering the brain not only paracellularly due to impaired tight junctions or vasculature but can also enter via transport or increased transcytosis [4]. Evans blue inhibits glutamate uptake via glutamate transporters and also is a kainate receptor antagonist [5, 6]. Thus, it should not be used in combination with electrophysiology experiments. Lastly, Evans blue has been reported to have toxic effects on the endothelium within hours after administration, suggesting that caution has to be exercised when designing and interpreting Evans blue experiments [3].

These limitations can be circumvented by using different tracers. Albumin coupled to fluorophores can assess changes in vascular and tight junctional integrity when these changes are so large as to allow entry of molecules 66.5 kDa and smaller. Yet, this approach can also indicate changes in transcytosis or transcellular transport because albumin is a substrate for transporters and can cross the endothelium via transcytosis (reviewed in [7]). Dextrans of varying molecular weights (3–2000 kDa) coupled to fluorophores are useful for assessment of the extent of paracellular BBB permeability (Fig. 1). Cadaverine, a molecule of similar size as free Evans blue dependent on the fluorophore that it is coupled to (640 Da when coupled to Alexa Fluor 488 and 950 Da when coupled to Alexa Fluor 555), enters the brain paracellularly when the vasculature or tight junctions are damaged. Importantly, cadaverine enters the healthy brain parenchyma in some brain areas, i.e., the hypothalamic median eminence, which has fenestrated capillaries and can thus serve as a positive control. In aging animals, cadaverine is leakier [8] and needs to be quantitatively compared to the experimental condition. Alexa Fluor 488, Alexa Fluor 555, and Alexa Fluor 647 contain carboxyl and/or sulfonic groups, which render the dextran or cadaverine conjugate PFA-fixable, so that there is no diffusion within the tissue after perfusion of the animal.

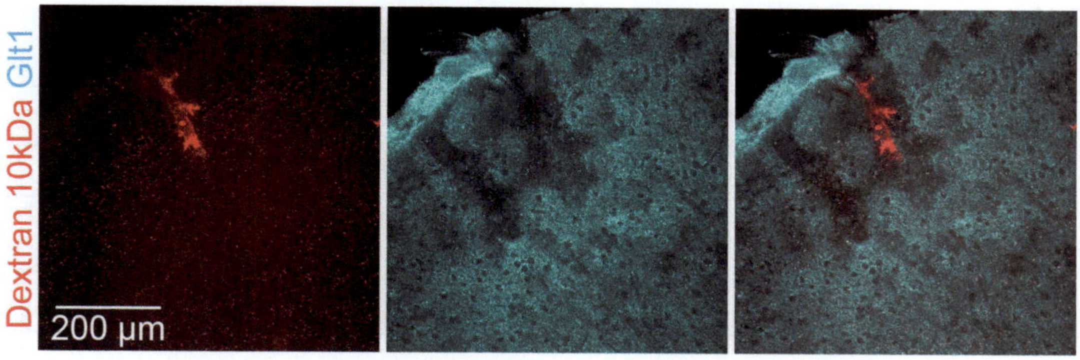

Fig. 1 Peripheral injection of dextran for assessing extent of paracellular BBB permeability. Fluorophore-conjugated dextran of 10 kDa size (red) permeating from vasculature into neural tissue in cortical regions overlapping with areas where glutamate transporter 1 (Glt1, cyan) exhibits decreased expression after a traumatic brain injury

In this protocol, we provide a quantitative method for assessing BBB permeability using the small molecule cadaverine that we have successfully optimized in the last years [9, 10].

2 Materials

1. 1× phosphate buffered saline (PBS), pH 7.3.
2. 4% Paraformaldehyde (PFA), diluted in 1× PBS, pH 7.3.
3. Surgical tools: large surgical scissors, small dissection scissors, curved forceps, spatula (with microspoon), serrated blunt forceps, and vaculet blood collection set (*see* **Note 1**).
4. 15 mL Falcon tubes.
5. Vibratome.
6. 24-Well plates.
7. 10× PO_4 buffer, pH 7.2–7.4: 65 g $NaH_2PO_4 \times H_2O$, 15 g NaOH, 2 N HCl, distilled H_2O to total volume of 500 mL.
8. Glycerol-based storing solution: 150 mL glycerol, 150 mL ethylene glycol, 50 mL 10× PO_4 buffer pH 7.2–7.4, 150 mL distilled H_2O.
9. Rocking platform at room temperature.
10. Paintbrushes.
11. Positive charged glass microscope slides.
12. Mounting media (DAPI-free).
13. Coverslips.
14. Clear nail polish.
15. Epifluorescence microscope.

Table 1
List of dyes required for the outlined methods for analysis of BBB dysfunction

Dyes				
Name	**Manufacturer**	**Catalog #**	**RRID**	**Concentration**
DAPI	ThermoFisher	D1306	AB_2629482	1:1000
AF555 Cadaverine	ThermoFisher	A30677	N/A	0.33 mg/mouse
AF488 Cadaverine	ThermoFisher	A30676	N/A	0.33 mg/mouse

16. Confocal microscope.
17. Image analysis software.
18. Fluorophore-conjugated cadaverine (*see* Table 1).
19. 1× PBS, sterile.
20. Isoflurane.
21. Vaporizer for anesthesia (for example, SomnoFlo® low-flow electronic vaporizer).
22. Anesthesia box.
23. Insulin syringes.

3 Methods

This protocol describes two methods to quantify leakage of cadaverine conjugated to a fluorophore (*see* **Note 2**).

3.1 Tissue Preparation and Imaging

1. Resuspend cadaverine conjugated to a fluorophore in sterile PBS solution at a concentration of 3.3 mg/1 mL.
2. Put the mouse under anesthesia with isoflurane 3% for 4 min in the anesthesia box.
3. Remove animal from anesthesia chamber, and retroorbitally inject 100 μL of resuspended cadaverine with an insulin syringe. Start 30 min timer (*see* **Notes 3–5**).
4. Place mouse back in home cage for 30 min.
5. When 30 min are up, transcardially perfuse the anesthetized mouse as described in [11].
6. Extract the brain from the skull, and immerse it in 15 mL Falcon tube with 10 mL of 4% PFA overnight at 4 °C.
7. The following day, wash the brain with filtered PBS, and store the washed brain in filtered PBS (*see* **Note 6**).

8. Section the brains using a vibratome in 50 μm slices. Section can be kept at 4 °C in PBS for short-term storage. For long-term storage, slices can be kept in 1.5 mL Eppendorf tubes in a glycerol-based storing solution at −20 °C (*see* **Note 7**).
9. This can be combined with co-staining with antibodies, avoiding the use of secondary fluorophores with the same color as cadaverine.
10. Using a paint brush, transfer brain slices into a well of a 24-well plate with DAPI 1:1000 in PBS.
11. Incubate sections in DAPI for 4–5 min on a rocking platform at room temperature.
12. Wash sections in PBS twice for 10 min on a rocking platform (*see* **Note 8**).
13. Check the staining on one section on a slide in a fluorescence microscope (*see* **Note 9**).
14. Mount sections on positively charged glass slides. When sections are dried, attach coverslip with water-based fluorescence-protective mounting media (*see* **Note 10**). Apply clear nail polish to coat the edges of the coverslip in order to seal it in place. Let it dry overnight.
15. After mounting the sections, acquire a stitched image of the cortex on a confocal microscope at 10× magnification and 1024 × 1024 resolution. Set blending at 15%. We recommend sampling multiple slices per animal. For an accurate estimation of BBB damage across the entire cortex, we most often collect a total of three to five sections.

3.2 Manual Assessment of Cadaverine Leakage in the Cortex (See Note 11) (Fig. 2)

1. In an image analysis software (for example, FIJI), draw a region of interest (ROI) around the entire cortex, using the corpus callosum as a boundary. The DAPI channel can also be used to determine this area.
2. Measure the area of the ROI. This is the "total cortex area" for this slice.
3. Draw freehand regions of interest around the cadaverine-positive cells. Selection criteria: cadaverine-positive cells are brighter than background, cadaverine-positive cells are clustered in reasonably close proximity to each other, and there are greater than ten cadaverine-positive cells within the ROI. Continue until all ROI areas are outlined.
4. Measure the total cortical area covered by these cadaverine-positive ROIs. This is going to be the "total leakage area" for this slice.
5. Repeat **steps 1–4** with all of the acquired images.

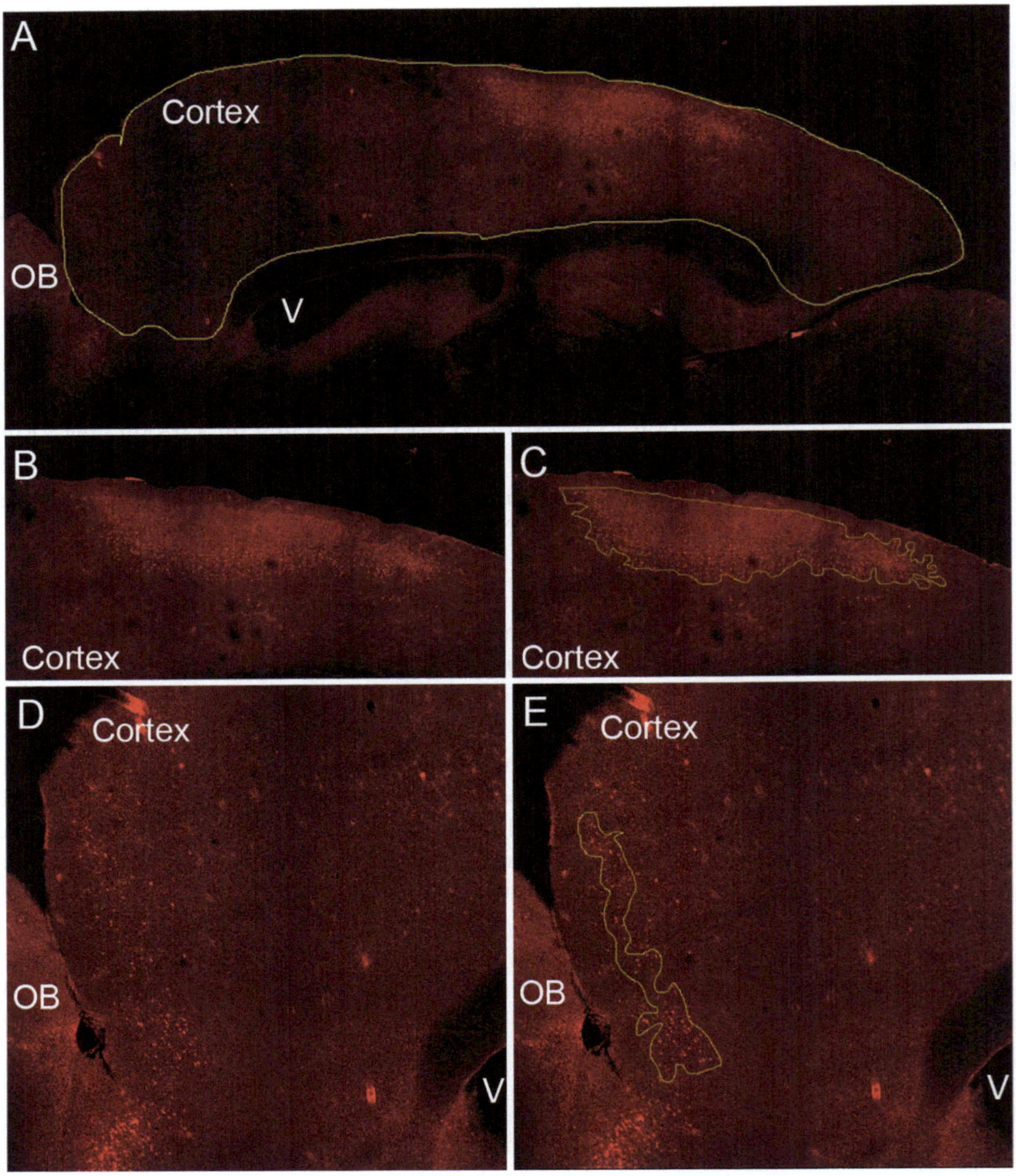

Fig. 2 Step-by-step procedure for manual quantification of cadaverine leakage in mouse brain cortex. (**a**) Selection of the whole cortex using freehand area selection. The selected area is encircled by a yellow line. This area will be measured. (**b**) Visual identification of large cadaverine-positive areas. Cadaverine (red) is located in neuronal soma of areas of leakage and identified by the experimenter. (**c**) Selection of the cadaverine-positive area using the freehand area selection. The selected area appears surrounded by a yellow line. This area will also be measured. (**d**) Visualization of small areas of cadaverine leakage in the cortex. In these areas, cells appear more disperse. (**e**) Selection of the disperse cadaverine area. The cadaverine-positive area is surrounded by a yellow line. The area encircled is measured. *OB* olfactory bulb, *V* ventricle

6. Calculate the percent of cadaverine leakage in the cortex using this equation:

$$\text{Percent leakage} = \frac{\text{Total leakage area of 5 slices}}{\text{Total cortex area of 5 slices}} \times 100$$

3.3 Automated Assessment of Cadaverine Leakage in the Cortex (Fig. 3)

1. This method selects leakage areas based on areas and shape. This method is more prone to omitting some smaller clusters of cadaverine-positive cells, but it reduces experimenter bias.
2. Draw a freehand ROI around the cortex. Use the corpus callosum as a boundary, and avoid including bright blood vessels on the edge of the slice. Measure the area of this ROI. This parameter will be "total cortex area."
3. Threshold the image such that cadaverine-positive cells are clearly highlighted. At this stage, some few blood vessels might be also selected. They will be excluded in further analysis (*see* **Note 12**).
4. Make a binary of the threshold image.

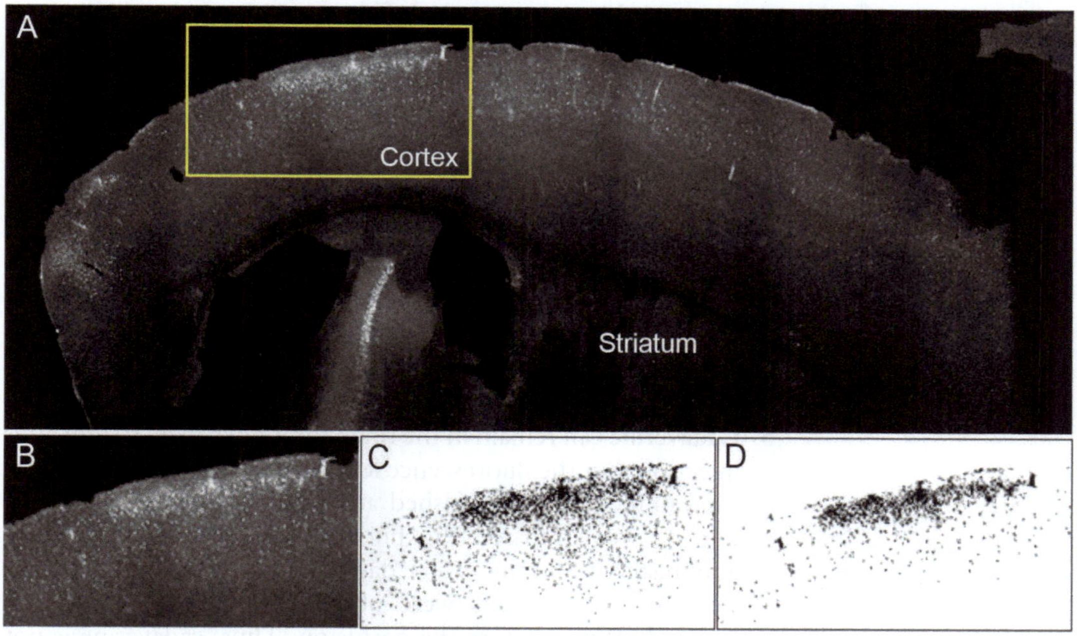

Fig. 3 Procedure for automated quantification of cadaverine leakage in mouse brain cortex. (**a**) Stitched imaged of the cadaverine signal in the cortex. A yellow square indicates the zoomed area. The cortex area will be encircled with a yellow line and the area measured. (**b**) Zoomed area of cortex showing a cadaverine-positive area. (**c**) Same area after binarizing the cadaverine-positive pixels. The area covered by these positive pixels will be quantified and divided by the total area of the cortex. (**d**) Same area as in (**c**) after adjustment by circularity for excluding blood vessels. Some information is lost, but no specific signal is eliminated in the process

5. Adjust size and circularity to include only cadaverine-positive cells. Examine the generated mask and compare to original image for accuracy. Adjusting circularity helps exclude blood vessels.
6. Measure the area covered by the mask. This parameter will be "total leakage area."
7. Repeat **steps 1–5** in five slices per animal.
8. Calculate the percent of cadaverine leakage in the cortex using this equation:

$$\text{Percent leakage} = \frac{\text{Total leakage area of 5 slices}}{\text{Total cortex area of 5 slices}} \times 100$$

4 Notes

1. Once tools have come in contact with PFA, they can no longer be used for in vivo procedures or with unfixed tissue and cells.
2. This methodology has specifically been validated with cadaverine conjugated to Alexa Fluor 488 and 555.
3. Injections into the retroorbital sinus are easier to learn and more reliable than tail vein injections. If the retroorbital injection worked well, a change of color can be observed in the mouse as the cadaverine is distributed in the blood. Urine changes color also. If the retroorbital injection is not performed well, bleeding and swelling around the eye can be observed. This can happen if the experimenter did not reach the retroorbital sinus with the syringe.
4. Ensure to not exceed the maximum volume that can be injected intravenously ($\leq$5 mL/kg or ~ 125 μL in a 25 g mouse). Higher amounts can lead to volume overload resulting in high blood pressure and vascular and organ damage that impact experimental results.
5. Cadaverine can remain in the brain for longer than 30 min after injection, but the fluorescence levels decrease with time and are more difficultly distinguished after 4 h after the injection. In Fig. 4, we compared the fluorescence levels of cadaverine administered 30 min, 4 h, 1 day, and 3 days before the perfusion, and best results were obtained when cadaverine was injected 30 min before the perfusion. Thus, cadaverine is not a great solution to label where BBB damage initially occured when tissue fixation takes place days later.
6. PBS should be filtered through a sterile 0.2 μm filter.
7. If brain slices were stored in glycerol-based storing solution before starting the staining, wash the slices in PBS twice for 10 min at room temperature in 24-well plates.

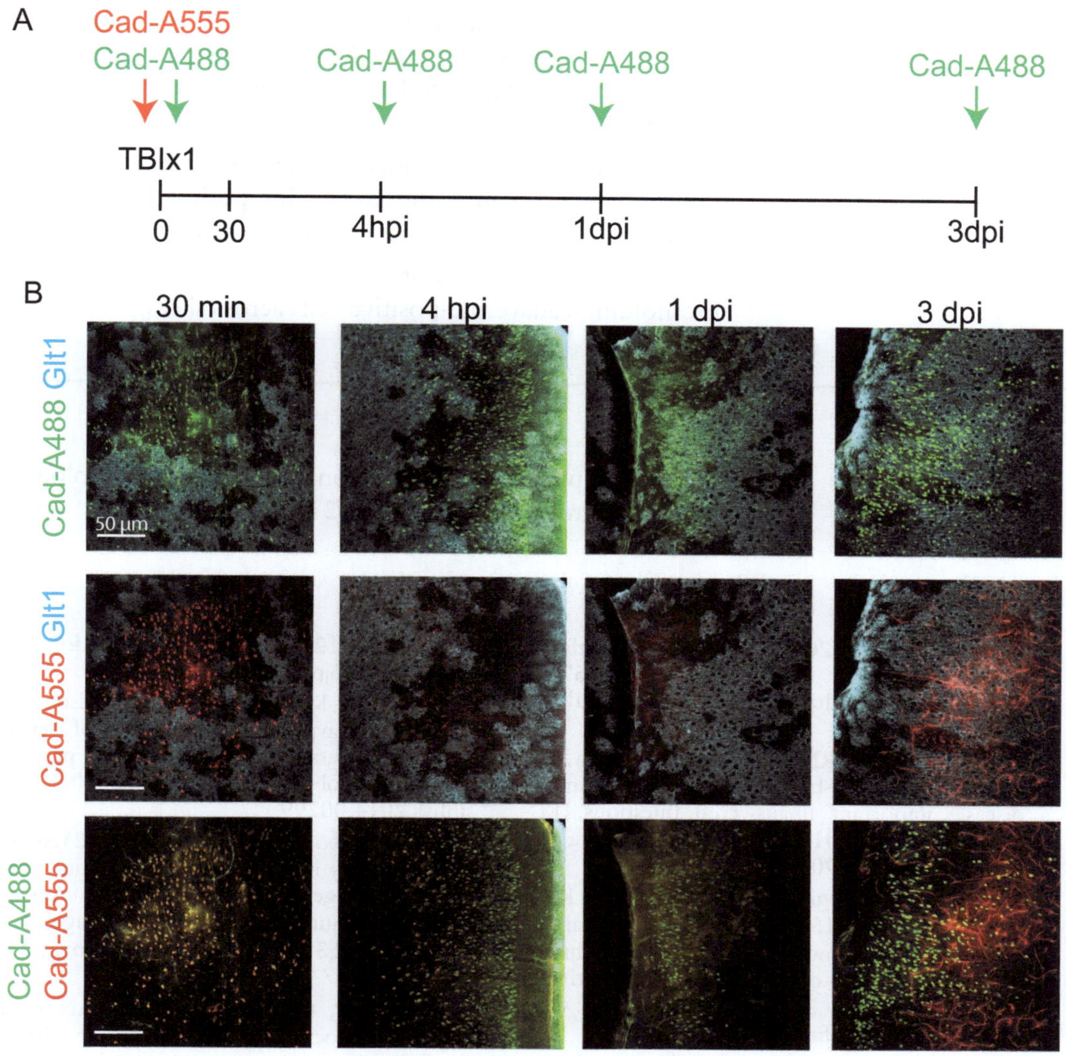

Fig. 4 Comparison of cadaverine fluorescence levels at several timepoints following retroorbital administration. (**a**) Experimental timeline: cadaverine conjugated to Alexa Fluor 555 (Cad-A555, red) was retroorbitally injected immediately after a single weight drop traumatic brain injury (TBIx1) followed by retroorbital injection of cadaverine conjugated to Alexa Fluor 488 (Cad-A488, green) immediately after TBIx1, 30 min post injury, 4 h post injury (4hpi), 1 day post injury (1dpi), or 3dpi. (**b**) Representative images of Cad-A555 (red) and Cad-A488 (green), injected at timepoints outline in (**a**), overlaid with staining for glutamate transporter 1 (Glt1, cyan)

8. For washing steps, transfer the sections to a different well with the washing solution using a paint brush.
9. If blood vessels appear fluorescent throughout the cortex (due to a suboptimal perfusion that did not flush all the dye out of the vasculature), exclude the animal from further analysis.

10. Make sure that the thickness of the coverslip is compatible with the objective that will be used. Most objectives require coverslips of number 1.5, which have a thickness between 0.16 and 0.19 mm.
11. Manual quantification of cadaverine leakage is fast but can be prone to bias because the cadaverine-positive area that is selected is subjectively chosen by the experimenter.
12. It may be appropriate to use different analysis parameters for thresholding cadaverine-positive cells across slices.

Acknowledgments

The work was supported by the National Institute of Neurological Disorders and Stroke (grant R01NS121145).

References

1. Ayloo S, Gu C (2019) Transcytosis at the blood-brain barrier. Curr Opin Neurobiol 57: 32–38. https://doi.org/10.1016/j.conb.2018.12.014
2. Profaci CP, Munji RN, Pulido RS, Daneman R (2020) The blood–brain barrier in health and disease: important unanswered questions. J Exp Med 217:e20190062. https://doi.org/10.1084/jem.20190062
3. Saunders NR, Dziegielewska KM, Møllgård K, Habgood MD (2015) Markers for blood-brain barrier integrity: how appropriate is Evans blue in the twenty-first century and what are the alternatives? Front Neurosci 9:385. https://doi.org/10.3389/fnins.2015.00385
4. Davis AA (2020) Plasma proteins have a ticket to ride the blood-brain barrier. Sci Transl Med 12:eabd3610. https://doi.org/10.1126/scitranslmed.abd3610
5. Roseth S, Fykse EM, Fonnum F (1995) Uptake of L-glutamate into rat brain synaptic vesicles: effect of inhibitors that bind specifically to the glutamate transporter. J Neurochem 65:96–103. https://doi.org/10.1046/j.1471-4159.1995.65010096.x
6. Schürmann B, Wu X, Dietzel ID, Lessmann V (1997) Differential modulation of AMPA receptor mediated currents by Evans blue in postnatal rat hippocampal neurones. Br J Pharmacol 121:237–247. https://doi.org/10.1038/sj.bjp.0701125
7. Merlot AM, Kalinowski DS, Richardson DR (2014) Unraveling the mysteries of serum albumin – more than just a serum protein. Front Physiol 5. https://doi.org/10.3389/fphys.2014.00299
8. Mills WA, Woo AM, Jiang S et al (2022) Astrocyte plasticity in mice ensures continued endfoot coverage of cerebral blood vessels following injury and declines with age. Nat Commun 13:1794. https://doi.org/10.1038/s41467-022-29475-2
9. Heithoff BP, George KK, Phares AN et al (2021) Astrocytes are necessary for blood-brain barrier maintenance in the adult mouse brain. Glia 69:436–472. https://doi.org/10.1002/glia.23908
10. George KK, Heithoff BP, Shandra O, Robel S (2022) Mild traumatic brain injury/concussion initiates an atypical astrocyte response caused by blood-brain barrier dysfunction. J Neurotrauma 39:211–226. https://doi.org/10.1089/neu.2021.0204
11. Marinković P, Godinho L, Misgeld T (2015) Generation of tissue sections for screening Thy1 mouse lines. Cold Spring Harb Protoc 2015:pdb.prot087684. https://doi.org/10.1101/pdb.prot087684

Chapter 11

Investigating the Pathology-Related Astroglial Plasticity in the Human Cerebral Cortex

Swetlana Sirko and Patrizia Della Vecchia

Abstract

Evaluation of astrocyte proliferation in human brain pathology plays a key role in the diagnostic and prognostic assessment of diseases, especially in cases of benign or malignant brain neoplasms. However, the proliferative potential of nonneoplastic astrocytes within the affected human brain parenchyma has not been well defined. Given the beneficial functions of proliferating reactive astrocytes for brain repair in experimental models of stroke and trauma, investigating the context-specific potential of human astrocytes to proliferate will be a major step forward in exploiting their roles in the genesis, progression, and outcome of neurological diseases in patients. Here, we describe a step-by-step protocol for immunofluorescent staining and neurosphere-forming assay tailored to uncover the astrocyte proliferation and neural stem cell properties in samples of human brain tissue obtained during neurosurgical resections or in a small brain biopsy. This protocol allows for reliable assessment and evaluation of adoptive astrocyte plasticity in the context with various neuropathological conditions in the human brain.

Key words Reactive astrocytes, Human brain, Cell proliferation, Stem cell properties, Neurosphere, Histopathology, Immunochemistry, Immunofluorescent staining

1 Introduction

Any form of pathophysiological processes in the brain, irrespective of their etiology, lead to the reactive astrogliosis that is generally characterized by an imbalance in the homeostatic properties of parenchymal astrocytes and the complex structural and functional changes that follow [1, 2]. These changes contribute to, or are even causative for, multiple states that astrocytes within affected brain parenchyma can assume depending on the type, location, and severity of the initial injury [3–7]. The emergence of distinct astrocyte phenotypes, both disease-specific and nonspecific, determines the resolution of secondary injury processes, such as alterations in blood–brain barrier (BBB) permeability (edema, hemorrhage), influx of neutrophils, metabolic dysregulation, hypoxia, cell damage (excitotoxicity, oxidative stress, apoptosis/necrosis), and tissue

Barbara Di Benedetto (ed.), *Astrocytes: Methods and Protocols*, Methods in Molecular Biology, vol. 2896,
https://doi.org/10.1007/978-1-0716-4366-2_11,

scarring [8–18] and is tightly related to the course and clinical outcomes of brain injury. This correlation is evident in the beneficial effects of proliferating reactive astrocytes on wound healing and functional recovery by reducing scar formation, helping to reseal the BBB, and restraining monocyte invasion in animal models of traumatic brain injury, hemorrhage, or stroke [8, 9, 12, 19, 20].

Of particular interest is that the transition of reactive astrocytes from a non-proliferative to a proliferative state, which is highly specific for disease conditions associated with BBB disruption, is accompanied by the acquisition of neural stem cell (NSC) properties in some of these proliferating astrocytes [4, 9, 21–24]. These particularly plastic astrocytes offer a valuable resource that can aid the development of wound healing therapies, yet much of our knowledge about this specific astrocyte subtype still relies on preclinical animal studies. In contrast to the relatively large amount of data available in experimental animals, only a few neuropathological studies have addressed the proliferation of human astrocytes outside the context of brain neoplasms [25–27]. Therefore, developing reliable detection methods to reveal the context-related proliferative potential of human non-neoplastic astrocytes might help identify this a highly specific astrocyte reaction within affected brain parenchyma for intervention at different stages of disease progression. This could support the development of therapeutic strategies to accelerate tissue regeneration (ideally to its original state) with minimal scarring.

Our recently published study discovered the disruption of BBB integrity as a common trigger for the proliferation of human parenchymal astrocytes and the emergence of NSC-like cells in patients with brain pathologies such as cerebral cavernoma, stroke, and medically refractory epilepsy with a higher load of cortical microbleeds or intraparenchymal hemorrhage [28]. We further confirmed the proliferative astrocyte phenotype in individual autopsy samples of the hemorrhage-affected human cerebral cortex due to brain trauma or stroke, as well as in cerebral tissue affected by prominent cerebrovascular edema upon severe acute respiratory syndrome coronavirus 2 infection (coronavirus disease 2019, COVID-19) [28].

In addition to revealing the crucial role of galectin-3 (GAL3) signaling in mediating such a therapeutically relevant phenotype of human astrocytes, we show that GAL3 is expressed in reactive human astrocytes entering the G_1 phase of cell cycle. This suggests that GAL3 predicts the proliferative status of reactive human astrocytes irrespective of the etiology of BBB disruption and at earlier cell cycle stages than MIB1(Ki67) and PCNA, which label cells in the S phase [29, 30]. These findings significantly extend previous analyses of astrocyte proliferation in humans and pave the way to modulate this highly specific astrocyte reaction for intervention at different stages of disease progression to accelerate tissue regeneration.

To investigate the potential of human astrocytes to proliferate in non-neoplastic conditions, we established a sensitive protocol for immunofluorescent staining that can be used to detect GAL3 and perform colocalization analysis of its expression in specimens obtained during neurosurgical resections or small brain biopsies, as well as in post-mortem brain tissue [28]. Furthermore, given that some of the proliferating reactive astrocytes at the injury site can acquire self-renewal and multipotential differentiation potential, the neutrosphere assay is an efficient in vitro model to investigate these NSC hallmarks in the reactive astrocytes in different experimental models of brain injury [21]. We used a modified protocol for culturing the cells dissociated from the affected human cerebral cortex under neurosphere-forming condition, enabling the examination of the injury-induced NSC capacity in reactive human astrocytes [28]. Here, we provide a detailed step-by-step protocol which allows the investigation of the context-related plasticity of human astrocytes for basic research as well as diagnostic assessment in various neuropathological states (Fig. 1).

2 Materials

2.1 General Considerations

1. Every tool or object used for tissue dissection and preparation should be sterilized beforehand by UV light, heat, or ethanol. Whenever ethanol is used for sterilization, it should be left for proper evaporation before usage to prevent the contact with the tissue samples, solutions, or the dissociated cells.
2. All procedures are performed at RT, unless otherwise stated.
3. All solutions/buffers/media are prepared on the day prior to surgery, unless otherwise stated.
4. For preparation of solutions/buffers/media, we use ultrapure Milli-Q water with a resistivity >18.2MΩ-cm, unless otherwise stated.

2.2 Human Tissue Samples

1. Research specimens of human cerebral cortex tissue were obtained from patients undergoing neurosurgery due to the nonneoplastic brain pathology, such as sporadic cerebral cavernous malformation, extra-axial benign (non-atypical/non-anaplastic) meningioma, or pharmacologically intractable epilepsy. Patients over 18 years of age were randomly selected according to their diagnosis/clinical state of disease, and each surgically treated patient had the same chance of being included in our study. No additional randomization was used during data collection. Brain surgery took place at the

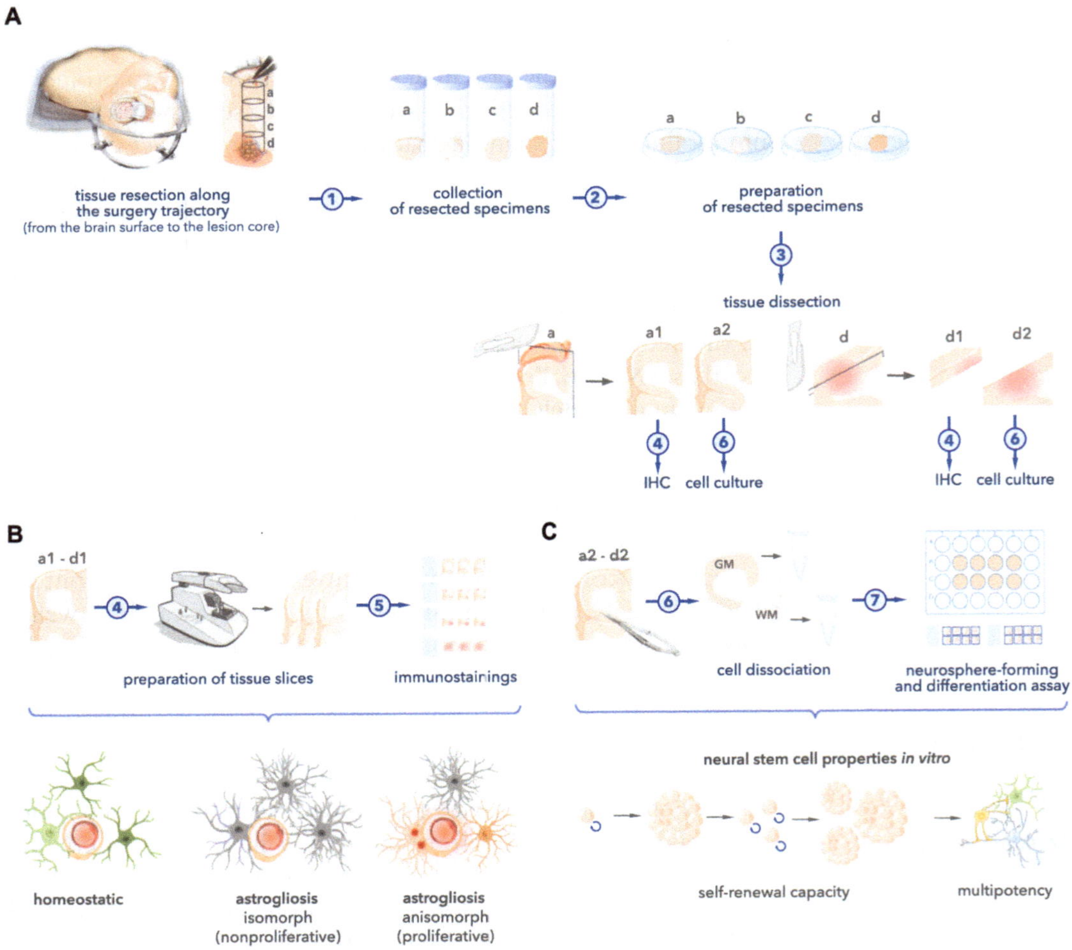

Fig. 1 Overview of the steps taken during the investigation of context-related astrocytes plasticity in affected human brain parenchyma

Department of Neurosurgery at the LMU Munich Hospital (Germany).

2. During the surgery, the stereotactic resected tissue was divided into a research specimen and a specimen that was used for histopathological diagnosis. All specimens for research purposes were collected in accordance to the Declaration of Helsinki and with the ethical guidelines approved by the Ethics Committee of the LMU Munich, Germany (Certificate of Compliance No. 17-263).

3. For analysis of individual autopsy samples, the post-mortem human cerebral cortex tissue affected by traumatic brain injury, stroke, or cerebrovascular edema in context with severe COVID-19 infection was collected (*see* **Note 1**). All post-mortem specimens were obtained and used for research in accordance to the legal guidelines of Government of Upper

Bavaria (BayKrG Art. 27 Abs. 4) and with the ethical guidelines approved by the Ethical Committee at the LMU Munich (Certificate of Compliance No. 225/20S, Declaration of No Objection No. 087-13).

4. The diagnosis, clinical features, and status of neurological disorder were defined and confirmed by neurologists at the Department of Neurosurgery, Clinic of Ludwig Maximilian University Munich (Germany), or neuropathologists at the Department of Neuropathology, Institute of Pathology, TU Munich (Germany). All research samples were blinded by irreversibly anonymization immediately after collection, thereby leaving no possibility to trace back specific subject, and except a neuropathological diagnosis, no individual participant data are available.

2.3 Dissection and Preparation of Tissue Samples

1. Phenol-free Hanks' Balanced Salt Solution (HBSS, e.g., Invitrogen 14025-050) for transportation of brain tissue samples from the surgery room to the laboratory.
2. Sterile surgical kit, tissue preparation tools (scalpel/scalpel blades, tissue forceps, micro-knives), sterile petri dishes, 15 mL/50 mL Falcon tubes, pipettes and pipette tips, etc.
3. Dissection buffer solution: 2 mM HEPES (e.g., Sigma-Aldrich H3375) in phenol-free Hanks' Balanced Salt Solution (HBSS, e.g., Invitrogen 14025-050).
4. Phosphate-buffered saline (PBS, 10×): 14.73 g $Na_2HPO_4 \times 2H_2O$ (e.g., Roth 4984.3), 2 g KH_2PO_4 (e.g., Merck 7778-77-0), 80 g NaCl (e.g., Sigma-Aldrich 7647-14-5), 2 g KCl (e.g., Merck 7447-40-7) in 1000 mL Milli-Q water and adjust pH to 7.4. Sterilize by autoclaving.
5. Stereomikroskop such as the Leica MZ6 ZOOM.

2.4 Preparation of Single Cell Suspension and Neurosphere Culture

1. Dissection buffer solution: phenol-free Hanks' Balanced Salt Solution (HBSS, e.g., Invitrogen 14025-050) with 2 mM HEPES (e.g., Sigma-Aldrich H3375).
2. Neurosphere medium: Dulbecco's modified Eagle medium/F12 (e.g., Gibco 21331-046) with 1% vol/vol Pen-Strep, 1% vol/vol B27, and 2 mM L-glutamine (e.g., all Invitrogen 15140-122, 17504044, and 25030081).
3. Differentiation medium: Dulbecco's modified Eagle medium/F12 (e.g., Gibco 21331-046) with 1% vol/vol Pen-Strep, 1% vol/vol B27, 2 mM L-glutamine (e.g., all Invitrogen 15140-122, 17504044, and 25030081), and 1% fetal calf serum (e.g., PAN Biotech P30-3302).

4. Fibroblast growth factor 2 (FGF2, e.g., PeproTech 100-18B) and epidermal growth factor (EGF; e.g., PeproTech 100-15), both at 20 ng mL^{-1} neurosphere medium.
5. 0.025% Trypsin-EDTA (e.g., Gibco 25200-056).
6. Ovomucoid (e.g., Roche).
7. 24-well plates (e.g., Sarstedt 83.3922.005).
8. With poly-L-ornithine- (PO, e.g., Sigma-Aldrich P3655) and laminin-coated (e.g., Sigma-Aldrich L2020) (both at 100 μg/mL) eight-chamber slides (e.g., ThGeyer 62001671).
9. Primary antibodies: βIII-tubulin (1:300; guinea pig polyclonal, e.g., Synaptic Systems 302304), GFAP (1:1000; rabbit polyclonal, e.g., Dako Z0334), and O4 (1:50; mouse monoclonal, e.g., Sigma O7139 clone O4). See secondary antibodies in Subheading 2.8.
10. 4′,6-Diamidino-2-phenylindole (DAPI; 0.1 mg mL^{-1}; e.g., Sigma D9542).
11. Eppendorf Safe-Lock tubes (2 mL, e.g., Eppendorf AG 0030120094).
12. Microscope cover glass (22 × 60 mm, e.g., Fisher Scientific, Roth H878.2).
13. Mounting medium: Aqua-Poly/Mount (e.g., Polysciences 18606).
14. Stereomikroskop such as the Leica MZ6 ZOOM.
15. Microscope Leica DMIL LED and LAS software (V4.6).
16. Laser-scanning confocal microscope such as the Zeiss LSM710 (Carl Zeiss) and software such as the ZEN (black edition, v.2.3 SP1).

2.5 Preparation of Free-Floating Sections for Immunofluorescence Staining

1. Paraformaldehyde (PFA, 20%): 67 g $Na_2HPO_4 \times 2H_2O$ (e.g., Roth 4984.1) and 200 g PFA (e.g., Roth 0335.4) dissolved in 800 mL Milli-Q water (heat to 60°C). Set pH to 7.4. Prepare 4% PFA in PBS 1×.
2. Agarose (4%): 40 g agarose (e.g., VWR Life Science 443666A) in 1000 mL PBS 1×.
3. Peel-A-Way embedding molds (22×30×20mm, e.g., Polysciences 18646B-1).
4. Vibrating blade microtome such as the Leica VT1000 S.
5. PBS 1× for the buffer tank of the vibratome.
6. 24-well plates (e.g., Sarstedt 83.3922.005).
7. Cryoprotective storage solution for tissue slices after cutting: 300 mL glycerol (e.g., Sigma-Aldrich G7893), 300 mL ethylene glycol pure (e.g., PanReac AppliChem—ITW Reagents

107-21-1), 100 mL 10× PO_4 buffer, and 300 mL Milli-Q water. For 10× PO_4 buffer: 137.99 g $NaH_2PO_4 \times H_2O$ (e.g., VWR International Ltd. 102454R), 37.5 g NaOH (e.g., Acros Organics 134070010) in 750 mL Milli-Q water. Adjust pH to 7.4.

8. Microscope glass slides (e.g., Epredia AA00000112E01MNZ20, ISO 8037/1).
9. Microscope cover glass (24 × 60 mm, e.g., Roth H878.2).
10. Fine brushes.

2.6 Preparation of Paraffin-Embedded Brain Tissue Sections for Immunofluorescence Staining

1. Sliding Benchtop Microtome such as the Leica SM2000 R.
2. Heating cabinet (Heraeus) at ~40 °C.
3. Water tank (for mounting) at ~60 °C.
4. Microscope glass slides (e.g., Epredia AA00000112E01MNZ20, ISO 8037/1).
5. Microscope cover glass (24 × 60 mm, e.g., Roth H878.2).
6. Fine brushes.

2.7 Immuno-fluorescence Staining of Free-Floating Brain Tissue Sections

1. Petri dish (90×16mm) and 12- or 24-well plates.
2. Pipettes, tubes, and fine brushes.
3. Microscope glass slides (e.g., Epredia AA00000112E01MNZ20, ISO 8037/1).
4. Microscope cover glass (24 × 60 mm, e.g., Roth H878.2).
5. Mounting medium: Aqua-Poly/Mount (e.g., Polysciences 18606).
6. Horizontal shaker such as the IKA-VIBRAX-VXR (speed ~50 min^{-1}).
7. Washing buffer: PBS 1×.
8. Antibody buffer solution: 20 g bovine serum albumin (BSA, e.g., Sigma-Aldrich A2153) and 5 mL Triton ×-100 (e.g., Sigma-Aldrich T9284) in 1000 mL PBS 1×.
9. Blocking solution: 20 g bovine serum albumin (BSA, e.g., Sigma-Aldrich A2153) and 5 mL Triton ×-100 (e.g., Sigma-Aldrich T9284) in 1000 mL PBS 1×.
10. Primary antibodies: GFAP (1:400; mouse IgG_1 monoclonal, Sigma G3893, clone G-A-5), GFAP (1:500; rabbit IgG polyclonal, e.g., Dako Z0334), GFAP (1:250; rabbit polyclonal, e.g., Sigma G9269), GAL3 (1:200; goat IgG polyclonal, e.g., R&D AF1197), CCND1 (1:200; rabbit IgG polyclonal, e.g., ThermoFisher RM-904-SO), MKI67 (1:200; mouse IgG_1 monoclonal, e.g., Dako M7240 clone MIB1), FN (1:250; rabbit IgG, e.g., Sigma-Aldrich F3648), and LAM1 (1:250; rabbit polyclonal, e.g., Bio-Techne NB300-144).

11. Secondary antibodies: anti-mouse IgG Alexa Fluor 488 (1: 1000; e.g., Invitrogen A21202), anti-mouse IgG Alexa Fluor 594 (1:1000; e.g., Invitrogen A21203), anti-mouse IgG Alexa Fluor 647 (1:1000; e.g., Invitrogen A32787), anti-mouse IgM Cy3-conjugated (1:1000; e.g., Jackson ImmunoResearch 715-165-020), anti-goat IgG Alexa Fluor 488 (1:1000; e.g., Invitrogen A11055), anti-goat IgG Cy3-conjugated (1:1000; e.g., Jackson ImmunoResearch 05-165-147), anti-rabbit IgG Alexa Fluor 488 (1:1000; e.g., Invitrogen A31573), and anti-rabbit IgG Cy3-conjugated (1:1000; e.g., Jackson ImmunoResearch 711-165-152).
12. 4′,6-Diamidino-2-phenylindole (DAPI; 0.1 mg mL^{-1}; e.g., Sigma D9542).
13. Eppendorf Safe-Lock tubes (2 mL, e.g., Eppendorf AG 0030120094).
14. Laser-scanning confocal microscope such as the Zeiss LSM710 (Carl Zeiss) and software such as the ZEN (black edition, v.2.3 SP1).

2.8 Immunofluorescence Staining of Paraffin-Embedded Brain Tissue Sections

1. Staining dishes with lids (8×).
2. Staining rack for microscope slides (1×).
3. Xylol (e.g., Fisher Scientific 1330-20-7).
4. 2-Propanol (e.g., Carl Roth 0733.2).
5. Ultrapure Milli-Q water.
6. Washing buffer: PBS 1×.
7. Citrate buffer (10 mM, pH 6; e.g., Carl Roth 6779.1).
8. Staining tray.
9. Heat-resistant pot.
10. PAP Pen (e.g., Kisker Biotech GmbH & Co. KG/Dutscher Group).
11. Pipettes, tubes.
12. Blocking solution: 20 g bovine serum albumin (BSA, e.g., Sigma-Aldrich A2153) and 5 mL Triton ×-100 (e.g., Sigma-Aldrich T9284) in 1000 mL PBS 1×.
13. Antibody buffer solution: 20 g bovine serum albumin (BSA, e.g., Sigma-Aldrich A2153) and 5 mL Triton ×-100 (e.g., Sigma-Aldrich T9284) in 1000 mL PBS 1×.
14. Eppendorf Safe-Lock tubes (2 mL, e.g., Eppendorf AG 0030120094).
15. Primary and secondary antibodies as described above.
16. 4′,6-Diamidino-2-phenylindole (DAPI; 0.1 mg mL^{-1}; e.g., Sigma D9542).

17. Microscope glass slides (e.g., Epredia AA00000112E01MNZ20, ISO 8037/1).
18. Microscope cover glass (22 × 60 mm, Fisher Scientific, e.g., Roth H878.2).
19. Mounting medium: Aqua-Poly/Mount (e.g., Polysciences 18606).
20. Laser-scanning confocal microscope such as the Zeiss LSM710 (Carl Zeiss) and software such as the ZEN (black edition, v.2.3 SP1).

3 Methods

3.1 Dissection and Preparation of the Human Cerebral Tissue Samples

1. Transfer each small tissue sample or en-block resected tissue immediately upon the resection into ice-cold phenol-free HBSS medium (*see* **Note 2**).
2. Place Falcon tubes with obtained tissue specimens in a Styrofoam container filled with ice for transport (*see* **Note 3**).
3. After arriving at the working bench, transfer specimens from transport solution in a sterile petri dish filled with ice-cold phenol-free HBSS medium (Fig. 2a1) (*see* **Note 4**).
4. Under a stereomicroscope, quickly but carefully remove the leptomeninges as completely as possible by grabbing them with forceps/microdissection scissors at the cutting edges and gently pulling to not damage the adjacent cerebral parenchyma (*see* **Note 5**).
5. Using a sterile scalpel, cut each tissue block/sample into two halves: one for slicing with vibratome and immunohistochemical analysis (*see* **Note 6**) and other one for in vitro culturing under neurosphere-forming condition (Fig. 2a2).
6. Place each of the "slicing" tissue blocks in individual Falcon tubes filled with 4% PFA for 24–36 h at 4 °C. After fixation, transfer tissue blocks in storage solution and place at −20 °C until ready for cutting.
7. Immediately transfer the tissue for cell culture into the sterile petri dish with ice-cold phenol-free HBSS medium.
8. Using angled forceps or a fine scissor, carefully separate the gray matter (GM) from white matter (WM). Gently transfer the dissected tissue into a 15 mL Falcon tube filled with 5 mL ice-cold dissection buffer solution. Keep the tube on ice during further dissection.

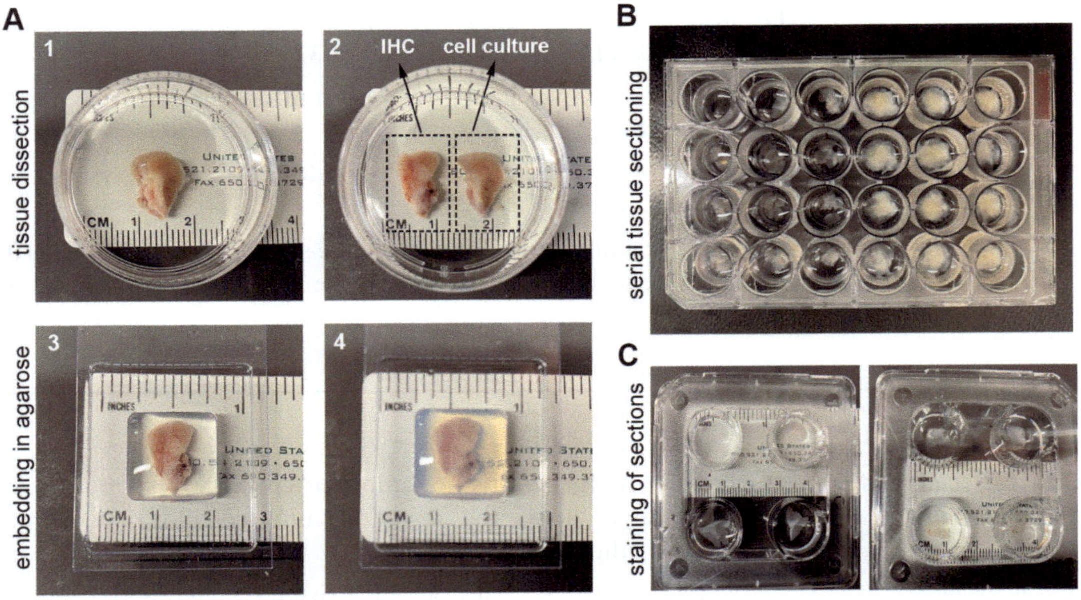

Fig. 2 Preparation of the human brain tissue sample for immunohistochemical analysis. Resected specimen was removed from transport solution and transferred in a sterile petri dish filled with ice-cold phenol-free HBSS medium (**a**1). Once the leptomeninges are completely removed, the tissue block was cut with a sterile scalpel into two halves (**a**2): one half was used for in vitro culturing under neurosphere-forming condition, and other one is embedded in agarose (**a**3) after fixation in 4% PFA. Once the agarose in the embedding mold has solidified (**a**4), the human brain tissue block was placed in the sample holder of the vibratome. The vibratome sections of the human brain tissue are collected serially into the 24-well plate filled with cryoprotective storing solution containing (**b**). For immunostaining, the selected sections are transferred into a four-well plate filled with PBS 1× (**c**) to equilibrate to the RT before further proceeding with the staining

3.2 Preparation of the Single Cell Suspension and Neurosphere-Forming Assay

1. After finishing the dissection of the GM and WM, mince the collected tissue with a fine scissor.
2. Spin down the tissue at 150 g for 1 min at RT.
3. In a laminar flow cabinet, aspirate the dissection buffer solution, add 1 mL neurosphere medium with 0.025% Trypsin-EDTA, and incubate at 37 °C in a water bath.
4. After 20 min, gently pipette three to five times up and down with a 1000 μL pipette and continue incubation at 37° C for the next 10 min.
5. To inhibit enzymatic tissue digestion, add 1 mL of ovomucoid and incubate at 37 °C for 10 min.
6. Triturate the cell suspension by pipetting up and down ten times using a 1000 μL pipette and then with 200 μL pipette (*see* **Note 7**).
7. Add 5 mL of neurosphere medium and pellet the cells at 200 g for 5 min at RT.

8. Slowly aspirate medium and then resuspend the cells in 1 mL fresh neurosphere medium.
9. Count cell number/viability and then plate the cells into single wells (24-well culture plate) at a density of one to five cells per μL in 1mL of neurosphere medium supplemented with FGF2 and EGF (both at 20 ng mL^{-1} neurosphere medium) per well.
10. After 7 days in vitro (div) incubation at 37 °C in a 5% CO_2 incubator, replace half of the medium with fresh growth factor–containing medium (*see* **Note 8**).
11. To passage the generated neurospheres (*see* **Note 9**), collect the generated neurospheres from each well into individual 15 mL tube using a 1000 μL pipette and spin them down at 200 g for 5 min at RT. Aspirate the media, add 1 mL of 0.05% Trypsin-EDTA, and incubate at 37 °C for 5 min in a water bath. Afterward, add 1 mL of ovomucoid, incubate for 5 min at 37 °C in a water bath and then triturate the neurospheres into single cell suspension. Spin the cells down at 200 g for 5 min and resuspend them in neurosphere medium. Seed cells into single wells (24-well culture plate) at a density of one to five cells per μL in 1mL of neurosphere medium supplemented with FGF2 and EGF (both at 20 ng mL^{-1} neurosphere medium) per well, and incubate the cell cultures in a 5% CO_2 37 °C incubator until the next passage (*see* **Note 10**).

3.3 Preparation of the Adherent Neurosphere Culture and Differentiation Assay

1. One day before seeding the neurospheres for differentiation assay, coat eight-chamber slides (ThGeyer 62001671) with poly-L-ornithine (100 μg mL^{-1}) overnight at 37 °C. On the day of seeding, wash chambers 3× with 300 μL PBS 1×, and coat with 100 μg mL^{-1} Laminin for 1 h at 37 °C. Afterward, wash chambers 3× with 300 μL PBS 1×. After the last wash, fill the chambers with 200 μL fresh prepared differentiation medium.
2. Seed the individual neurospheres (at 50–150 μm in diameter) onto coated eight-chamber slides and incubate at 37 °C in a 5% CO_2 incubator for 7 days.
3. After 7 days, assess the type of differentiated cells in the migration area by triple immunolabeling for βIII-tubulin (as a neuronal marker), GFAP (as an astrocyte marker), and O4 (as an oligodendrocyte marker). Therefore, gently aspirate the differentiation medium and wash the chambers 3× with ice-cold PBS 1×.
4. Fix the cells with ice-cold 4% PFA for 5 min at RT and then wash 3× with PBS 1×.
5. Incubate fixed cells with the primary antibodies against βIII-tubulin (1:300), GFAP (1:1000), and O4 (1:50), all diluted in antibody buffer solution for 2 h at RT.

6. Wash the cell cultures 3× 5 min in PBS 1× and then incubate with secondary fluorescence conjugated antibodies for 1 h at RT. After 3× washing in PBS 1×, incubate cells with DAPI for 5 min to label cell nuclei.
7. After final washing in PBS 1×, mount the chamber slides and analyze using fluorescence microscopy (*see* examples in Fig. 3d, e).

3.4 Preparation of the Free-Floating Vibratome Sections for Immunofluorescence Staining

1. Take out tissue samples from the freezer and allow them to equilibrate to the room temperature (RT) for about 20–30 min.
2. Right before use, prepare solidified agarose at a concentration of 4% according to standardized protocol. Briefly, mix agarose powder with a given volume of PBS 1× in a microwavable flask and microwave for 2 min at 900 W and then at 750 W until the agarose is completely dissolved (1-2 min). Let agarose solution cool down to about 37–40 °C (~5 min).
3. Embed the block in agarose to stabilize brain tissue during the slicing process. An individual block can be positioned in the center of a Peel-A-Way embedding mold and pipette warm agarose solution over top of block in a mold until tissue is entirely covered (Fig. 2a3). Agarose should solidify in ~5 min (Fig. 2a4) (*see* **Note 11**).
4. Once the agarose has solidified, glue the face of the embedded block with a drop of superglue directly to the specimen disc of the vibratome (*see* **Note 12**).
5. Fill the buffer tray of the vibratome with PBS 1×.
6. Adjust settings on the vibratome depending on the consistency and integrity of the tissue and start cutting at 50 μm thickness.
7. As soon as the slice dissociates from the block (*see* **Note 13**), collect it with a thin brush, and put them into the fresh prepared 24-well plate with storage solution following the series (Fig. 2b) and then put the plate at −20 °C until further proceeding with the staining (Fig. 2c).

3.5 Preparation of Paraffin-Embedded Brain Tissue Slices for Immunofluorescence Staining

1. Prepare a sufficiently large water tank at constant temperature of ~60 °C for collection and mounting the tissue sections. Fix the paraffin block on the specimen holder of the microtome.
2. Fix the paraffin block on the specimen holder of the microtome and adjust the height and angle of the sample holder for precise alignment of the tissue sample.
3. Start cutting manually at 14 μm thickness at an individual pace.
4. Collect the sections with a thin brush and transfer them carefully to the water tank to unfold.

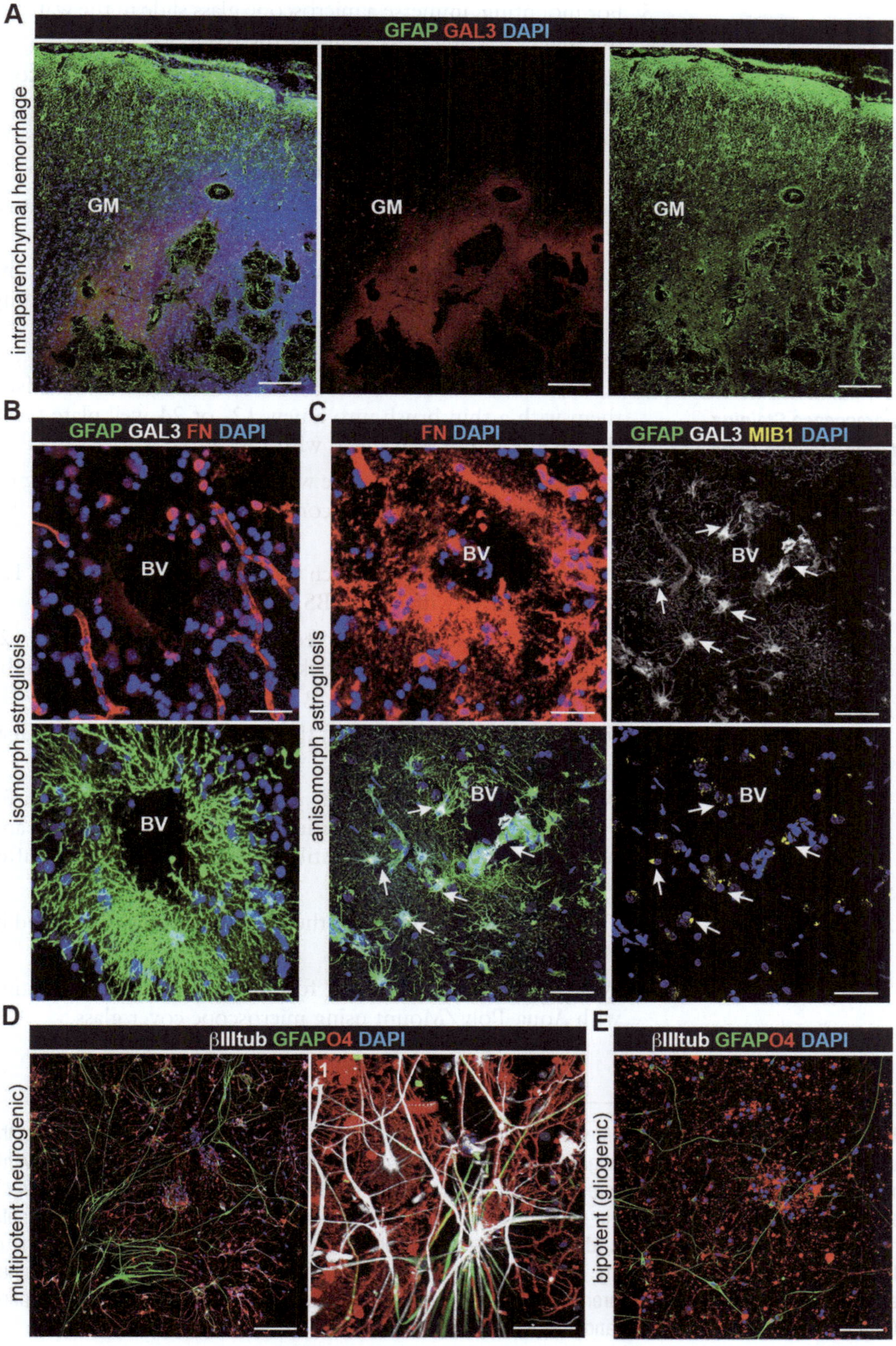

Fig. 3 Astroglial expression of GAL3 within the hemorrhage-affected human cerebral cortex is tightly correlated with acquisition of NSC-like properties at the injury site. Representative GAL3 and GFAP immunolabeling in the gliotic cerebral cortex areas affected by intraparenchymal hemorrhage (**a**). Confocal images

5. For mounting, immerse a microscope glass slide in the water of the water tank and place it under the tissue section.
6. Transfer sections with a thin brush in sequence on microscope glass slides (each on a different slide) without creases.
7. The paraffin tissue sections start to adhere to the glass slides as soon as they are removed from the water.
8. Put the slides with tissue sections in a heater overnight with a constant temperature of ~42 °C.
9. Remove the slides with the tissue sections from the heater the next day and store the slides at RT until further proceeding with the staining (*see* **Note 14**).

3.6 Immunofluorescence Staining of Free-Floating Brain Tissue Slices

1. Take out stored tissue sections from the freezer and transfer them with a thin brush into a new 12- or 24-well plate (each slice in a different well) filled with PBS 1×.
2. Place the 12- or 24-well plate with the sections for staining on the horizontal shaker at a constant speed of 50 min^{-1} for 10 min at RT.
3. Wash with PBS 1× for 3× (each time using new wells of the 12- or 24-well plate with fresh PBS 1×).
4. Afterward, transfer the tissue sections into new wells of the 12- or 24-well plate filled with blocking solution and place it on the horizontal shaker for 45 min at RT.
5. Incubate the tissue slices with the primary antibody for 48 h at 4 °C.
6. Wash the sections with PBS 1× on the horizontal shaker for 3× 10 min at RT before incubating the slices with species- and subclass-specific secondary antibody diluted in antibody buffer solution for 4 h at RT.
7. To label cell nuclei, incubate the sections with DAPI diluted in PBS 1× for 10 min.
8. Wash the sections in PBS 1× for 3× 10 min before mounting with Aqua-Poly/Mount using microscope cover glass.

Fig. 3 (continued) showing representative examples of the isomorph astrogliosis in epileptogenic region of the human cerebral cortex without any signs of vasogenic edema or bleeding (**b**). Anisomorphic astrogliosis within the hemorrhage-affected human cerebral parenchyma and appearance of proliferating $GAL3^{+}MIB1^{+}GFAP^{+}$-reactive astrocytes (white arrows) in the vicinity of disrupted vascular walls (**c**). Fluorescence micrographs show examples of the neurogenic (**d** and a 3D reconstruction of maximum intensity projection in **d**1) or gliogenic (**e**) differentiation capacity of neurospheres formed from the cells dissociated from the human cerebral cortex affected by intraparenchymal hemorrhage. Gray matter, GM; blood vessel, BV. Scale bars: 100 μm (**a**), 25 μm (**b**, **c** and **d**1), and 50 μm (**d** and **e**)

3.7 Immunofluorescence Staining of Paraffin-Embedded Brain Tissue Sections

1. For immunofluorescence staining of paraffin-embedded tissue sections, put staining dishes with lids (8×) under a hood.
2. Use Milli-Q water to dilute isopropanol and prepare the following dilution series (isopropanol: 100%, 90%, 80%, 70%).
3. For deparaffinization and rehydration of the paraffin sections, pour xylol, isopropanol, and PBS 1× into the staining dishes in specific order (2× xylol, 2× isopropanol (100%), 1× isopropanol (90%), 1× isopropanol (80%), 1× isopropanol (70%), 1× PBS 1×). Close the lids directly to prevent evaporation.
4. Place the microscope glass slides with the paraffin sections in the slide rack and immerse the rack in the solutions for 3 min each, following the sequence described above.
5. Wash the sections with PBS 1× once for 10 min.
6. Transfer the slide rack with the paraffin sections to a heat-resistant pot filled with citrate buffer and microwave for 15 min at 800 W and 10 min at 400 W and then let them cool down for 20 min at RT (~ 40 °C).
7. Place the paraffin sections in a staining tray.
8. Outline the staining area on the microscope glass slides using a PAP Pen (*see* **Note 15**).
9. Wash the sections 3× 10 min with PBS 1× using transfer pipettes to apply.
10. Incubate the paraffin sections in blocking solution for 45 min at RT and then with the primary antibody (diluted in antibody buffer solution) overnight at 4°C.
11. After washing the slides with PBS 1× for 3× 10 min at RT, incubate the paraffin sections with species- and subclass-specific secondary antibody diluted in antibody buffer solution for 90 min at RT.
12. To label cell nuclei, dilute DAPI in PBS 1× and incubate sections for 10 min.
13. Wash the paraffin sections with PBS 1× for 3× 10 min at RT and mounting with Aqua-Poly/Mount.
14. Analyze the immunostainings using fluorescence microscopy.

3.8 Image Acquisition

1. To acquire confocal fluorescent microscopic images, we use laser-scanning confocal microscope Zeiss LSM710 (Carl Zeiss) and ZEN software (black edition, v.2.3 SP1).
2. For optimal visualization and evaluation of immunolabeling, adjust the microscope settings for a digital image processing Z-stacking (e.g., an interval of optical sections at 1.5 μm, the thickness/distance of the plane of focus) to combine multiple images taken at different focal distances to provide a composite

image with a greater depth of field. For accurate analysis, use consistent settings.

3. Fluorescence imaging of immunolabeled cell cultures can be performed either with a laser-scanning confocal microscope or an epifluorescence microscope such as the Zeiss Axio Imager M2 (Carl Zeiss) using the ZEN 2 pro software.
4. For quantification of neurosphere-forming capacity, we use phase-contrast images acquired with Leica DMIL LED (Leica) using Leica Application Suite (LAS) software (v.4.6).

4 Notes

1. Postmortem tissue samples were collected from individuals over 18 years of age and only when the autolytic changes within the brain parenchyma were minimal.
2. Encapsulate the post-mortem tissue samples for fixation in ice-cold 4% PFA directly after resection.
3. Tissue samples should not be squished or exhibited to any kind of mechanical stress during transfer.
4. Depending on the size of the tissue block, it can be directly dissected into smaller blocks by cutting perpendicular to the tissue surface using a scalpel.
5. The final cortex surface should be looking smooth and homogenous, without any vessels left.
6. To obtain a clear orientation of the slices later on, cut the cortical tissue straight perpendicular from the surface toward the WM.
7. Avoid generating excess air bubbles during the triturating and wash steps as this can affect cell viability.
8. Replace half of the media with fresh growth factor-containing neurosphere medium every second day.
9. We recommend quantification of the generated neurospheres per well after 14 div and to calculate the incidence of neurosphere-forming cells per 10,000 cells. Nonspherical cell clusters as well as neurospheres smaller than 50 μm in diameter should be excluded from counts.
10. We recommend passaging of neurospheres every 14 days after plating.
11. Multiple tissue blocks can also be embedded in agarose using one Peel-A-Way embedding mold instead of in individual molds.
12. Ensure that agarose block is in contact with the adhesive.

13. Do not exhibit any kind of mechanical stress on the slices by pipette blowing or handling them with other objects.
14. We recommend waiting 1 day after removing the paraffin section from the heater before further proceeding with the staining.
15. Ensure an intact outline around the paraffin section on the microscope glass slide during staining. In case of interruption, use absorbent tissue to dry the slide around the paraffin section and apply a new one using a PAP Pen.

References

1. Escartin C, Galea E, Lakatos A, O'Callaghan JP, Petzold GC, Serrano-Pozo A et al (2021) Reactive astrocyte nomenclature, definitions, and future directions. Nat Neurosci 24(3): 312–325
2. Verkhratsky A, Butt A, Li B, Illes P, Zorec R, Semyanov A et al (2023) Astrocytes in human central nervous system diseases: a frontier for new therapies. Signal Transduct Target Ther 8(1):396
3. Sirko S, Neitz A, Mittmann T, Horvat-Brocker A, von Holst A, Eysel UT et al (2009) Focal laser-lesions activate an endogenous population of neural stem/progenitor cells in the adult visual cortex. Brain 132 (Pt 8):2252–2264
4. Sirko S, Behrendt G, Johansson PA, Tripathi P, Costa M, Bek S et al (2013) Reactive glia in the injured brain acquire stem cell properties in response to sonic hedgehog. Cell Stem Cell 12(4):426–439
5. Mattugini N, Merl-Pham J, Petrozziello E, Schindler L, Bernhagen J, Hauck SM et al (2018) Influence of white matter injury on gray matter reactive gliosis upon stab wound in the adult murine cerebral cortex. Glia 66(8): 1644–1662
6. Sofroniew MV (2020) Astrocyte reactivity: subtypes, states, and functions in CNS innate immunity. Trends Immunol 41(9):758–770
7. Bardehle S, Kruger M, Buggenthin F, Schwausch J, Ninkovic J, Clevers H et al (2013) Live imaging of astrocyte responses to acute injury reveals selective juxtavascular proliferation. Nat Neurosci 16(5):580–586
8. Frik J, Merl-Pham J, Plesnila N, Mattugini N, Kjell J, Kraska J et al (2018) Cross-talk between monocyte invasion and astrocyte proliferation regulates scarring in brain injury. EMBO Rep 19(5):e45294
9. Lange Canhos L, Chen M, Falk S, Popper B, Straub T, Gotz M et al (2021) Repetitive injury and absence of monocytes promote astrocyte self-renewal and neurological recovery. Glia 69(1):165–181
10. Koupourtidou C, Schwarz V, Aliee H, Frerich S, Fischer-Sternjak J, Bocchi R et al (2024) Shared inflammatory glial cell signature after stab wound injury, revealed by spatial, temporal, and cell-type-specific profiling of the murine cerebral cortex. Nat Commun 15(1):2866
11. Burda JE, O'Shea TM, Ao Y, Suresh KB, Wang S, Bernstein AM et al (2022) Divergent transcriptional regulation of astrocyte reactivity across disorders. Nature 606(7914):557–564
12. Wanner IB, Anderson MA, Song B, Levine J, Fernandez A, Gray-Thompson Z et al (2013) Glial scar borders are formed by newly proliferated, elongated astrocytes that interact to corral inflammatory and fibrotic cells via STAT3-dependent mechanisms after spinal cord injury. J Neurosci 33(31):12870–12886
13. Sirko S, Irmler M, Gascon S, Bek S, Schneider S, Dimou L et al (2015) Astrocyte reactivity after brain injury-: the role of galectins 1 and 3. Glia 63(12):2340–2361
14. Robel S, Sontheimer H (2016) Glia as drivers of abnormal neuronal activity. Nat Neurosci 19(1):28–33
15. Robel S, Buckingham SC, Boni JL, Campbell SL, Danbolt NC, Riedemann T et al (2015) Reactive astrogliosis causes the development of spontaneous seizures. J Neurosci 35(8): 3330–3345
16. Munoz-Ballester C, Robel S (2023) Astrocyte-mediated mechanisms contribute to traumatic brain injury pathology. WIREs Mech Dis 15(5):e1622
17. Bush TG, Puvanachandra N, Horner CH, Polito A, Ostenfeld T, Svendsen CN et al (1999) Leukocyte infiltration, neuronal degeneration, and neurite outgrowth after ablation of

scar-forming, reactive astrocytes in adult transgenic mice. Neuron 23(2):297–308

18. Troscher AR, Gruber J, Wagner JN, Bohm V, Wahl AS, von Oertzen TJ (2021) Inflammation mediated epileptogenesis as possible mechanism underlying ischemic post-stroke epilepsy. Front Aging Neurosci 13:781174
19. Heimann G, Canhos LL, Frik J, Jager G, Lepko T, Ninkovic J et al (2017) Changes in the proliferative program limit astrocyte homeostasis in the aged post-traumatic murine cerebral cortex. Cereb Cortex 27(8): 4213–4228
20. Robel S, Bardehle S, Lepier A, Brakebusch C, Gotz M (2011) Genetic deletion of cdc42 reveals a crucial role for astrocyte recruitment to the injury site in vitro and in vivo. J Neurosci 31(35):12471–12482
21. Gotz M, Sirko S, Beckers J, Irmler M (2015) Reactive astrocytes as neural stem or progenitor cells: in vivo lineage, in vitro potential, and genome-wide expression analysis. Glia 63(8): 1452–1468
22. Magnusson JP, Zamboni M, Santopolo G, Mold JE, Barrientos-Somarribas M, Talavera-Lopez C et al (2020) Activation of a neural stem cell transcriptional program in parenchymal astrocytes. elife 9:9
23. Buffo A, Rite I, Tripathi P, Lepier A, Colak D, Horn AP et al (2008) Origin and progeny of reactive gliosis: a source of multipotent cells in the injured brain. Proc Natl Acad Sci USA 105(9):3581–3586
24. Zamanian JL, Xu L, Foo LC, Nouri N, Zhou L, Giffard RG et al (2012) Genomic analysis of reactive astrogliosis. J Neurosci 32(18):6391–6410
25. Colodner KJ, Montana RA, Anthony DC, Folkerth RD, De Girolami U, Feany MB (2005) Proliferative potential of human astrocytes. J Neuropathol Exp Neurol 64(2): 163–169
26. Dowling P, Husar W, Menonna J, Donnenfeld H, Cook S, Sidhu M (1997) Cell death and birth in multiple sclerosis brain. J Neurol Sci 149(1):1–11
27. Schonrock LM, Kuhlmann T, Adler S, Bitsch A, Bruck W (1998) Identification of glial cell proliferation in early multiple sclerosis lesions. Neuropathol Appl Neurobiol 24(4): 320–330
28. Sirko S, Schichor C, Della Vecchia P, Metzger F, Sonsalla G, Simon T et al (2023) Injury-specific factors in the cerebrospinal fluid regulate astrocyte plasticity in the human brain. Nat Med 29(12):3149–3161
29. Whitfield ML, George LK, Grant GD, Perou CM (2006) Common markers of proliferation. Nat Rev Cancer 6(2):99–106
30. Brown DC, Gatter KC (2002) Ki67 protein: the immaculate deception? Histopathology 40(1):2–11

Chapter 12

Super-resolution Analysis of Astrocyte Morphology Using Expansion Microscopy

Gerald Seifert, Erik Sommer, Stefan Passlick, and Christian Henneberger

Abstract

Analyzing the structure of astrocytes, their specific morphological features, and their remodeling is important for understanding how this cell type fulfils its many functions. This is because cell shape determines the propagation of intracellular signals and their subcellular compartmentalization. At the same time, it determines which other cells in the neuropil an astrocyte is closely in contact with and can most effectively exchange signals with. One experimental challenge has been that the most abundant small astrocytic processes cannot be resolved with diffraction-limited microscopy. Typically, this obstacle was overcome by using electron microscopy, but the continuous development of super-resolution microscopy has produced many alternative techniques. One is expansion microscopy (ExM) [1]. ExM, in principle, expands the tissue while preserving the relative positioning of labels that mark structures of interest (e.g., fluorescent labels), which increases the effective spatial resolution of light microscopy without improving the spatial resolution of the microscope itself. The advantage of ExM is that it requires only a little more laboratory infrastructure than immunohistochemistry combined with confocal fluorescence microscopy. We have previously applied this universal technique to the analysis of the structure of astrocytes and of their fine processes and their perisynaptic arrangement. Here, we present a comprehensive protocol for visualizing and localizing astrocytes, synaptic structures, and synaptic and astrocytic proteins in fixed brain tissue.

Key words Super-resolution microscopy, Expansion microscopy, Astrocyte morphology, Perisynaptic astrocytic processes

1 Introduction

Astrocytes, neurons, and other glial cells in the brain interact and mutually modulate each other's activity by many mechanisms (e.g., neurotransmitter uptake, bi-directional synaptic signaling, metabolite supply). It is thought that the distinct morphology of astrocytes enables them to be particularly effective modulators and supporters of neuronal and synaptic function: They occupy clearly delineated and mostly nonoverlapping tissue domains, in which their fine processes can approach many thousands of synapses [2]. They also tightly contact blood vessel with their so-called

Barbara Di Benedetto (ed.), *Astrocytes: Methods and Protocols*, Methods in Molecular Biology, vol. 2896, https://doi.org/10.1007/978-1-0716-4366-2_12,

endfeet (for review [3]). Because of their close contact with synapses, perisynaptic astrocytic processes are in an ideal position to sense and modify synaptic activity and plasticity. Therefore, a major question has been if astrocytes modulate and support all synapses in their territory similarly and indiscriminately or whether they do so selectively in a synapse-specific manner and supported by subcellular synapse-specific specializations [4].

Such hypotheses can be tested by identifying the rules that govern whether and to what extent synapses are approached and covered by perisynaptic astrocytic processes and by analyzing the proteins expressed by astrocytes in a given synaptic environment. However, this cannot be fully resolved using diffraction-limited microscopy, because of the small size of synaptic structures and perisynaptic astrocytic processes [5]. Traditionally, this obstacle is overcome by high-resolution techniques like electron microscopy, which has been used to analyze astrocyte morphology in many studies (for example, [6]). However, despite the excellent spatial resolution of electron microscopy and its ability to precisely localize membranes and organelles, it is relatively labor-intensive, requires highly specialized equipment, and labeling specific proteins and structures can be difficult. Many super-resolution microscopy techniques such as structured illumination microscopy, single-molecule localization microscopy, and stimulated emission depletion microscopy have been successfully used as alternative approaches for studying astrocytes and their configuration at synapses [5, 7]. A particularly interesting alternative is expansion microscopy (ExM) [1], which, instead of increasing the spatial resolution of the microscopy method itself, expands the tissue while preserving the relative positioning of labels marking structures of interest (e.g., a fluorescent label of a protein), thus increasing the effective spatial resolution of, for instance, confocal microscopy. Updated and newer protocols allow the experimenter to use conventional antibodies and fluorescent proteins and the widely available laboratory infrastructure and expertise for immunohistochemistry and confocal fluorescence microscopy to achieve a manyfold higher effective spatial resolution compared to confocal microscopy on its own [8].

We have successfully used this technique for analyzing the synapse-specific distribution of astrocytic processes and glutamate transporters at excitatory synapses [9], for studying the structure of GABAergic tripartite synapses [10], for validating morphological analyses based on two-photon excitation fluorescence microscopy [11], and for visualizing astrocytic endfeet and their gap junction protein expression [12]. Because ExM is not designed with a particular cell type in mind, we have applied it also to the analysis of NG2-glia with minimal changes [12–14].

The protocol described here is for fixed tissue slices and is also called protein-retention ExM [15]. It is based on crosslinking endogenous proteins, genetically encoded fluorescent proteins, and/or fluorophore-labeled antibodies to an expandable polyelectrolyte hydrogel. Overall, the presented procedure consists of four main steps. After performing the necessary immunohistochemistry to label the structures of interest, the first step of protein-retention ExM is the attachment of anchors to proteins, i.e., to fluorescently labeled secondary antibodies in our case. In this anchoring step, a molecular anchor containing polymerizable carbon–carbon double bonds is added to proteins, which enables them to be bound to a polymer gel synthesized in the next step. Second, the tissue slices are incubated in a sodium acrylate–based monomer solution, and in situ polymerization is initiated. In this step, proteins and labels are integrated via their anchors into the polymer gel. As a result, secondary antibodies labeled with fluorescent dyes, for instance, are now linked to the polymer. Third, protein digestion degrades proteins, and a transparent hydrogel is obtained. In a fourth step, the hydrogel is expanded by incubation in water.

The presented protocol is mostly based on [8], which typically expands the tissue about fourfold and thus increases the effective spatial resolution by the same factor. As mentioned, ExM is rapidly evolving, and new applications and labeling and expansion techniques are constantly developed (for review [16–18]). Examples are the improvement of the magnification [19, 20] and the visualization of proteins along with cellular ultrastructure [21].

2 Materials

2.1 Animals

1. Adult (10–12 weeks old) mice (e.g., GFAP-EGFP mice [22] for the examples presented in Figs. 1 and 2b or Thy1-YFP [23] for examples in Fig. 2a). Licenses for perfusion fixation (see below) are typically required. All our experiments involving animals were performed in full compliance with regulations of the European Union, the University of Bonn, and the Landesamt für Natur, Umwelt und Verbraucherschutz (LANUV) North Rhine-Westphalia, Germany. Licenses for animal experiments were granted by the LANUV where needed.

2.2 Preparation of Brains for Immunohistochemistry

1. Syringe pump for transcardial perfusion of small animals.
2. Phosphate-buffered saline (PBS) stock solution (10×, total volume 1 L, deionized water): 1370 mM NaCl, 27 mM KCl, 100 mM Na_2HPO_4, and 17.6 mM KH_2PO_4. (adjust pH to 7.4).
3. 4% Paraformaldehyde (PFA) in PBS (1×), pH 7.4.
4. Laboratory dissection tools for brain preparation and extraction.
5. Ketamine.

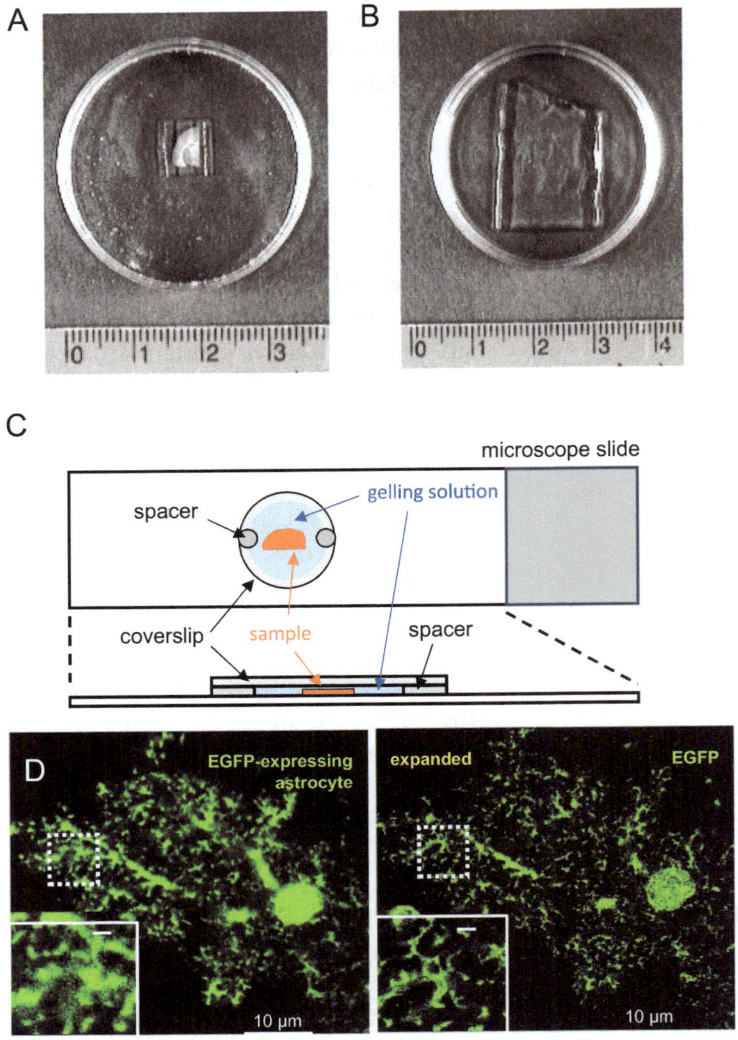

Fig. 1 Expansion microscopy (ExM) for visualization of astrocytes. Example of brain slices before (**a**) and after (**b**) expansion. Note the enlargement of the section and the transparent appearance (ruler in cm). (**c**) Illustration of the incubation chamber used for gelling from the top (upper panel) and from the side (bottom panel). (**d**) Comparison of confocal microscopy of the same individual astrocyte expressing EGFP [22] before (left) and after expansion (right) (modified from [9]). The scale bars correspond to pre-expansion dimensions (i.e., size within sample divided by the expansion factor for the right panel). Insets represent higher magnification examples as indicated (scale bar 1 μm) and reveal the more clearly defined astrocyte branches in ExM and the disappearance of out-of-focus structures

6. Xylazine.
7. Perfusion syringe.
8. Butterfly needle green 21G.

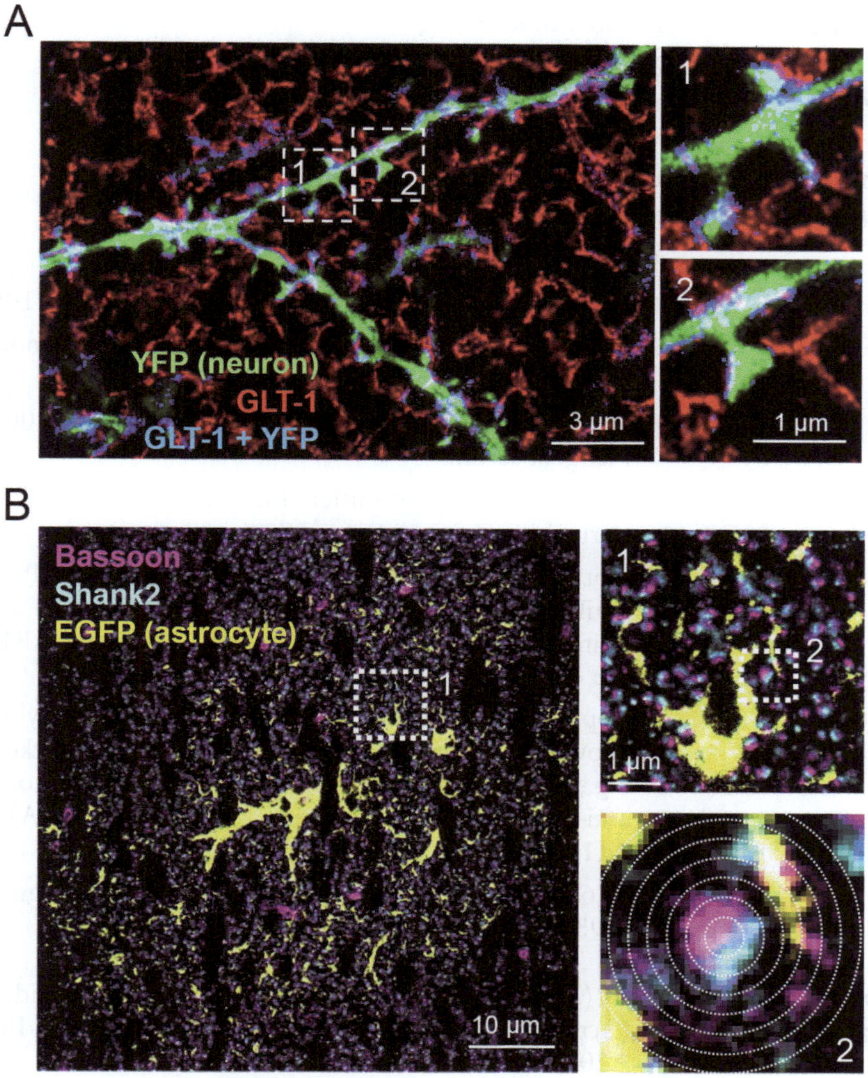

Fig. 2 Examples of ExM of excitatory synaptic connections and nearby astrocyte processes. Modified from [9]. (**a**) Left panel: ExM of postsynaptic spines of a YFP-expressing radial oblique dendrite of a CA1 pyramidal neuron (green) and the surrounding GLT-1 (glutamate transporter 1)-positive (red) astrocyte processes (left panel). Regions of immediate juxtaposition, i.e., appearing co-localized, are displayed in blue. Right panels: Numbered regions of interest corresponding to magnified image sections outlined in the left panel (dashed boxes). Note the thin astrocyte processes approaching synaptic spines. (**b**) ExM of EGFP-expressing astrocytes and pre- and postsynaptic sites (single focal plane). Left panel: Example of a triple label ExM experiment (astrocyte: cytosolic EGFP, yellow; pre-synaptic label: bassoon, magenta; postsynaptic density (PSD) label: shank2, cyan). Top right panel: Enlarged section from boxed region in the left panel (white, "1"). Bottom right panel: Top right panel further magnified (white box, "2") corresponding to $1.25 \times 1.25\ \mu m^2$. Example of ring/shell-like regions of interest for the analysis of astrocytes near excitatory synapses

2.3 Preparation of Slices for Immunohistochemistry

1. Vibratome.
2. 24-well plates.
3. Glue for fixating brain.

2.4 Immunofluorescence

1. 24-well plates.
2. Horizontal shaker.
3. Washing buffer: PBS (1×).
4. Permeabilization buffer: PBS (1×), 0.1% Triton X-100.
5. Blocking buffer: PBS (1×), 1% Triton X-100, 10% normal goat serum (also, *see* Subheading 3.1).
6. Primary antibody buffer: PBS (1×), 1% Triton X-100, 5% normal goat serum.
7. Secondary antibody buffer: PBS (1×), 1% Triton X-100, 2% normal goat serum.
8. Suitable primary antibodies: e.g., chicken anti-GFP polyclonal antibody (1:500, Abcam, ab13970) for Figs. 1d and 2b or guinea pig anti-GLT-1 (1:500, e.g., Millipore, AB1783) for the example in Fig. 2a.
9. Appropriate secondary antibodies: for the primary antibodies above, for example, Alexa Fluor 488 goat anti-chicken (1:200, e.g., ThermoFisher, A11039) and Alexa Fluor 568 or 594 goat anti-guinea pig (1:200; e.g., ThermoFisher, A11075 or A11076).
10. 4′,6-Diamidino-2-phenylindole (DAPI, Sigma-Aldrich D9542, 1:5000) to label nuclei.

2.5 Anchoring and Polymerization

1. 6-((Acryloyl)amino)hexanoic acid, succinimidyl ester (Acryloyl-X SE or AcX; 10 mg/mL in DMSO, e.g., Invitrogen, A20770).
2. Microscope cover glass for spacer (5 mm, circular, e.g., Fischer Scientific, 50-949-439).
3. Microscope cover glass (13 mm, circular, e.g., VWR, 631-0148).
4. Microscope slides with frosted edge.
5. Incubator.
6. Fine brush.
7. Pipettes and other general laboratory equipment.
8. 10% ammonium persulfate (APS; 1 g/10 mL water).
9. 10% N,N,N′,N′-tetramethylethane-1,2-diamine (TEMED; 1 g/10 mL water).
10. 0.5% 4-hydroxy-2,2,6,6-tetramethyl-piperidine-1-oxyl (TEMPO or 4HT; 0.05 g/1 mL water).
11. Monomer solution (see Subheading 2.8).

2.6 Digestion

1. Cell culture dishes, 35 mm diameter.
2. Razor blade.
3. Digestion buffer (see Subheading 2.8).

2.7 Expansion and Mounting

1. Dissection microscope.
2. Cell culture dishes, 35 mm diameter.
3. Water.
4. Microscope cover glass 22 mm × 50 mm for transferring the gel.
5. Microscope cover glass 15 mm × 15 mm.
6. Ibidi chamber μ-slide two-well glass bottom (e.g., Ibidi, 80287) (*see* **Note 1**).
7. 0.1% Poly-L-lysine, 150–300 kDa (e.g., Sigma-Aldrich, P8920).

2.8 Solutions

1. Monomer solution (final volume of 9.4 mL). The solvent for all stock solutions below is water.
   (a) Sodium acrylate (2.25 mL of a stock solution of 38 g/100 mL, final concentration 0.97 M).
   (b) Acrylamide (0.50 mL of a stock solution of 50 g/100 mL, final concentration 0.37 M).
   (c) N,N′-Methylenbisacrylamide (0.75 mL of a stock solution of 2 g/100 mL, final concentration 10 mM).
   (d) Sodium chloride (4.0 mL of a stock solution of 29.2 g/100 mL, final concentration 2.13 M).
   (e) PBS (1.0 mL of a 10× stock solution, final concentration 1.06×).
   (f) Deionized water (0.9 mL).
2. Digestion buffer (final volume of 100 mL).
   (a) Triton X-100 (0.5 g, final concentration 8 mM).
   (b) EDTA, disodium (0.2 mL of a stock solution of 0.5 M and pH 8.0, final concentration 1 mM).
   (c) Tris-Cl (5 mL of a stock solution of a 1 M aqueous solution, pH 8.0, final concentration 50 mM).
   (d) NaCl (4.68 g, final concentration 0.8 M).
   (e) Add deionized water to obtain a volume of 100 mL.
   (f) For digestion, proteinase K (800 U/mL) is diluted 1:100 in an aliquot of 2 mL digestion buffer per slice in a cell culture dish (final concentration 8 U/mL). Instead of NaCl, an equimolar solution of guanidine HCl can be used.

3 Methods

3.1 Preparation of Brain Tissue Slices and Immunohistochemistry

1. Tissue fixation by transcardial perfusion fixation. It typically requires a license for performing this animal experiment.
   (a) Anaesthetize mice (for instance, with xylazine and ketamine) according to the license.
   (b) Transcardially perfuse with ice-cold PBS (30 mL) and ice-cold 4% PFA in PBS (30 mL) at a constant flow of 6 mL/min using a syringe pump.
   (c) Extract the whole brain and incubate in 4% PFA in PBS overnight at 4 °C. Then, store in PBS at 4°C until cutting. For long-time storage, PBS is supplemented with 0.1% NaN_3.
2. Cut tissue slices with a thickness of 60 μm using a vibratome, and transfer slices into PBS in a 24-well chamber.
3. For immunohistochemistry (*see* **Note 2**), incubate slices for 2 h at room temperature in blocking/permeabilization solution (e.g., PBS containing 0.1% Triton X-100 and, if compatible with the antibodies used, 10% normal goat serum (NGS) or an alternative).
4. Incubate with primary antibodies in blocking buffer containing 1% Triton X-100 and 5% NGS (or alternative, see above) overnight at 4 °C by gently shaking.
5. Next day, wash slices 3× in PBS for 10 min.
6. Incubate with secondary antibodies in blocking buffer containing 1% Triton X-100 and 2% NGS (or similar, see above) overnight at 4 °C by gently shaking.
7. Next day, wash slices 3× in PBS for 10 min.
8. Separate the hemispheres using a razor blade on a microscope slide, and/or trim the slices under microscope control retaining the region of interest.
9. Stain and inspect nuclear staining for determining the expansion factor later (*see* **Note 3**).
   (a) Incubate slices in water containing DAPI (1:5000 dilution of a stock of 20 mg/mL (in water)) for 10 min at room temperature under gentle shaking.
   (b) Wash slices once with water for 10 min and twice with PBS for 10 min.
   (c) Take sample images of nuclei staining by inspection on a confocal scanning microscope under UV excitation using a suitable objective (e.g., 20× water immersion objective) for determining the expansion factor later.

(d) Place tissue section on a microscope slide using a fine brush. Cover with a coverslip, and inspect (microscope slide upside down) on an inverted confocal microscope.

(e) In the hippocampus, nuclei of dentate granule cells from the tip of the granular cell layer are suitable for determining the expansion factor because the structure is easy to find before and after expansion and because their nuclei are fairly circular. Determine the diameter of ~20 nuclei of each slice (e.g., using FIJI [24]). The mean diameter in each slice will be used later to calculate the expansion factor.

3.2 Anchoring of Proteins

1. Prepare the solution of the molecular anchor AcX by 1:100 dilution of 10 mg/mL AcX/DMSO stock solution in PBS.
2. Incubate the tissue slices in AcX/PBS overnight at room temperature in the dark (no shaking) (*see* **Note 4**).
3. On the next day, discard the AcX solution, and wash the slices 3× for 15 min in PBS.

3.3 Monomer Incubation and Polymerization

1. Prepare the gelling solution by mixing the monomer solution containing sodium acrylate (final concentration 0.97 M), acrylamide (0.37 M), and N,N′-methylenebisacrylamide as cross-linker (10 mM) in NaCl/PBS with the stock solutions of TEMED (polymerization accelerator), TEMPO (radical scavenger and mediator of controlled radical polymerization), and APS (radical polymerization initiator) at the ratio 47:1:1:1. Prepare the gelling solution on ice to prevent undesired early gelation. Preparation of corresponding stock solutions and monomer solution was reported in detail [17] (also see Subheadings 2.5 and 2.8 and **Note 5**).
2. Incubate the tissue slices in gelling solution for 30 min at 4 °C in a fridge (no shaking).
3. Perform the polymerization, as explained in the following steps, in a liquid layer between a microscope slide and a coverslip. As a spacer between them, we use two small cover slips of 5 mm diameter positioned at the sides of the tissue slice. *See* Fig. 1c for an illustration. The thickness of the spacer coverslips is 130–160 μm compared to an initial slice thickness of 60 μm.
4. Pipette 5–10 μL chilled gelling solution between the two spacers on the microscope slide. Place the tissue slice soaked with gelling solution inside this drop. Pipette another 5–10 μL of gelling solution on top of the tissue slice. Place the coverslip (13 mm diameter) on top of spacers.
5. Using a small pipette, add some gelling solution at the edge of the 13 mm coverslip to avoid crystallization of the gelling

solution. This also makes sure that the slice does not dry and remains completely embedded within the gelling solution.

6. Place the microscope slide with the tissue into a humidified box (easily built from an empty pipet box filled with a bit of water at the bottom) to avoid the evaporation of water and crystallization. For the polymerization of monomers into a hydrogel, put the box in an incubator for 2 h at 37 °C.
7. After polymerization, remove, and discard the coverslip on top of the tissue slice now embedded in gel by first carefully wetting the edges of the gel with PBS using a fine brush or, if necessary, with the help of a razor blade.
8. Trim the gel using a razor blade as necessary.
9. Gently transfer the gel with the embedded tissue slice to a 35 mm cell culture dish filled with digestion buffer (2 mL, see below).

3.4 Protein Digestion and Tissue Expansion

1. Transfer the gel with the embedded tissue slice into the digestion buffer containing Tris-Cl (pH 8.0, 50 mM), EDTA, disodium (pH 8.0, 1 mM), Triton X-100 (8 mM), and NaCl (0.8 M).
2. Start protein digestion by adding 8 U/mL proteinase K to the digestion buffer.
3. Incubate the gel overnight in digestion buffer supplemented with proteinase K at 25 °C in the dark. Because of slight expansion of gels in digestion buffer, the gels should be incubated in sufficiently large cell culture dishes.
4. Next day, discard the digestion buffer, and replace by water to start the expansion of the hydrogel. Expansion should be performed at least for 2 h at room temperature while exchanging water every 20 min.
5. Control the swelling and volume increase after each water exchange. After expansion, the sample should be expanded and transparent without remaining slice contours (*see* Fig. 1).
6. For microscopy of expanded gels, use a μ-slide two-well glass bottom chamber, a so-called Ibidi chamber (*see* **Note 1**). Coat the bottom of the Ibidi chamber with 0.1% poly-L-lysine (150–300 kDa).
7. Trim the gels as needed, and place them in wells of the Ibidi chamber. Carefully transfer the gels from cell culture dishes into the wells of the Ibidi chamber using a 22 mm × 50 mm cover glass for the transfer. Cover the gel in the Ibidi chamber by a 15 mm × 15 mm coverslip to prevent desiccation and shrinking.

3.5 Post-expansion Confocal Microscopy, Deconvolution, and Analysis

1. An epifluorescence microscope can be used for getting a first impression. We routinely use confocal microscopy for all further steps. Use the DAPI-stained nuclei (see above) as a landmark for orientation (e.g., dentate gyrus granular cells and pyramidal cell layers in the hippocampus).
2. Determine the average diameter of nuclei using the same technique as above (also, *see* **Note 3**). Calculate the expansion factor for each experiment from the average pre- and post-expansion diameters. In our experience, the linear expansion is ~4–5-fold, which corresponds to a tissue volume increase by a factor of ~64–125. This should be considered when choosing the confocal microscope for analyzing the post-ExM samples (e.g., inverted vs. upright microscope, working distance of objective) and when inspecting the post-ExM samples.
3. For data acquisition from expanded gels, we routinely use an inverted laser scanning confocal fluorescence microscope (Leica SP8) for acquiring image stacks. The acquisition settings will depend on the exact brand and type of microscope used, and the usual considerations regarding confocal fluorescence microscopy apply (e.g., choosing an objective with a high numerical aperture suitable for fluorescence microscopy, carefully setting the excitation and acquisition parameters, avoiding undersampling).
4. Perform a deconvolution of the acquired image stacks if needed. The optimal procedure will depend on many parameters (e.g., brand and type of the confocal microscope, used objective). Manufacturers typically provide instructions/information/software for image deconvolution.
5. When setting up the method, test if expansion is isotropic (i.e., equal expansion factor in all spatial dimensions). It may also be necessary to establish or estimate the effective spatial resolution and to assess possible tissue distortions introduced by ExM (see, for instance, [9]). Also *see* **Note 6** regarding further considerations about tissue volume and its changes due to tissue fixation.
6. Many different types of analysis can be applied to the acquired data. An example is shown in Fig. 2b, in which the distribution of astrocytic processes was analyzed around excitatory synapses by investigating the occupancy of spherical shells with increasing diameter centered on excitatory synapses. See [9] for further details regarding analysis and [10] for a similar type of analysis for GABAergic synapses.

4 Notes

1. The Ibidi chambers are designed for cell cultures, but we found that they are very useful for expansion microscopy when using an inverted microscope. One needs to test if they are compatible with the microscope that will be used, because, for instance, the microscope objective could be too wide when using an upright microscope.
2. It is crucial to firmly establish the immunohistochemical protocol first. This includes testing the specificity of primary and secondary antibodies. A robust and intense staining is typically needed for two reasons. First, the many experimental steps can lead to a loss of fluorophores. Second, the volume density of the fluorophores is greatly reduced by tissue expansion (e.g., 64-fold for a 4-fold expansion in each dimension). The choice of secondary antibodies and fluorophores is important and requires some testing because, for instance, the retained fluorescence after ExM can vary considerably between fluorophores [16]. In our experience, it is also advisable to include a specific label for the cell type of interest (e.g., EGFP-expressing astrocytes [22], Fig. 1), which provides a reference for the interpretation of data. Typically, the fluorescence intensity of endogenously expressed fluorescent proteins like EGFP is not sufficient and needs to be amplified (e.g., using a primary antibody against EGFP and a suitable secondary antibody).
3. Obtaining the factor by which the tissue was expanded is critical for converting distances/areas/volumes in the expanded tissue to their pre-expansion values. In our experience, the expansion factor varies between experiments and slices (typically four to five). Therefore, the expansion factor must be determined in each experiment. There are multiple approaches for that. For the hippocampus, we found that the diameter of nuclei of dentate gyrus granular cells at the tip of the dentate gyrus granular cell layer imaged before and after expansion is a reliable and easily obtainable measure of the expansion factor of the entire specimen because the nuclei are fairly circular and the region is easy to find before and after expansion.
4. To attach anchors to proteins in tissue slices, N-hydroxysuccinyl esters (NHS) are used, for example, methacrylic acid N-hydroxysuccinimide ester (MA-NHS) and acryloyl-X, SE (AcX). Both NHS-esters have been used as anchors for antibodies to be bound to the hydrogel polymerized throughout tissue slices. For AcX at a concentration of 0.1 mg/mL in PBS (0.35 mM), incubation overnight was

recommended, while MA-NHS at 1 mM was successfully used with only 1 h incubation of fixed tissue slices, both at room temperature [9–12].

5. It was reported that decreasing the concentration of the cross-linker in the monomer solution can increase the expansion factor at the cost of greater gel fragility [16]. In our hands, the stability of the gel also critically depended on the concentration of sodium acrylate in the monomer solution. In the preparation of monomer solution and in particular of the sodium acrylate stock, it should be taken into account that the sodium acrylate crystals have a significant volume and, therefore, water should be added last to obtain the desired volume (100 mL in this protocol). It is frequently mentioned that the quality of the batch of sodium acrylate should be high to obtain a colorless and easily soluble stock solution in water. Prior to polymerization, the tissue slices are incubated in monomer solution with a polymerization initiator, accelerator, and radical quencher for 30 min at 4 °C. The addition of polymerization initiator and accelerator should be performed in a chilled monomer solution and on ice to avoid undesired polymerization. Using a monomer solution composed of N, N-dimethylacrylamide, sodium acrylate and acrylamide was reported to allow expansion of hydrogels by the factor 10 [19, 20].
6. Chemical fixation can collapse the extracellular space and thereby alter the relative proportions of tissue volume occupied by cells. Regarding astrocytes, chemical fixation has been demonstrated to affect the morphology of perisynaptic astrocytic processes [25]. However, the fraction of tissue volume occupied by astrocytes is similar when obtained from acute hippocampal slices and from electron microscopy after chemical fixation when the latter is corrected for tissue shrinkage [26]. Also, see [27] for a recent study of the effect of chemical fixatives on extracellular space and astrocytes.

Acknowledgments

We thank Thomas Erdmann for support with animal breeding and laboratory maintenance. This work was supported by intramural funding from the Medical Faculty of the University of Bonn and by the German Research Council (DFG SPP1757 HE6949/1, HE6949/3, and FOR2795 P2 to C.H.).

References

1. Chen F, Tillberg PW, Boyden ES (2015) Expansion microscopy. Science 347:543–548. https://doi.org/10.1126/science.1260088
2. Bushong EA, Martone ME, Jones YZ, Ellisman MH (2002) Protoplasmic astrocytes in CA1 stratum radiatum occupy separate anatomical domains. J Neurosci 22:183–192
3. Díaz-Castro B, Robel S, Mishra A (2023) Astrocyte endfeet in brain function and pathology: open questions. Annu Rev Neurosci 46: 101–121. https://doi.org/10.1146/annurev-neuro-091922-031205
4. Rusakov DA, Bard L, Stewart MG, Henneberger C (2014) Diversity of astroglial functions alludes to subcellular specialisation. Trends Neurosci 37:228–242. https://doi.org/10.1016/j.tins.2014.02.008
5. Heller JP, Rusakov DA (2017) The nanoworld of the tripartite synapse: insights from super-resolution microscopy. Front Cell Neurosci 11: 374. https://doi.org/10.3389/fncel.2017.00374
6. Ventura R, Harris KM (1999) Three-dimensional relationships between hippocampal synapses and astrocytes. J Neurosci 19: 6897–6906. https://doi.org/10.1523/JNEUROSCI.19-16-06897.1999
7. Henneberger C, Bard L, Panatier A et al (2020) LTP induction boosts glutamate spillover by driving withdrawal of Perisynaptic Astroglia. Neuron 108:919–936. https://doi.org/10.1016/j.neuron.2020.08.030
8. Chozinski TJ, Halpern AR, Okawa H et al (2016) Expansion microscopy with conventional antibodies and fluorescent proteins. Nat Methods 13:485–488. https://doi.org/10.1038/nmeth.3833
9. Herde MK, Bohmbach K, Domingos C et al (2020) Local efficacy of glutamate uptake decreases with synapse size. Cell Rep 32: 108182. https://doi.org/10.1016/j.celrep.2020.108182
10. Brunskine C, Passlick S, Henneberger C (2022) Structural heterogeneity of the GABAergic tripartite synapse. Cells 11:3150. https://doi.org/10.3390/cells11193150
11. Minge D, Domingos C, Unichenko P et al (2021) Heterogeneity and development of fine astrocyte morphology captured by diffraction-limited microscopy. Front Cell Neurosci 15:669280. https://doi.org/10.3389/fncel.2021.669280
12. Deshpande T, Li T, Herde MK et al (2017) Subcellular reorganization and altered phosphorylation of the astrocytic gap junction protein connexin43 in human and experimental temporal lobe epilepsy. Glia 65:1809–1820. https://doi.org/10.1002/glia.23196
13. Patt L, Tascio D, Domingos C et al (2023) Impact of developmental changes of GABAA receptors on interneuron-NG2 glia transmission in the hippocampus. Int J Mol Sci 24: 13490. https://doi.org/10.3390/ijms241713490
14. Timmermann A, Tascio D, Jabs R et al (2023) Dysfunction of NG2 glial cells affects neuronal plasticity and behavior. Glia 71:1481–1501. https://doi.org/10.1002/glia.24352
15. Tillberg PW, Chen F, Piatkevich KD et al (2016) Protein-retention expansion microscopy of cells and tissues labeled using standard fluorescent proteins and antibodies. Nat Biotechnol 34:987–992. https://doi.org/10.1038/nbt.3625
16. Wassie AT, Zhao Y, Boyden ES (2019) Expansion microscopy: principles and uses in biological research. Nat Methods 16:33–41. https://doi.org/10.1038/s41592-018-0219-4
17. Asano SM, Gao R, Wassie AT et al (2018) Expansion microscopy: protocols for imaging proteins and RNA in cells and tissues. Curr Protoc Cell Biol 80:e56. https://doi.org/10.1002/cpcb.56
18. Wen G, Leen V, Rohand T et al (2023) Current progress in expansion microscopy: chemical strategies and applications. Chem Rev 123: 3299–3323. https://doi.org/10.1021/acs.chemrev.2c00711
19. Truckenbrodt S, Maidorn M, Crzan D et al (2018) X10 expansion microscopy enables 25-nm resolution on conventional microscopes. EMBO Rep 19:e45836. https://doi.org/10.15252/embr.201845836
20. Klimas A, Gallagher BR, Wijesekara P et al (2023) Magnify is a universal molecular anchoring strategy for expansion microscopy. Nat Biotechnol 1–12. https://doi.org/10.1038/s41587-022-01546-1
21. M'Saad O, Bewersdorf J (2020) Light microscopy of proteins in their ultrastructural context. Nat Commun 11:3850. https://doi.org/10.1038/s41467-020-17523-8
22. Nolte C, Matyash M, Pivneva T et al (2001) GFAP promoter-controlled EGFP-expressing transgenic mice: a tool to visualize astrocytes and astrogliosis in living brain tissue. Glia 33: 72–86
23. Feng G, Mellor RH, Bernstein M et al (2000) Imaging neuronal subsets in transgenic mice

expressing multiple spectral variants of GFP. Neuron 28:41–51. https://doi.org/10.1016/s0896-6273(00)00084-2

24. Schindelin J, Arganda-Carreras I, Frise E et al (2012) Fiji: an open-source platform for biological-image analysis. Nat Methods 9:676–682. https://doi.org/10.1038/nmeth.2019
25. Korogod N, Petersen CC, Knott GW (2015) Ultrastructural analysis of adult mouse neocortex comparing aldehyde perfusion with cryo fixation. elife 4:e05793. https://doi.org/10.7554/eLife.05793
26. Medvedev N, Popov V, Henneberger C et al (2014) Glia selectively approach synapses on thin dendritic spines. Philos Trans R Soc B 369:20140047. https://doi.org/10.1098/rstb.2014.0047
27. Idziak A, Inavalli VVGK, Bancelin S et al (2023) The impact of chemical fixation on the microanatomy of mouse organotypic hippocampal slices. eNeuro 10. https://doi.org/10.1523/ENEURO.0104-23.2023

Chapter 13

Astrocyte Morphological Adaptations in a Mouse Model of Social Anxiety Disorder

Laura Boi, Rohit Menon, and Barbara Di Benedetto

Abstract

The social fear conditioning (SFC) paradigm, which is an animal model of social anxiety disorder, has been extensively used to investigate the cellular and molecular mechanisms underlying the psychopathology of this disorder. Studies using the SFC paradigm have suggested that the lateral septum (LS) plays a pivotal role in regulating social fear. While several lines of evidence have indicated the involvement of various neuronal mechanisms within the LS in modulating social fear, the role played by LS astrocytes in the tripartite synapses in regulating social fear expression is not yet fully understood. Over the years, astrocytes have emerged as key players in regulating synaptic transmission and a plethora of animal behaviors. Given their dynamic morphological and molecular response to various stimuli, studying astrocyte morphology remains an excellent tool for understanding their role in brain function and disease. By applying immunofluorescent-immunohistochemical methods, here we describe a protocol to characterize changes in the morphology of astrocytes in response to social fear acquisition using confocal microscopy.

Key words Astrocytes, Astrocyte morphology, Social fear, Lateral septum, Mice

1 Introduction

The septal area has been linked to social behavior in both humans and animals [1]. Specifically, the lateral septum (LS) has been identified as a central hub connecting various brain regions implicated in the acquisition and expression of social fear and avoidance [1, 2]. Located in the subcortical area and delimited by the lateral ventricles, LS serves as a convergence point of a rich network of inputs from diverse brain regions such as the hypothalamus, ventral tegmental area, and hippocampus [1, 3, 4]. Additionally, it projects to executive downstream nuclei within the hypothalamus to drive an array of defensive and motivational social behaviors [5–7]. While previous animal studies on social fear have primarily focused on the neuronal component within the LS [6], the pivotal role of astrocytes in the tripartite synapse has been grossly overlooked.

Barbara Di Benedetto (ed.), *Astrocytes: Methods and Protocols*, Methods in Molecular Biology, vol. 2896, https://doi.org/10.1007/978-1-0716-4366-2_13,

In addition to their well-established roles in the neuronal support and blood–brain barrier functionality [8], astrocytes also actively participate in the regulation of emotional behaviors, including anxiety and depression [9, 10]. As the understanding of astrocytic contribution to behavioral mechanisms and disorders grows [9, 10], their characteristic plasticity in response to various stimuli has garnered a lot of interest. To gain insights into the stimuli-specific astrocyte responses and functions, various approaches including the study of astrocyte morphology have been employed [11, 12]. Astrocytes display a characteristic morphology with multiple branching processes that depart from the cell body, giving them the typical star-like shape [8]. Changes in astrocyte morphology often reflect alterations in their function in response to stimuli [8, 12]. Therefore, studying astrocyte morphology is important for understanding how they respond to specific stimuli and their role in neural circuits, behavioral function, and disease [12]. Alterations of glial fibrillary acidic protein (GFAP) expression, a major component of astrocyte cytoskeletal crucial for maintaining their structure, has been found to be related with diverse neurodegenerative diseases, anxiety, and depressive behaviors [12–15].

To provide a better and more holistic understanding of the underlying cellular mechanism of social fear, here, we present a method to analyze astrocyte morphology in an animal model of social anxiety disorder [2]. In this approach, we employed an immunofluorescent-immunohistochemical protocol to detect GFAP, combined with confocal microscopy, to achieve high-resolution visualization of the astrocyte structure. By analyzing various morphological parameters, this method allows us to unravel astrocyte responses and gain insights into their involvement in the learning process of social fear acquisition.

2 Materials

2.1 Animals

1. Wild-type CD1 male mice (Charles River, Sulzfeld, Germany, 8–9 weeks of age at the start of experiments) are housed under standard laboratory conditions (12/12 h light/dark cycle, lights on at 06:00, 22°C, 60 humidity, food and water ad libitum) in polycarbonate cages (16 × 22 × 14 cm). All experimental procedures are performed between 08:00 and 14:00 h in accordance with the Guide for the Care and Use of Laboratory Animals of the Government of Oberpfalz and the guidelines of the NIH.

2.2 Preparation of Brains for Immunohistochemistry

1. Ketamine (10%, 1 mL/kg) and xylazine (2%, 0.5 mL/kg).
2. Mini-peristaltic pump for intracardial perfusion of small animals.

3. Phosphate-buffered saline (PBS, 10×): 137 mM NaCl, 27 mM KCl, 14 mM Na_2HPO_4, and 43 mM KH_2PO_4. Bring total volume to 1 L with deionized water (*see* **Note 1**). Adjust pH to 7.4. Sterilize by autoclaving.
4. 4% Paraformaldehyde (PFA) in PBS 1×, pH 7.4.
5. Cryoprotective solution I (for brains after perfusion): 30% sucrose in PBS 1×.
6. N-methylbutane.
7. Laboratory dissection tools for brain preparation and extraction.

2.3 Preparation of Slices for IF-IHC

1. Cryostat.
2. 12-well plates (*see* **Note 2**).
3. Optimal cutting temperature (OCT) compound to fix brains for cutting in the cryostat.
4. Cryoprotective solution II (for slices after cutting): 25% ethylene glycol and 25% glycerol in PBS 1× pre-cooled to −20 °C.

2.4 Immunofluorescent Staining

1. 24-well plates.
2. Washing buffer: PBS 1×.
3. Blocking/permeabilization buffer: PBS 1×, 0.5% Triton ×-100, and 5% normal goat serum (NGS).
4. Primary antibody buffer solution: PBS 1×, 0.5% Triton ×-100, and 5% normal goat serum (*see* **Note 3**).
5. Secondary antibody solution: PBS 1×.
6. Primary antibody.
   (a) Rabbit monoclonal anti-GFAP antibody (1:200, Cell Signaling D1F4Q XP) to label the cytoskeletal structure of astrocytes.
7. Secondary antibody.
   (a) Goat anti-rabbit Alexa Fluor 488 (1:1000, Invitrogen).
8. Round-shaped glass plate to mount slices on slides.
9. Pre-cleaned microscope slides (Superfrost, e.g., Fisher Scientific).
10. Roti®Mount FluorCare DAPI.
11. Microscope cover glass (22 × 60 mm, e.g., Fisher Scientific, Pittsburgh, PA).
12. Pipettes, brushes, and other general laboratory equipment.
13. Horizontal shaker.
14. Confocal microscope (CLSM, Zeiss LSM 980).
15. ImageJ Software (http://www.imagescience.org).

3 Methods

3.1 Preparation of Brains for Immunofluorescent-Immuno-histochemistry (IF-IHC)

1. Following behavioral experiment (2), anaesthetize the animal by administering intraperitoneally (i.p.) a mixture of ketamine and xylazine (*see* **Note 4**).
2. Transfer the animal on a perfusion tray, and fix their four paws with tape or pin so that nothing moves during the procedure.
3. From this step on, operate within a fume hood (*see* **Note 5**).
4. With the help of forceps, hold the skin of the animals above the chest, and with scissors, open it by cutting a small hole below sternum. Carefully cut laterally just under the diaphragm. Then, carefully cut the diaphragm along the entire length of the ribcage. Clamp the sternum with a pair of hemostats, and lift it up to fully expose the chest cavity with heart and lung.
5. Carefully tear off pericardial sac to enable to have a clear view of the heart.
6. Take the apex of the heart with your thumb and forefinger or a good pair of tissue forceps. Insert the perfusion needle into left ventricle, and hold it there with a pair of hemostats, and make a small incision on the right atrium using fine scissors.
7. Let the cold (4 °C) PBS 1× flow at a steady drip of 19 mL/min (*see* **Note 6**) until the liver turns to a pale color, which indicates a good flush of arteries/veins from blood with PBS 1×.
8. Change the PBS 1× to cold PFA 4%, and continue perfusing the animal for 3–4 min (*see* **Note 7**) until all paws and liver become whitish and blenched, respectively.
9. Remove needle and clamps (hemostats), and decapitate the head with a pair of scissors.
10. After decapitation, open the skull and remove the brain.
11. Place each brain in a 50 mL Falcon tube for post-fixation overnight (o.n.) in 4% PFA at 4 °C.
12. After 1 day, transfer the brains to cryoprotective solution I. Let the brains in this solution at 4 °C until they sink.
13. After sinking of brains, wipe them from excess of liquid and flash-frozen N-methylbutane pre-cooled in dry ice before placing them in a piece of aluminum foil (or other foil) labelled with the number of the respective animal.
14. Place brains at −20 or −80 °C until ready for cutting and staining (*see* **Note 8**).

3.2 Preparation of Free-Floating Sections for IF-IHC

1. Take out brains from the freezer, and allow them to equilibrate to the cutting temperature of the cryostat for about 20–30 min (*see* **Note 9**).

2. Prepare a 12-well plate with about 2 mL of cryoprotective solution II at −20 °C per well.
3. Prepare the brain holder of the cryostat with a thin layer of OCT compound, and immerge the brain into it in a vertical position (olfactory bulbs toward you) (*see* **Note 10**); surround it at the base with additional OCT, and keep it in this position until the OCT is completely frozen.
4. Place the block in the correct position in the cryostat, and start cutting at 30–40 μm (*see* **Note 11**).
5. Collect the slices using a thin brush that has been previously dampened with a small drop of cryoprotective solution II to ease the peeling of the slice (*see* **Note 12**). Place the slice into the 12-well plate following the series for each brain region.
6. Once the brain has been cut to the desired extent, remove it from the block, and put it back into the foil for further cutting, if necessary.
7. Put the plate at −20 °C until proceeding with the staining (*see* **Note 13**).
8. Procedure for the double IF-IHC.
9. Remove the plates from the freezer, and, using a thin brush, select a series of LS region for staining from each brain.
10. Transfer slices in a 24-well plate with about 1 mL of PBS 1× per well, and let them wash on a horizontal shaker for 20 min at room temperature (RT). Discard the buffer, and repeat this step other two times to allow a wash-through of the cryoprotective solution II (*see* **Note 14**).
11. Remove the buffer and incubate slices in the blocking/permeabilization solution for 1 h at RT.
12. Remove the blocking solution and incubate slices in the antibody solution containing the primary antibody o.n. on a horizontal shaker at 4 °C (*see* **Note 15**).
13. Remove the antibody solution and wash slices three times with PBS 1× for 20 min each, on a horizontal shaker at RT.
14. Remove the antibody primary solution (*see* **Note 16**) and incubate slices with secondary antibody solution for 2 h on a horizontal shaker at RT (in a closed box or covered with aluminum foil to keep the fluorescent dyes protected from light).
15. To keep the fluorescent dyes protected from light, from this step on, operate in a dimly lit condition.
16. Remove the secondary antibody solution, and wash slices for three times with PBS 1× for 20 min each at RT (keeping always the plate protected from light).

17. Put slices from one series in a glass petri dish filled with PBS 1×, and mount them on a slide with the help of a thin brush to gently drift them.
18. Let slices to air-dry inside a closed box or under an aluminum foil.
19. Put few drops of mounting medium containing DAPI on slides, and cover them with a microscope cover glass (*see* **Note 17**).
20. Keep sections in a refrigerator until microscopic analysis.

3.3 Confocal Microscopy

1. Visualize and select the region of interest for image capture with a confocal laser scanning microscope using a 20× objective.
2. Switch to a 63× oil-immerged objective and capture three z-sections per each brain (Fig. 1a–c), each 30 μm thick, with a 0.5 μm/z section interval, and acquired at 1024 × 1024 resolution.
3. Save files in the existing proprietary CZI file format (.czi) file format.

3.4 Morphological Analysis

1. To analyze astrocyte morphology, open the file using ImageJ software with the FIJI update.
2. Emphasize details within the stack, by "flattening" the 3D stack through Z-projection at maximum intensity.
3. To facilitate visualization of the astrocyte soma, generate a merge channel with GFAP and DAPI, represented in green and blue, respectively.
4. Within the picture frame, select astrocytes whose structure is fully captured. Analyze the number of primary processes, the length of the longest process, and the area covered by each astrocyte (*see* **Note 18**).
5. To do this, first visualize the soma of the astrocyte, and count the number of primary processes. Then, use the segmented line tool to measure the length of one of the longest processes. Finally, measure the total area covered by each astrocyte using the polygon selection tool.
6. The measurements of the length and the area coverage are expressed as arbitrary units (a.u.).

3.5 Statistical Analysis

1. For each brain (animal) and parameter considered, calculate the average among the three pictures. Perform a normality test using Shapiro–Wilk test to ensure the data follows a normal distribution.
2. If the data is normally distributed, use an unpaired t-test of the two groups of interest. Use the Mann–Whitney test if the data is not normally distributed.

4 Notes

1. All solutions are prepared in deionized water.
2. Cutting in a three-series step-size is ideal for a complete representation of the lateral septum along the rostral–caudal axis.
3. Ideally, the serum used for the blocking or dilution should come from the same species in which the secondary antibody is raised. This ensures compatibility, reduces the risk of nonspecific binding, and improves the reliability of the IF-IHC assay.
4. Allow a few minutes for the mouse to reach a surgical plane of anesthesia, then test for the loss of response by gently pinching the tail or toe.
5. To minimize inhalation of formaldehyde vapor, operate always within a fume hood and wear a suitable protective equipment.
6. In order to utilize the circulatory system effectively, the flow should be adjusted to the physiological pressure of the species.
7. Efficient perfusion is crucial for obtaining high-quality tissue suitable for morphological analysis, with reduced background staining and improved clarity of the images. When brains are well perfused and fixed, they show a whitish color, without any evident blood vessels on the surface.
8. If you intend to store the brains at −80 °C, it would be ideal to transfer them to −20 °C the day before slicing. This gradual thawing process ensures a smoother slicing procedure.
9. Based on our experience, we have found that for obtaining slices with a thickness of 30–40 μm, a cryostat temperature −20 °C is ideal for soft tissues such as the adult animal brain tissue.
10. To guarantee uniform and accurate slicing, consider making a straight cut through the cerebellum and positioning the brain on the holder without any angle.
11. To ensure the preservation of morphology as much as possible during sectioning, it is ideal to use a minimum of slice size of 30 μm. This allows for a greater possibility of analyzing extensive branching processes and structure within the same section.
12. Ensure that the bristle of the brush are not damaged, as this could potentially harm the tissue quality.
13. Since the section will float above the solution and potentially become stuck at the walls of well, make sure they are fully immersed in the cryoprotective solution before placing them in the freezer.

14. During the change of solutions, slices should never get dried. Additionally, handle the process gently, avoiding any pressure on the slices to prevent damage to the tissue.
15. Once again, before placing the plate on the shaker, ensure the slices are fully immersed in the solution and that the agitation is not too strong.
16. You may store the antibody primary solution at −20 °C for future use, typically for one to two additional times.
17. Take care of avoid forming bubbles in the mounting medium, as they may interfere with imaging on the confocal microscope. If bubbles do form, gently remove them using a pipette tip pressed against the cover slip.
18. For easier analysis, you might opt to keep the GFAP channel grayscale and invert the colors to depict the astrocyte cytoskeleton in black (*see* Fig. 1d, e).

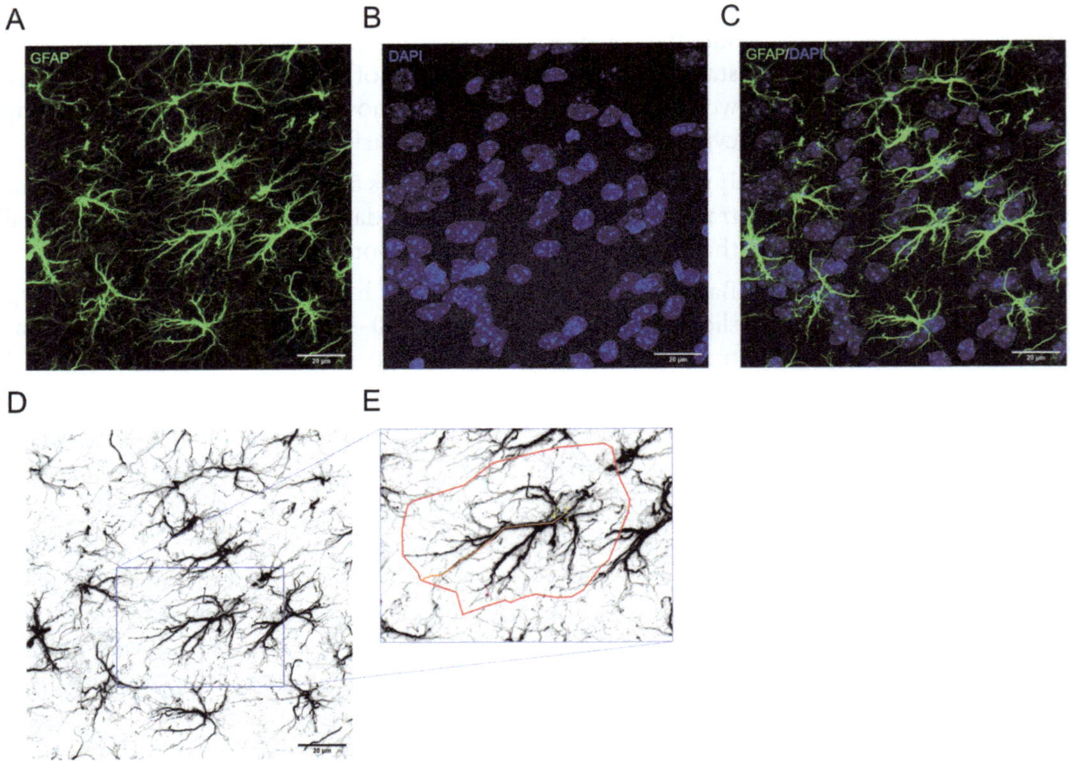

Fig. 1 Analysis of astrocyte morphology with GFAP and DAPI staining. Confocal images of astrocytes from the lateral septum (LS) of either social fear conditioned (SFC$^+$) or unconditioned (SFC$^-$) mice are taken containing GFAP (green) and DAPI (blue), which label respectively the astrocyte cytoskeleton and nucleus (**a**, **b**, **c**). Using ImageJ, astrocyte morphology is analyzed (**d**, **e**) by counting the number of primary processes (number in yellow) and measuring the length of the longest process (orange) and the area covered by each astrocyte (red). Scale bar = 20 μm

Acknowledgments

The authors wish to thank Dr. Carl P. Meinung for providing his support for the initial establishment of the protocol, later adapted and presented in this study. This work was supported by grants from the German Research Council (DFG, GRK2174) to RM and BDB.

References

1. Menon R, Süß T, Oliveira VEM et al (2022) Neurobiology of the lateral septum: regulation of social behavior. Trends Neurosci 45:27–40. https://doi.org/10.1016/j.tins.2021.10.010
2. Toth I, Neumann ID, Slattery DA (2012) Social fear conditioning: a novel and specific animal model to study social anxiety disorder. Neuropsychopharmacology 37:1433–1443. https://doi.org/10.1038/npp.2011.329
3. Rizzi-Wise CA, Wang DV (2021) Putting together pieces of the lateral septum: multifaceted functions and its neural pathways. eNeuro 8. https://doi.org/10.1523/ENEURO.0315-21.2021
4. Wirtshafter HS, Wilson MA (2021) Lateral septum as a nexus for mood, motivation, and movement. Neurosci Biobehav Rev 126:544–559. https://doi.org/10.1016/j.neubiorev.2021.03.029
5. Wong LC, Wang L, D'Amour JA et al (2016) Effective modulation of male aggression through lateral septum to medial hypothalamus projection. Curr Biol 26:593–604. https://doi.org/10.1016/j.cub.2015.12.065
6. Menon R, Grund T, Zoicas I et al (2018) Oxytocin signaling in the lateral septum prevents social fear during lactation. Curr Biol 28:1066–1078. https://doi.org/10.1016/j.cub.2018.02.044
7. Carus-Cadavieco M, Gorbati M, Ye L et al (2017) Gamma oscillations organize top-down signalling to hypothalamus and enable food seeking. Nature 542:232–236. https://doi.org/10.1038/nature21066
8. Verkhratsky A, Nedergaard M (2018) Physiology of Astroglia. Physiol Rev 98:239–389. https://doi.org/10.1152/physrev.00042.2016
9. Shigetomi E, Koizumi S (2023) The role of astrocytes in behaviors related to emotion and motivation. Neurosci Res 187:21–39. https://doi.org/10.1016/j.neures.2022.09.015
10. Kofuji P, Araque A (2021) Astrocytes and behavior. Annu Rev Neurosci 44:49–67. https://doi.org/10.1146/annurev-neuro-101920-112225
11. Yu X, Nagai J, Khakh BS (2020) Improved tools to study astrocytes. Nat Rev Neurosci 21:121–138. https://doi.org/10.1038/s41583-020-0264-8
12. Zhou B, Zuo Y-X, Jiang R-T (2019) Astrocyte morphology: diversity, plasticity, and role in neurological diseases. CNS Neurosci Ther 25:665–673. https://doi.org/10.1111/cns.13123
13. Banasr M, Duman RS (2008) Glial loss in the prefrontal cortex is sufficient to induce depressive-like behaviors. Biol Psychiatry 64:863–870. https://doi.org/10.1016/j.biopsych.2008.06.008
14. Rajkowska G, Hughes J, Stockmeier CA et al (2013) Coverage of blood vessels by astrocytic endfeet is reduced in major depressive disorder. Biol Psychiatry 73:613–621. https://doi.org/10.1016/j.biopsych.2012.09.024
15. Miguel-Hidalgo JJ, Baucom C, Dilley G et al (2000) Glial fibrillary acidic protein immunoreactivity in the prefrontal cortex distinguishes younger from older adults in major depressive disorder. Biol Psychiatry 48:861–873. https://doi.org/10.1016/s0006-3223(00)00999-9

Chapter 14

An In Vitro Assay to Measure Astrocyte-Dependent Synaptic Phagocytosis in Health and Major Depressive Disorder

Diego Julian Vesga-Jimenez, Eugenia Vivi, and Barbara Di Benedetto

Abstract

Immunofluorescent methods in vitro have been widely used to characterize the distribution and extent of colocalization of molecules located in various intracellular compartments. These are key processes to better understand the role of those molecules in physiological and pathological conditions. In the brain, interactions of astrocytes with neurons are fundamental for the establishment and proper functioning of neuronal circuits. Among these, the phagocytosis of synapses that need to be removed is a process known as synaptic pruning. When this process is dysregulated, it might lead to the development of several brain disorders. Here, we describe the adaptation of a method to characterize the astrocyte-mediated synaptic phagocytosis and evaluate astrocytic (dys)functions in healthy cells and cells derived from an animal model of major depressive disorder (MDD).

Key words Confocal microscopy, Astrocytes, Phagocytosis, Prefrontal cortex, Major depressive disorder, Synaptic pruning

1 Introduction

Astrocytes display several functions essential for the brain's well-functioning, such as supplying the structural support to the blood–brain barrier (BBB), giving metabolic aid to neurons, regulating inflammatory processes in the brain, maintaining ionic homeostasis, and releasing neurotrophic factors [1–6].

Major depressive disorder (MDD) is a heterogeneous disease with a wide variety of symptoms and causes [7], among which the reduction in astrocyte numbers and impaired morphology in most areas of the central nervous system (CNS) are very peculiar characteristics [8–10]. In recent years, astrocytes have got more relevance as putative targets for pharmacological interventions in many psychiatric disorders due to their capability of remodeling synapses [10, 11]. Alterations in the homeostasis of this process will lead to

Barbara Di Benedetto (ed.), *Astrocytes: Methods and Protocols*, Methods in Molecular Biology, vol. 2896,
https://doi.org/10.1007/978-1-0716-4366-2_14,

imbalances between excitatory and inhibitory outputs, which might lead to the development of MDD [12–14].

Therefore, developing reliable detection methods to reveal deficits in astrocyte-mediated synaptic pruning might help to improve our understanding of the disease and support the development of alternative therapeutic strategies to reverse such deficits.

Recent studies using a combination of histological and software-based quantification methods showed that MDD models display a reduction in astrocytes' phagocytosis of synapses [11, 15, 16]. Here, we describe the details of our method established to quantify in vitro the astrocyte-dependent synaptic phagocytosis in cells derived from two animal models of MDD, the high-anxiety behavior (HAB) rats and the Wistar Kyoto rats (WKY) [17].

To this aim, we established a sensitive immunofluorescent-immunohistochemical (IF-IHC) staining which allows the simultaneous detection of synaptophysin, a marker of presynaptic puncta, and the lysosomal-associated membrane protein 1 (LAMP1), a marker of lysosomes, inside astrocytes. Moreover, we developed a computer-based tool to measure the amount of co-localizing signals between synaptophysin and LAMP1. This method is useful to diagnose differences in astrocyte-dependent phagocytosis of presynaptic puncta between healthy cells and cells derived from an animal model of disease, further helping to propose a reduced phagocytosis as a cause (or consequence) of the depressive-like behavior in these animals.

2 Materials

2.1 Animals for Cell Culture

1. Cortical astrocytes were obtained from the brains of newborn pups (postnatal day (P)1–P4) from either Wistar rats or high-anxiety behavior (HAB) rats (from the laboratory of Prof. Inga Neumann, University of Regensburg) or Wistar Kyoto rats (WKY). Procedures follow institutional guidelines of the University of Regensburg, Germany, as well as the European Community Council Directive (86/609/EEC). The Animal Welfare Office at the Animal Care and Use Facility of the University of Regensburg approves all experiments. In accordance with the German Animal Welfare Act (Tierschutzgesetz, Articles 4 and 7), no additional approval for the post-mortem removal of brain tissue is necessary.

2.2 Media

1. Dissection Solution (DS).
   (a) Hank's Balanced Salt Solution (HBSS) 500 mL, e.g., Thermo Scientific 14025092.
   (b) Pen-Strep, e.g., Thermo Scientific 10378016.

   - (c) 1 M HEPES, e.g., Thermo Scientific 15630080.
   - (d) 200 mM L-Glutamine, e.g., Thermo Scientific™ 25030081.
2. Neuron Growth Medium (NGM).
   - (a) Neurobasal™ Plus Medium 500 mL, e.g., Gibco 21103049.
   - (b) B27 Supplement, e.g., Gibco 17504044.
   - (c) 200 mM L-Glutamine or GlutaMAX™ Supplement, e.g., Thermo Scientific™ 35050061.
   - (d) Antibiotics/antimitotic, e.g., Sigma A5955.
3. Astrocyte Full Medium.
   - (a) DMEM, Dulbecco's Modified Eagle +4.5 g/L D-glucose, L-glutamine, e.g., Gibco 11965092.
   - (b) Heat-inactivated fetal calf serum (FCS), e.g., Gibco 26010074.
   - (c) Sodium pyruvate 100 mM (100×), e.g., Sigma-Aldrich 11360070.
   - (d) Antibiotics/antimitotic.
   - (e) MEM Non-Essential Amino Acids (NEAA) (100×), 100 mL, e.g., Gibco 11140050.
   - (f) 1 M HEPES.
4. Low medium.
   - (a) DMEM, Dulbecco's Modified Eagle +4.5 g/L D-glucose, L-glutamine.
   - (b) Sodium pyruvate 100 mM (100×).
   - (c) Antibiotics/antimitotic.
   - (d) MEM Non-Essential Amino Acids (NEAA) (100×).
   - (e) 1 M HEPES.
5. 24-well plate coated w/ PDL and laminin, e.g., Corning Costar®.
6. Trypsin-EDTA solution, e.g., Sigma T4049.
7. Millipore H_2O.
8. Culture one.
9. Dissection tools.
10. Petri dishes.
11. T75 flasks.
12. Neubauer chamber.
13. Trypan blue.

2.3 Preparation of Slides for Immunofluorescence-Immunocytochemistry (IF-IHC)

1. Round-shaped glass (coverslips) coated with poly-D-lysine (PDL) to plate cells.
2. Phosphate-buffered saline (PBS, 10×): 137 mM NaCl, 27 mM KCl, 14 mM Na_2HPO_4, and 43 mM KH_2PO_4. Bring total volume to 1 L with deionized water.
3. Adjust pH to 7.5. Sterilize by autoclaving.
4. 4% Paraformaldehyde (PFA) in PBS 1×, pH 7.5.
5. PBS 1×.
6. 24-well plates.
7. Pre-cleaned microscope slides (e.g., Superfrost/Plus, Fisher Scientific).
8. Pipettes and other general laboratory equipment like dissection kit.
9. Horizontal shaker.

2.4 Immunofluorescent Staining

1. 24-well plates.
2. Washing buffer: PBS 1×.
3. Normal goat serum (NGS).
4. Blocking/permeabilization buffer: PBS 1×, 0.1% Triton ×-100, and 2% normal goat serum.
5. Antibody buffer solution: PBS 1×, 0.1% Triton ×-100, and 2% normal goat serum.
6. Primary antibodies:
   (a) Mouse monoclonal anti-glial fibrillary acidic protein (GFAP) antibody (1:400, e.g., Sigma-Aldrich Chemicals G3893) to label astrocytes.
   (b) Mouse monoclonal anti-S-100 (β-subunit) antibody (1: 1000, e.g., Sigma-Aldrich Chemicals S2532) to label astrocytes.
   (c) Rabbit polyclonal anti-lysosome associated membrane protein 1 (LAMP1) antibody (1:100, e.g., Abcam ab24170) to label lysosomes.
   (d) Chicken polyclonal anti-synaptophysin1 antibody (1:500, e.g., SySy 101006) to label presynaptic puncta.
7. Secondary antibodies:
   (a) Cy3-conjugated anti-mouse IgG antibody (1:400, e.g., Thermo Scientific A10521).
   (b) Biotin-conjugated anti-rabbit IgG antibody (1:500, e.g., Jackson Lab 11-065-003).
   (c) Alexa Fluor™ 488-conjugated goat anti-chicken IgG antibody, (1:500, e.g., Invitrogen A11035).

8. Tertiary antibodies:
   (a) Alexa Fluor™ 647-conjugated streptavidin (1:1000, e.g., Invitrogen S21374).
   (b) 4′,6-Diamidino-2-phenylindole (DAPI, 1:1000, e.g., Sigma D9542) to label nuclei.
9. Mounting medium (e.g., DAKO S3023).
10. Round-shaped glass plate to mount slices on slides.
11. Pre-cleaned microscope slides (e.g., Superfrost/Plus, Fisher Scientific).
12. Confocal microscope.

3 Methods

3.1 Preparation of Astrocyte–Neuron Co-culture

3.1.1 Preparation of Rat Primary Astrocytes (All Steps Are Executed at Room Temperature (RT) Except When Stated Otherwise)

1. Primary astrocytes are isolated from your region of interest from pup brains between P1 and P4.
2. Decapitate pups, open the skull, carefully extract the brains, and remove meninges.
3. Isolate your region of interest by opening the two hemispheres under a light microscope, and cut it out in Dissection Solution (we keep male and female brains separated).
4. Mince the tissue in small pieces.
5. Transfer the minced tissues into a 15 mL Falcon in dissociation buffer.
6. Carefully discard buffer.
7. Wash tissues with 2 mL Trypsin-EDTA.
8. Incubate tissue with 1,5 mL Trypsin-EDTA for approximately 20 min at 37 °C in the water bath (invert the Falcon tube every 5 min).
9. Discard trypsin.
10. Wash three times with 3–4 mL with Astrocyte Full Medium.
11. Add 1.5 mL of Astrocyte Full Medium, and homogenize the tissue with a fire-polished Pasteur pipette until suspension is homogenous.
12. Centrifuge at 1200 rpm for 3 min.
13. Discard the supernatant.
14. Resuspend the cells in Astrocyte Full Medium (~10 mL).
15. Plate cells on poly-D-lysine (PDL)-coated T75 cm^2 flask at 37 °C and 5% CO_2 in oxygen-reduced conditions (5% O_2).
16. Let cells grow in culture for approximately 7–10 days in vitro (changing the medium every ~4 days).

17. After reaching approximately 90% confluency, shake flasks for 6 h on a rotating shaker at circa 300 rpm in order to remove non-astrocytic cells from the culture.
18. Discard medium.
19. Wash cells with warm PBS.
20. Pipette 3 mL of warm Trypsin-EDTA and incubate for 4 min at 37 °C.
21. Add 7 mL of Astrocyte Full Medium in the flask to neutralize trypsin.
22. Collect cells in a 15 mL Falcon tube by pipetting up and down several times.
23. Centrifuge cells at 1200 rpm for 3 min.
24. Discard supernatant.
25. Resuspend cells in 10 mL freezing medium, and aliquot them (approximately 1×10^6 cells/vial).
26. Gradually freeze cells:
    (a) −20 °C for 24 h
    (b) −80 °C for 24 h
    (c) Liquid nitrogen for long-term storage

3.1.2 Culture and Maintenance of Rat Primary Astrocytes

1. Cell aliquots containing approximately 1×10^6 cells (stored in liquid nitrogen) are quickly thawed in the water bath and plated on a PDL-coated T75 flask in Astrocyte Full Medium at 37 °C and 5% CO_2 in oxygen-reduced conditions (5% O_2).
2. Change medium every 3–4 days.

3.1.3 Splitting Cells for Experiments/Culture Maintenance

1. Once confluent (~80–90%), discard medium and wash once with warm PBS.
2. Add 3 mL of Trypsin-EDTA in the flask and incubate for 4 min at 37 °C.
3. Add 7 mL of Astrocyte Full Medium and collect cells in a 15 mL Falcon.
4. Centrifuge at 1200 rpm for 3 min.
5. Discard supernatant and resuspend cells in ~10 mL of astrocyte medium.
6. Primary astrocyte cells are used until passage 3 (*see* **Note 1**).

3.2 Preparation of Rat Primary Neurons

1. Use late embryonic stage (E18) rat fetuses, euthanize them with CO_2, remove uterus, free individual fetuses from the embryonic sack, and collect them in T75 flask with 50 mL DS.
2. Place fetuses into a Petri dish and decapitate them, open the skull, carefully extract the brains, and remove meninges.

3. Dissect the two hemispheres and place brains in a new Petri dish with DS.
4. Isolate the cortex and mince the tissue.
5. Transfer the minced tissue to a 15 mL Falcon with DS.
6. Carefully discard DS.
7. Wash tissues with 2 mL Trypsin-EDTA.
8. Incubate tissue with 1.5 mL Trypsin-EDTA for approximately 20 min at 37 °C in the water bath (invert the Falcon tube every 5 min).
9. Discard Trypsin.
10. Wash three times with 3–4 mL with Neuron Growth Medium.
11. Add 1.5 mL of Neuron Growth Medium, and homogenize the tissue with a fire-polished Pasteur pipette until suspension is homogenous (circa 20×).
12. Centrifuge at 900 rpm for 5 min.
13. Discard supernatants. Resuspend in 1 mL of Neuron Growth Medium.
14. Counting cells in Neubauer chamber.
15. Cell plating: 45,000 cells/well for single cell culture and 90,000 cells/well for co-culture.
16. Renew half of the Neuron Growth Medium once per week.
17. Let neurons grow 2–3 weeks, depending on the maturational stage required.

3.2.1 Fixation

1. Remove medium.
2. Wash three times for 5 min with PBS 1×.
3. Incubate with PBS 1× with PFA 4% for 40 min.
4. Wash three times for 10 min with PBS 1×.
5. Store at 4 °C.

3.3 Immunofluorescent-Immunocytochemistry-GFAP/S100ß/LAMP1/Synaptophysin

1. Take out the coverslips and place them in a fresh 24-well plate.
2. Wash carefully two times for 10 min with PBS 1× (all incubation and washing steps require the plate to be on the shaker at around 100 rpm; *see* **Note 2**).
3. Add blocking/permeabilization solution: PBS 1× with 0.1% Triton ×-100/2% NGS, and incubate for 1 h (*see* **Note 3**).
4. Wash carefully three times for 10 min with PBS 1×.
5. Incubate in primary antibody solution: PBS 1× with 0.1% Triton ×-100 and 2% NGS O/N at 4 °C protected from light. Make sure to cover fully the wells; the recommended minimum amount is 150 μL.

6. The next day, wash carefully three times with PBS 1× and keep it protected from light.
7. Incubate in secondary antibody solution: PBS 1× with 2% NGS for 1.5 h protected from light.
8. Wash carefully three times with PBS 1× and keep protected from light.
9. Incubate the tertiary staining solution: PBS 1× with 2% NGS for 1.5 h protected from light.
10. Wash carefully three times with PBS 1× and keep protected from light.
11. Put a single drop of mounting medium on slides for each coverslip and place the coverslip downward on slides (*see* **Notes 4**).

3.4 Confocal Microscopy

1. Choose the suitable lasers to detect the fluorochromes and adjust settings to ensure optimal fluorescent saturation.
2. Capture a minimum of 6–7 images per coverslip, each consisting of 14 Z-stack slices with a thickness of 0.30 μm, using an Olympus confocal microscope equipped with a 40× oil immersion objective (*see* **Note 5** and **6**).
3. Save the acquired images (Fig. 1a, b) in the Olympus image file format.

3.5 Analysis of Astrocyte-Mediated Phagocytosis in Astrocyte–Neuron Co-cultures

1. Open ImageJ.
2. Drag&Drop to ImageJ.
   Color Mode: Default.
3. Change to 8-bit: Image → Type → 8Bit; for every channel.
4. Open ROI manager: Analyze → Tool → ROI manager.
5. Merge Channels → Image → Color →Merge Channel.
   (a) Preset is C1 (red): (C2) please change it according to colors on the images.
   (b) Preset is C2 (green): (C1) please change it according to colors on the images.
   (c) Preset is C3 (blue): (C3) please change it according to colors on the images.
6. Find a cell that is not overlapping with others.
7. Image → Stacks → Z Projekt → Projekttype: Max Intensity (for every astrocyte, use five to six slices with the highest intensity).
8. Image → Color → Channel tool → choose astrocyte channel.
9. Freehand selections: mark the astrocyte → Add (to ROI manager).

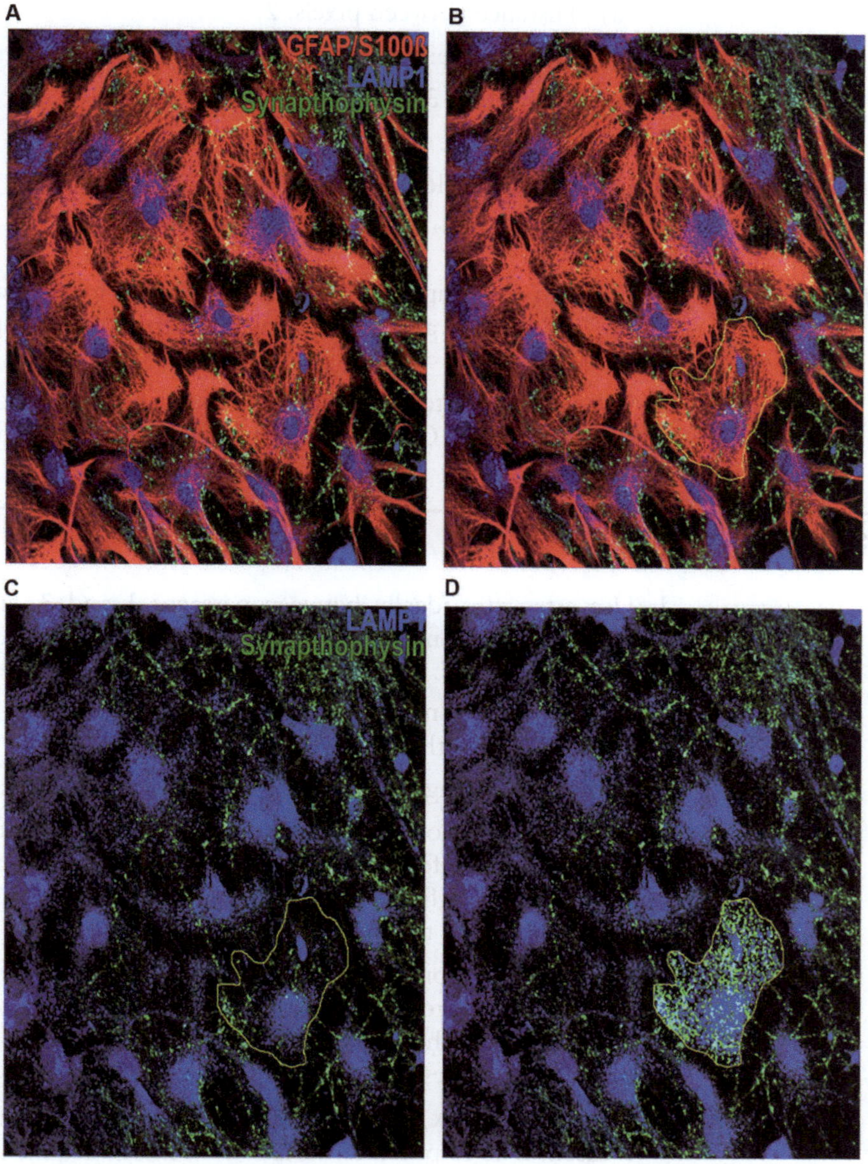

Fig. 1 Graphic example of the process for the ImageJ analysis. (**a**) Merging of the channels. (**b**) Generating the region of interest (ROI) in a target astrocyte. (**c**) Choosing the channels to analyze (e.g., synaptophysin/LAMP1) for every ROI. (**d**) Quantifying the colocalization spots inside any selected ROI

10. Image → Color → Split Channels → remove astrocyte channel.
11. Image → Color → Merge Channels.
12. Plugins → ComDet v.0.5.5 → Detect particles.
13. Select pixels and intensity for every channel, and keep them identical for every image.
14. Suggested starting parameters (*see* **Notes 7** and **8**):

   (a) Distance between pixels: 2
   (b) Particle size: 5
   (c) Threshold: 3
15. Summary opens.
16. Copy to Excel file.
17. Also save ROI → rename: e.g., NAB Pic1 Astro1.

3.6 Statistical Analysis

1. For each coverslip: calculate the mean value for "Synaptophysin," "LAMP1," and colocalizing spots from the average of ten astrocytes.
2. Use unpaired t-test for the comparison between two groups and one-way ANOVA when comparing more than two groups.

4 Notes

1. Use astrocytes ideally between passages 1 and 3, and avoid excessive passaging to preserve their morphological and functional properties.
2. All the incubation and washing steps require the plate to be on a shaker on less than 110 rpm.
3. During the change of solutions, slices should never get dried.
4. Be careful not to form bubbles in the mounting medium as they might interfere with imaging at the confocal microscope. In case bubbles form, you can gently pull them out with the help of a pipette tip pressed on the coverslip.
5. Choose the appropriate image frame size, considering the subsequent image analysis. Ensure that one or more astrocytes are captured in the image, avoiding isolation or excessive overlap with dendrites.
6. Utilize five to six Z-stack slices where the targeted astrocyte is located. Due to potential variations in image acquisition influenced by experimental procedures related to immunostaining, the condition of the co-culture, and the status of the fluorescent fluorophore, it is advisable to conduct a pilot immunostaining using coverslips from various co-cultures to establish the most suitable settings for image analysis.
7. Particle size and threshold is the same for both channels.
8. Parameters need adjustment based on the pictures and staining.

Acknowledgments

The authors wish to thank Susanne Wallner for the support in the execution of some experiments used to prepare the images for the protocol presented here. This work was supported by intramural funding from the University of Regensburg and by the German Research Council (DFG GRK2174) to BDB.

References

1. Khakh BS, Sofroniew MV (2015) Diversity of astrocyte functions and phenotypes in neural circuits. Nat Neurosci 18:942–952. https://doi.org/10.1038/nn.4043
2. Sofroniew MV (2020) Astrocyte reactivity: subtypes, states, and functions in CNS innate immunity. Trends Immunol 41:758–770. https://doi.org/10.1016/j.it.2020.07.004
3. Abbott NJ, Rönnbäck L, Hansson E (2006) Astrocyte–endothelial interactions at the blood–brain barrier. Nat Rev Neurosci 7:41–53. https://doi.org/10.1038/nrn1824
4. Allen NJ (2014) Astrocyte regulation of synaptic behavior. Annu Rev Cell Dev Biol 30:439–463. https://doi.org/10.1146/annurev-cellbio-100913-013053
5. Mahmoud S, Gharagozloo M, Simard C, Gris D (2019) Astrocytes maintain glutamate homeostasis in the CNS by controlling the balance between glutamate uptake and release. Cells 8:184. https://doi.org/10.3390/cells8020184
6. Oberheim NA, Goldman SA, Nedergaard M (2012) Heterogeneity of astrocytic form and function. In: Milner R (ed) Astrocytes: methods and protocols. Humana Press, Totowa, pp 23–45
7. Arnaud AM, Brister TS, Duckworth K et al (2022) Impact of major depressive disorder on comorbidities: a systematic literature review. J Clin Psychiatry 83. https://doi.org/10.4088/JCP.21r14328
8. Cobb JA, O'Neill K, Milner J et al (2016) Density of GFAP-immunoreactive astrocytes is decreased in left hippocampi in major depressive disorder. Neuroscience 316:209–220. https://doi.org/10.1016/j.neuroscience.2015.12.044
9. Rajkowska G, Hughes J, Stockmeier CA et al (2013) Coverage of blood vessels by astrocytic endfeet is reduced in major depressive disorder. Biol Psychiatry 73:613–621. https://doi.org/10.1016/j.biopsych.2012.09.024
10. Rajkowska G, Stockmeier CA (2013) Astrocyte pathology in major depressive disorder: insights from human postmortem brain tissue. Curr Drug Targets 14:1225–1236
11. Lee J-H, Kim J, Noh S et al (2021) Astrocytes phagocytose adult hippocampal synapses for circuit homeostasis. Nature 590:612–617. https://doi.org/10.1038/s41586-020-03060-3
12. Ortinski PI, Dong J, Mungenast A, Yue C, Takano H, Watson DJ, Haydon PG, Coulter DA (2010) Selective induction of astrocytic gliosis generates deficits in neuronal inhibition. Nat Neurosci 13(5):584–591
13. Antoine MW, Langberg T, Schnepel P, Feldman DE (2019) Increased excitation-inhibition ratio stabilizes synapse and circuit excitability in four autism mouse models. Neuron 101:648–661. https://doi.org/10.1016/j.neuron.2018.12.026
14. Dejanovic B, Wu T, Tsai M-C et al (2022) Complement C1q-dependent excitatory and inhibitory synapse elimination by astrocytes and microglia in Alzheimer's disease mouse models. Nat Aging 2:837–850. https://doi.org/10.1038/s43587-022-00281-
15. Park J, Choi Y, Jung E et al (2021) Microglial MERTK eliminates phosphatidylserine-displaying inhibitory post-synapses. EMBO J 40:e107121. https://doi.org/10.15252/embj.2020107121
16. Lee E, Chung W-S (2019) Glial control of synapse number in healthy and diseased brain. Front Cell Neurosci 13:42
17. Neumann ID, Wegener G, Homberg JR et al (2011) Animal models of depression and anxiety: what do they tell us about human condition? Prog Neuro-Psychopharmacol Biol Psychiatry 35:1357–1375. https://doi.org/10.1016/j.pnpbp.2010.11.028

Chapter 15

Cortical Organotypic Brain Slice Cultures to Examine Sex- and Age-Dependent Astrocyte-Mediated Synaptic Phagocytosis

Anastasiia Nagumanova, Lea R. Seeholzer, and Barbara Di Benedetto

Abstract

Astrocytes, the most abundant glial cells in the brain, are an integral part of the synaptic compartment and contribute to synaptic pruning, a key process for refining neural circuits during early postnatal development (PND). Dysregulations in this process are implicated in various neuropsychiatric disorders, including major depressive disorder (MDD). To investigate astrocyte functions in a physiologically relevatpdelnt context, organotypic brain slice cultures (OBSCs) offer a powerful model, reproducing more closely in vivo conditions than traditional cell cultures and preserving complex brain architecture and interactions. Here, we present OBSCs as an ex vivo culturing method to provide a platform to explore astrocyte-mediated synaptic pruning dynamics in the rat prefrontal cortex (PFC) during PND. Our approach is based on assessing the role of MEGF10, a key protein involved in synaptic pruning, alongside the synaptic markers synaptophysin and PSD95, using Western blotting to analyze the expression levels of these markers in the cortex of developing rat pups. Additionally, we combine immunofluorescence staining with confocal imaging and IMARIS 9.8 software-assisted analysis to investigate the colocalization of the lysosomal marker LAMP1 with synaptic and astrocytic markers to evaluate the precise rate of synaptic engulfment. The methods presented here allow a deeper examination of an astrocyte-mediated synaptic remodeling in healthy and pathophysiological conditions.

Key words Organotypic brain slice culture, Astrocyte, Phagocytosis, Synaptic elimination, Prefrontal cortex, Critical period, Major depressive disorder

1 Introduction

Astrocytes, the predominant subtype of glial cells in the central nervous system (CNS), present diverse functions crucial for brain homeostasis and function. They support endothelial cells at the blood–brain barrier (BBB), regulate ion homeostasis around synapses, provide metabolic support to neurons, and play a crucial role in synaptogenesis [1–3]. During the critical period of postnatal brain development in rodents, astrogenesis goes along with active

Barbara Di Benedetto (ed.), *Astrocytes: Methods and Protocols*, Methods in Molecular Biology, vol. 2896,
https://doi.org/10.1007/978-1-0716-4366-2_15,

synaptogenesis, leading to the formation of functional neuronal networks [4–6].

Altered properties of astrocytes have been shown in various neuropsychiatric disorders, including Alzheimer's disease [7], Parkinson's disease [8], epilepsy [9], and major depressive disorder (MDD) [10]. In particular, astrocytes play an important role in synaptic pruning, a process necessary for refining neural circuits during development [11]. The prefrontal cortex (PFC) is especially susceptible to environmental and endogenous influences during early postnatal development, affecting its structural and functional maturation and potentially shaping neural and behavioral long-term outcomes [12, 13]. Astrocyte-mediated synapse elimination involves the activation of signaling systems such as the MEGF10 phagocytic pathway. MEGF10 is a protein predominantly expressed in astrocytes and is essential for maintaining synaptic homeostasis [14].

While primary dissociated cultures facilitate the study of single-cell populations, organotypic brain slice cultures (OBSCs) offer a more comprehensive model to analyze interactions between various brain cell types. OBSCs preserve the structural and synaptic organization of the brain, making them valuable for ex vivo investigations of cellular and molecular processes [15–17]. Moreover, OBSCs present a more ethical and efficient alternative to in vivo studies, reducing the need for animal models.

In this chapter, we describe two approaches to investigate astrocyte-mediated phagocytosis of synapses in the rat PFC during postnatal development using OBSCs. Adapted from Stoppini et al. [18] and Humpel [19], our methodology incorporates sensitive immunofluorescent-immunohistochemical staining to detect glial fibrillary associated protein (GFAP, an astrocytic marker), along with synaptophysin (a presynaptic marker) and lysosomal associated membrane protein 1 (LAMP1, a lysosomal marker). Additionally, this protocol includes Western blot analysis to characterize the expression of synaptic markers PSD95 and synaptophysin, alongside the phagocytic receptor MEGF10, in lysates from cortical tissue.

Together, these methods are helpful to get a better understanding of astrocytic mechanisms underlying brain disorders such as MDD while preserving the complexity of the cellular environment and minimizing animal use.

2 Materials

2.1 Animals

1. Male and female non-selectively bred Wistar (NAB) or Wistar Kyoto (WKY) rats at between postnatal (P) days 4 and 7 were used (purchased from Charles River Laboratories). Procedures follow institutional guidelines of the University of Regensburg,

Germany, as well as the European Community Council Directive (86/609/EEC). All experiments are approved by the Animal Welfare Office at the Animal Care and Use Facility of the University of Regensburg. In accordance with the German Animal Welfare Act (Tierschutzgesetz, Articles 4 and 7), no additional approval for the post-mortem removal of brain tissue is necessary. All efforts were taken to minimize animal pain or discomfort.

2.2 Preparation of Organotypic Brain Slice Cultures (OBSCs)

1. Dissection tools: surgical scissors (sharp-blunt, straight), dissection scissors (sharp), scalpel, standard pattern forceps (serrated), narrow pattern forceps (curved and straight), and spatula (with microspoon).
2. 3.5 cm dissection dish.
3. Sterivex filters (0.22 μm).
4. Dissection medium with $CaCl_2$ (1 mM), D-glucose (10 mM), KCl (4 mM), $MgCl_2$ (5 mM), $NaHCO_3$ (26 mM), sucrose (234 mM), phenol red (0.1%), and penicillin–streptomycin (100 U/mL).
5. OBSC culture medium with Minimum Essential Medium (MEM), EBSS (Earle's balanced salt solution), 25 mM HEPES, 25% heat-inactivated horse serum, 5 mg/mL D-glucose, 1 mM GlutaMAX, and 1% penicillin–streptomycin.
6. Six-well plates.
7. 30 mm culture plate inserts (0.4 μm pore).
8. Vibratome.
9. Carbogen (95% O_2 + 5% CO_2)/-cell culture incubator.
10. Pasteur pipette.

2.3 OBSC Fixation

1. Phosphate-buffered saline (PBS, 1×), pH to 7.5 (*see* **Note 1**).
2. 4% Paraformaldehyde (PFA) in PBS 1×, pH 7.5.
3. 0.1% sodium azide in PBS 1×.
4. Parafilm.

2.4 Immuno-fluorescent-Immunohisto-chemistry (IF-IHC)

1. 24-well plate.
2. Scalpel.
3. PBS 1×.
4. Blocking/permeabilization buffer with PBS 1×, 0.1% Triton -×-100, and 2% normal goat serum.
5. Antibody buffer solution: PBS 1×, 0.1% Triton ×-100, and 2% normal goat serum (for primary antibodies)/2% NGS in PBS 1× (for secondary antibodies and tertiary staining solution) (*see* **Note 2**).

6. Primary antibodies:
   (a) Mouse monoclonal anti-glial fibrillary acidic protein (GFAP) antibody (1:400, e.g., Sigma-Aldrich Chemicals G3893) to label the astrocytes.
   (b) Rabbit anti-lysosomal-associated membrane protein 1 (LAMP1) antibody (1:100, e.g., Abcam ab24170) to label lysosomes within the cells.
   (c) Chicken anti-synaptophysin antibody (1:500, e.g., SySy 101006) to label presynaptic puncta.
7. Secondary antibodies:
   (a) Cy3-conjugated anti-mouse IgG antibody (1:400, e.g., Thermo Fisher Scientific A10521).
   (b) Biotin-conjugated anti-rabbit IgG antibody (1:500, e.g., Jackson Lab 11-065-003).
   (c) Alexa Fluor™ 488-conjugated goat anti-chicken IgG antibody, (1:500, e.g., Invitrogen A11035).
8. Tertiary staining molecules:
   (a) Alexa Fluor™ 647-conjugated streptavidin (1:1000, e.g., Invitrogen S21374).
   (b) 4′,6-Diamidino-2-phenylindole (DAPI, 1:1000, e.g., Sigma D9542) to label nuclei
9. Mounting medium (e.g., DAKO S3023).
10. Narrow pattern forceps (straight).
11. Microscope slides.
12. Microscope cover glass.
13. Pipettes and other general laboratory equipment.
14. Horizontal shaker.
15. Confocal microscope.

2.5 Purification of the Proteins from OBSCs

1. RIPA buffer with Tris–HCl pH 7.5 (50 mM), NaCl (150 mM), SDS (0.1%), Triton ×-100 (1%), sodium deoxycholate (0.5%), protease inhibitors (10 mL), and Millipore H_2O.
2. PBS 1×.
3. 1.5 mL Eppendorf tubes.
4. Narrow curved forceps.
5. Sonicator/homogenizer.
6. Centrifuge.

2.6 Immunoblotting

1. SDS-PAGE gels:
   (a) Millipore H_2O.
   (b) Acrylamide.

   (c) APS.
   (d) TEMED.

2. 5× Laemmli buffer with Trizma base (TRIS, 0.3 M), sodium dodecyl sulphate (SDS, 10%), glycerin (5 M), ethylenediaminetetraacetic acid (EDTA, 25 mM), DL-dithiothreitol (DTT, 0.5 M), and bromphenol blue (0.5 mM).
3. 5× Running buffer with glycine (959 mM), TRIS (123 mM), SDS (17 mM), and Millipore H_2O. Adjust pH to 8.3.
4. Transfer buffer with glycine (150 mM), TRIS (20 mM), methanol (20%), and Millipore H_2O. Adjust pH to 8.3.
5. TBST with TRIS (10 mM), NaCL (150 mM), Tween 20 (0.1%), and Millipore H_2O. Adjust pH to 7.4.
6. Protein marker (pre-stained).
7. Western blotting equipment: chambers, power source, filter papers, foam pads, cassettes, and polyvinylidene fluoride (PVDF) membranes.
8. Methanol.
9. 5% Milk/BSA (bovine serum albumin) in TBST.
10. 1.5 mL Eppendorf tubes.
11. Primary antibodies:
    (a) Rabbit anti-multiple EGF-like-domains 10 (MEGF10) antibody (1:500, e.g., Thermo Fisher).
    (b) Mouse anti-postsynaptic density protein 95 (PSD95) antibody (1:2500, e.g., Sigma-Aldrich) as a marker of phagocytic astrocytes.
    (c) Rabbit anti-synaptophysin antibody (1:5000, e.g., Abcam).
    (d) Rabbit anti-β-actin antibody (1:10,000, e.g., Abcam) as a loading control.
    (e) Rabbit anti-cofilin monoclonal antibody (1:1000, e.g., Cell Signaling) as a loading control.
12. Secondary antibodies:
    (a) Horseradish peroxidase (HRP)-conjugated anti-mouse (1: 2500, e.g., Dianova).
    (b) Biotinylated anti-rabbit IgG antibody (1:500/1:2500/1: 5000/1:10000, e.g., Jackson Lab).
13. Enhanced chemiluminescence (ECL) substrate.
14. ImageJ software.

3 Methods

3.1 Preparation of Organotypic Brain Slice Cultures (OBSCs)

1. Pre-incubate inserts in culture medium at least 2 h in advance (or late the day before). Add 1 mL of culture medium into each well, and transfer inserts using forceps into the six-well plate (*see* **Note 3**).
2. For the preparation of brain slices, decapitate rat pups of 4–7 PND with a rough cut using sharp scissors, keeping the head submerged in carbogenated ice-cold dissection medium (*see* **Note 4**).
3. Expose and open the skull, placing the brain into dissection medium, and carefully remove meninges and blood vessels of the brain using narrow pattern forceps. Using scalpel, remove the olfactory bulb and cerebellum with a rough cut, creating a flat surface for embedding the brain to cutting disk. Mount the brains with a thin layer of superglue.
4. Cut coronal slices containing the PFC (400 μm thickness, nine slices of both hemispheres) using a vibratome platform submerged in cold carbogenated dissection medium at a speed of 0.06 mm/s (Fig. 1). Transfer three slices per well into six-well plate filled with culture medium (*see* **Note 5**).
5. Under sterile conditions, place the slices onto the pre-incubated inserts with the culture medium six-well plate, ensuring slices are not submerged but exposed to air at the surface (maximum three slices per insert) (*see* **Note 6**). Incubate at 37 °C in 5% CO_2-cell culture incubator.

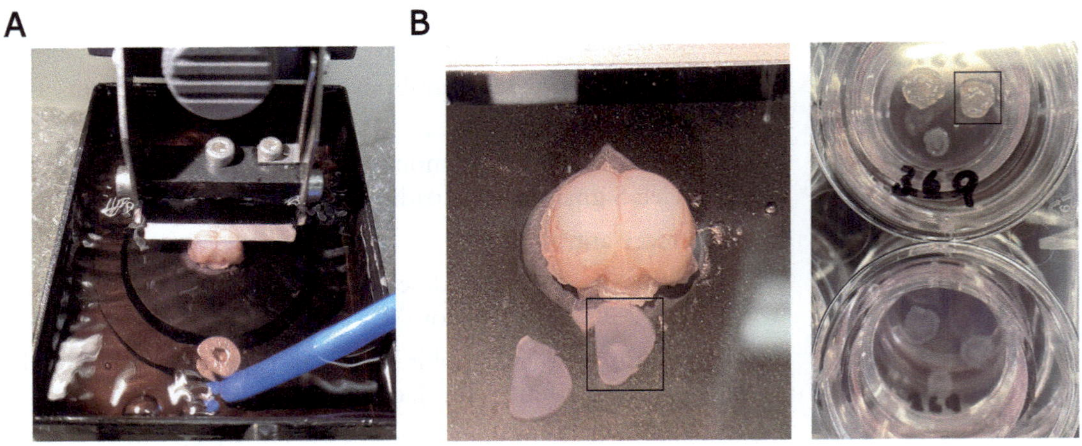

Fig. 1 Experimental setup of organotypic brain slice culture (OBSCs) preparation (**a**). The Brain is sectioned coronally using a vibratome to obtain 400-μm-thick slices. Transferring of brain slices to sterile culture inserts with semipermeable membranes (**b**). Inserts are placed in six-well culture plates containing sufficient culture medium to nourish the slices by diffusion through the membrane and incubated at 37 °C in a humidified atmosphere with 5% CO_2 for optimal growth

6. To keep OBSCs in culture up to 21 days in vitro (DIV), replace half of the OBSC culture medium (500 μL removed, replaced with 800 μL due to evaporation) every 2–3 days.

3.2 Preparation of OBSCs for Immunofluorescent-Immunohisto chemistry (IF-IHC)

1. Transfer membranes containing brain slices into a six-well plate, and wash with 1–2 mL PBS 1× for 10 min. Separate each slice by cutting the membrane using scalpel, and place in wells filled with 1.2 mL 4% paraformaldehyde in PBS 1×; leave gently shaking overnight (ON) at 4 °C.
2. Place the slices into new six-well plate washing them three times with PBS 1× at room temperature (RT). Finally, transfer membranes with the slices into wells with 1.5 mL 0.1% sodium azide in PBS solution. Seal the plate with parafilm and store at 4 °C.
3. Procedure for the IF-IHC.
4. Remove the plates from the fridge, place slices from each brain in 1 well of a 12-well plate with about 2 mL of PBS 1× per well, and let them swim on a horizontal shaker for 10 min (repeat three times).
5. Incubate slices in "blocking/permeabilization buffer" for 2 h at RT.
6. Remove the "blocking/permeabilization buffer," and place slices in 200 μL "antibody buffer solution" with primary antibodies ON horizontally shaking at 4 °C (*see* **Note** 7).
7. Wash slices with PBS 1× three times for 20 min each.
8. Remove PBS solution, and incubate slices in 200 μL "antibody buffer solution" with secondary antibodies for 2 h on a horizontal shaker at RT (*see* **Note 8**).
9. Remove the antibody solution, and wash slices with PBS 1× three times for 10 min each.
10. Remove the "washing buffer" and incubate slices in "antibody buffer solution" with tertiary antibodies for 2 h on a horizontal shaker at RT.
11. Remove the antibody solution, and wash slices with PBS 1× three times for 10 min each.
12. Using forceps (grab the membrane, avoid destruction of the slice), place slices from one brain on the top of the microscope slides facing the same direction and let slices to air-dry for a few min under an aluminum foil.
13. Put few drops of mounting medium on slides, and cover them with a microscope cover glass (*see* **Note 9**).
14. Keep sections in a refrigerator until microscopic analysis.

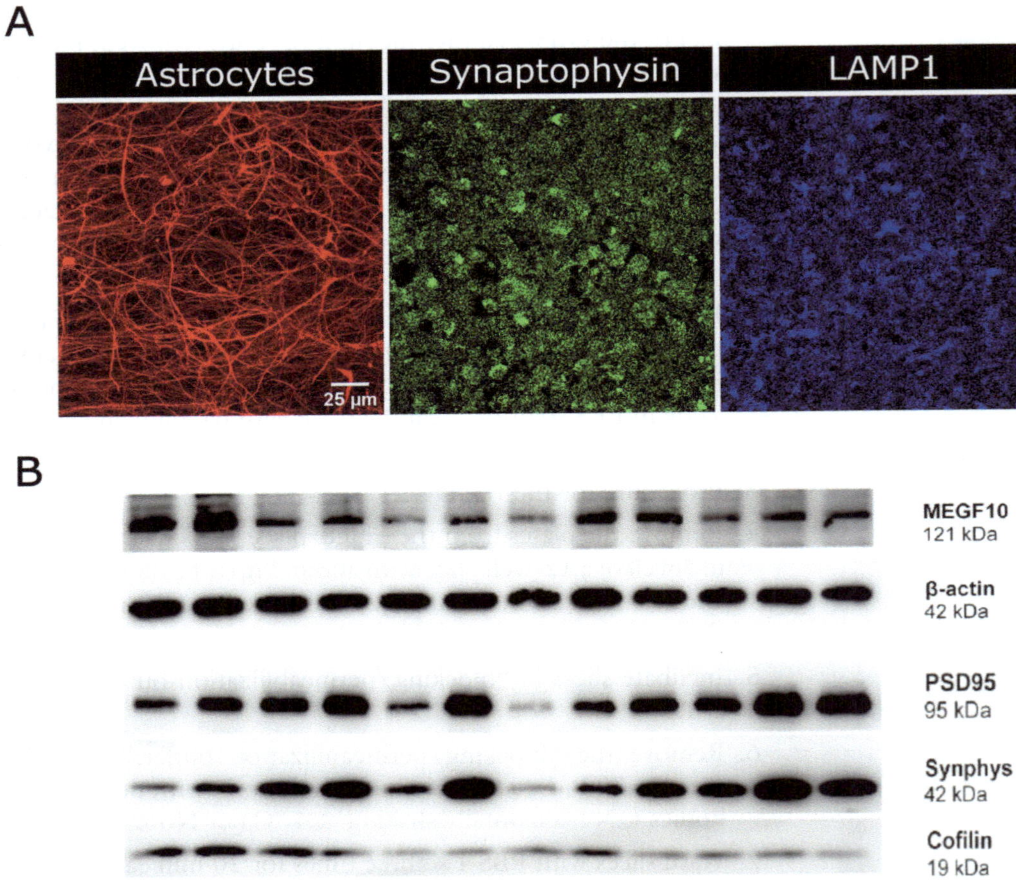

Fig. 2 Representative confocal images of OBSCs labelled with GFAP (astrocytes, red), synaptophysin (green), and LAMP1 (blue) in male-derived OBSCs (**a**). The composite of overlaying single-channel images is considered as the "Merge." Scale bar = 25 μm. Representative bands from immunoblots showing protein expression of MEGF10, β-actin, PSD95, synaptophysin, and cofilin in the cortex isolated from female-derived OBSCs at different time points (**b**)

3.3 Confocal Microscopy

1. Select the region of interest for image capture: prelimbic and infralimbic cortex of the PFC.
2. Using a 63× oil objective, capture a minimum of 6 images per OBSC (30 optical sections, 0.5 μm Z-step size for each image, GFAP as a reference) using the Zeiss LSM 880 Airyscan confocal microscope (Carl Zeiss Microscopy GmbH, Jena, Germany) (Fig. 2a).
3. Save files in the Imaris file format (.ims).
4. Perform analysis on whole image area for every picture with IMARIS 9.8 software (Bitplane, Zurich, Switzerland). For that, create a colocalization channel of LAMP1+ and synaptophysin+; mask with the GFAP+ channel. Astrocyte phagocytosis is determined as the ratio of "synphys+/LAMP1+/GFAP+" to "synphys+/LAMP1+" colocalized voxels and is further used for statistical analysis.

3.4 Protein Isolation from OBSCs

1. Remove the medium and wash in and out the inserts with ice-cold PBS 1×.
2. Use forceps to scrape OBSCs from the membrane, collecting prefrontal cortex (PFC) tissue (*see* **Note 10**). Transfer them into tubes with 500 μL of ice-cold RIPA buffer and incubate on ice for 15 min.
3. Sonicate tissue homogenates (2 × 20 s, 20% intensity), and centrifuge at 15,000 rcf for 25 min at 4 °C. Collect the supernatant containing proteins and store at −20 °C.

3.5 Immunoblotting

1. Dilute samples in Millipore H_2O to an equal amount of total protein and denaturate with 5× Laemmli buffer at 95 °C for 5 min.
2. Load samples with the total amount of 12 μg of protein and the marker into the gel (either 10% for MEGF10 or 12% separation gel for PSD95 and synaptophysin analysis—*see* **Note 11**). Separate them by the molecular size via electrophoresis at a constant 80 V for 15 min and 100 V for approximately 2 h in 1× running buffer. Transfer the resolved proteins to nitrocellulose membranes for 20 min at 70 V and then 2 h at 110 V.
3. Block the membranes either in 5% milk (β -actin, cofilin) or BSA (MEGF10, PSD95, synaptophysin) in TBST for 1 h at RT and incubate with primary antibodies (*see* **Note 12**) ON at 4 °C.
4. Wash ×3 membranes with TBST and incubate with respective secondary antibodies for 2 h at RT (*see* **Note 12**). Wash ×3 membranes with TBST again.
5. Detect and visualize protein samples by enhanced chemiluminescence, adjusting the brightness, focus, exposure, and interval time (Fig. 2b). Analyze the intensity of the bands quantitatively using ImageJ software. The immunoreactivity of the proteins of interest should be normalized to β-actin or cofilin for relative comparisons of results.

3.6 Statistical Analysis

1. Use ANOVA test with repeated measures, as slices derived from the same brain will be analyzed at different time points. Follow the analysis by Tukey's post hoc test, and present data as the mean ± standard deviation (SD).
2. Statistical significance is to be considered when the p value is equal or less than 5% ($p < 0.05$).

4 Notes

1. All solutions are prepared in deionized water with a resistivity of 18.2 MΩ-cm.
2. Ideally, the serum should be derived from the same species in which the secondary antibody is raised. In case of IF-IHC with multiple secondary antibodies derived from different species, we have got good experiences with the normal goat serum.
3. While transferring culture inserts into the wells filled with OBSC culture medium, avoid trapped bubbles beneath (by moving the insert with forceps).
4. Dissection medium must be ice-cold during dissection and sectioning to prevent nucleic acid degradation, and maintain the tissue as much as possible [20].
5. The explants consist of 400 μm/coronal slices of the rostral end of the rodent brains, giving a high resolution. We suggest cutting nine slices per each hemisphere, distributing them into three wells in a following manner that would be ideal for a complete representation of the different regions of the brain:
   (a) Well A1: slice 1, 4, 7
   (b) Well A2: slice 2, 5, 8
   (c) Well A3: slice 3, 6, 9

   Keep the six-well-plates with slices submerged in the culture medium in the incubator no longer than 3 h, while cutting the next brain.
6. While replacing the slices onto inserts, make sure no bubbles are trapped under and no medium covers the slices.
7. During the change of solutions, slices should never get dried.
8. As secondary/tertiary antibodies are light-sensitive, plates must be covered with an aluminum foil to keep the fluorescent dyes protected from light.
9. Be careful not to form bubbles in the mounting medium as they might interfere with imaging at the confocal microscope. In case bubbles form, you can gently pull them out with the help of a pipette tip pressed on the cover slip.
10. It is doubtful to selectively isolate only PFC from other cortical regions; therefore, the tissue may contain proteins from neighboring areas.
11. The choice of the concentration of the gel is based on the target's molecular weight in which protein would resolve better.
12. Antibodies are diluted in 2,5% BSA-TBST (MEGF10), 5% BSA-TBST (PSD95, synaptophysin), and 5% milk-TBST

(β-actin, cofilin). The concentration of antibody collusion buffer depends on the specific antibody and can be adjusted.

Acknowledgments

The authors would like to thank Britta Wenske and Susanne Wallner for their constant support during the establishment of the OBSCs, the execution of the experiments, and critical technical advancements. This work was supported by the Department of Psychiatry and Psychotherapy of the University of Regensburg and by the German Research Council (DFG-GRK2174 "Neurobiology of Emotion Dysfunction" (P1)) to BDB. The FV3000 confocal microscope was funded by a grant (INST 89/506-1 FUGG, 91b GG) from the Deutsche Forschungsgemeinschaft (DFG).

References

1. Liu CY, Yang Y, Ju WN, Wang X, Zhang HL (2018) Emerging roles of astrocytes in neurovascular unit and the tripartite synapse with emphasis on reactive gliosis in the context of Alzheimer's disease. Front Cell Neurosci 10(12):193
2. Sofroniew MV, Vinters HV (2010) Astrocytes: biology and pathology. Acta Neuropathol (Berl) 119(1):7–35
3. Bosworth AP, Allen NJ (2017) The diverse actions of astrocytes during synaptic development. Curr Opin Neurobiol 47:38–43
4. Allen NJ (2013) Role of glia in developmental synapse formation. Curr Opin Neurobiol 23(6):1027–1033
5. Stogsdill JA, Eroglu C (2017) The interplay between neurons and glia in synapse development and plasticity. Curr Opin Neurobiol 42: 1–8
6. Eroglu C, Barres BA (2010) Regulation of synaptic connectivity by glia. Nature 468(7321):223–231
7. Paumier A, Boisseau S, Jacquier-Sarlin M, Pernet-Gallay K, Buisson A, Albrieux M (2022) Astrocyte–neuron interplay is critical for Alzheimer's disease pathogenesis and is rescued by TRPA1 channel blockade. Brain 145(1):388–405
8. Bosson A, Boisseau S, Buisson A, Savasta M, Albrieux M (2015) Disruption of dopaminergic transmission remodels tripartite synapse morphology and astrocytic calcium activity within substantia nigra pars reticulata. Glia 63(4):673–683
9. Crunelli V, Carmignoto G, Steinhäuser C (2015) Novel astrocyte targets: new avenues for the therapeutic treatment of epilepsy. Neuroscientist 21(1):62–83
10. Jia X, Gao Z, Hu H (2021) Microglia in depression: current perspectives. Sci China Life Sci 64(6):911–925
11. Chung WS, Allen NJ, Eroglu C (2015) Astrocytes control synapse formation, function, and elimination. Cold Spring Harb Perspect Biol 7(9):a020370
12. Kolb B, Mychasiuk R, Muhammad A, Li Y, Frost DO, Gibb R (2012) Experience and the developing prefrontal cortex. Proc Natl Acad Sci 109(supplement_2):17186–17193
13. Drzewiecki CM, Juraska JM (2020) The structural reorganization of the prefrontal cortex during adolescence as a framework for vulnerability to the environment. Pharmacol Biochem Behav 199:173044
14. Chung WS, Clarke LE, Wang GX, Stafford BK, Sher A, Chakraborty C et al (2013) Astrocytes mediate synapse elimination through MEGF10 and MERTK pathways. Nature 504(7480): 394–400
15. Bak A, Schmied K, Jakob ML, Bedogni F, Squire OA, Gittel B, Jesinghausen M, Schünemann KD, Weber Y, Kampa B, van Loo KM (2024) Temporal dynamics of neocortical development in organotypic mouse brain cultures: a comprehensive analysis. J Neurophysiol 132(3):1038–1055
16. Ullrich C, Daschil N, Humpel C (2011) Organotypic vibrosections: novel whole sagittal

brain cultures. J Neurosci Methods 201(1): 131–141
17. Vivi E, Seeholzer LR, Nagumanova A, Di Benedetto B (2023) Early age- and sex-dependent regulation of astrocyte-mediated glutamatergic synapse elimination in the rat prefrontal cortex: establishing an organotypic brain slice culture investigating tool. Cells 12(23):2761
18. Stoppini L, Buchs PA, Muller D (1991) A simple method for organotypic cultures of nervous tissue. J Neurosci Methods 37(2):173–182
19. Humpel C (2018) Organotypic brain slice cultures. Curr Protoc Immunol 123(1):e59
20. Thompson D, Odufuwa AE, Watt JA (2023) Microdissection of the supraoptic nucleus and posterior pituitary gland from fresh unfixed tissue. MethodsX 1(11):102388

Chapter 16

Development and Validation of Stress Paradigms to Study Astrocytes in Rodent Models of Stress Disorders

Lewis Depaauw-Holt, Sarah Peyrard, Anthony Bosson, and Ciaran Murphy-Royal

Abstract

There is growing appreciation for astrocytes as essential partners in the central response to stress. Insight from post-mortem human brain tissue has revealed extensive astrocyte remodeling in people who suffered from psychiatric disorders including major depressive disorder. Rodent models of stress disorders align with human data, showing that astrocyte dysfunction directly contributes to cellular and behavioral maladaptation following intense or chronic stress. Despite these links, more research is required to determine astrocyte-specific stress signalling mechanisms and whether these pathways can be targeted in a therapeutic strategy. Here, we detail an early-life stress paradigm that induces lifelong stress susceptibility associated with astrocyte dysfunction. In addition, we provide a detailed guideline to validate a stress model.

Key words Astrocytes, Early-life stress, Neuroendocrinology, Stress paradigm, Glucocorticoids

1 Introduction

Astrocytes play key roles in regulating synaptic transmission, plasticity, and behavior [1]. Through decades of research, it has become apparent that astrocytes not only fine-tune neuronal activity but also provide context to neural circuits, including satiety, arousal, and stress, among many others [2–8]. This concept has been formalized in a contextual guidance hypothesis which proposes astrocytes as contextual integrators responding to state changes in a brain region–specific manner [9].

One particular context to which astrocytes are acutely responsive is stress, via the stress hormone receptors, glucocorticoid and mineralocorticoid receptors [6, 10, 11]. Indeed, there is an abundance of literature surrounding the profound impact of stress on astrocyte function using various methods including genetic as well as behavioral models of stress. These models induce a variety of

Barbara Di Benedetto (ed.), *Astrocytes: Methods and Protocols*, Methods in Molecular Biology, vol. 2896,
https://doi.org/10.1007/978-1-0716-4366-2_16,

behavioral phenotypes including anhedonia, anxiety-like behavior, cognitive impairments, and sleep dysfunction [6, 11–22].

Despite extensive research investigating how astrocytes are affected by stress [10, 23–26], the mechanisms underlying astrocyte sensitivity to stress remain poorly defined. One hypothesis is that their physical proximity to the vasculature renders these cells particularly sensitive to blood-borne signals including peripheral hormones such as glucocorticoids [12, 27–30]. The strategic location of astrocytes, bridging the vasculature and synapses, is likely to endow these cells with the distinct capacity to link peripheral changes with central adaptations and vice versa [9]. As such, understanding how astrocytes respond to stress hormones, participate in the central stress response, and whether these cells play a role in maladaptive stress responses following extensive stress demands further investigation.

Here, we outline implementation of an early-life stress paradigm and how to validate the hormonal stress response.

We use a protocol developed by Peña and colleagues that has been shown to enhance susceptibility to stress later in life [21]. This stressor alone, without any subsequent challenge, may not induce an overt phenotype but may require subsequent intervention or challenge. Other more intense or prolonged early-life stress paradigms could be used and compared [31], as the precise timing of stress during distinct developmental windows could have distinct effects on how astrocytes integrate into brain circuits [32, 33].

2 Materials

2.1 Animals and Materials for Early-Life Stress

1. Jax laboratory C57BL/6 J mice are used for all experiments (P45–P90). All experiments were approved and performed in accordance with guidelines for maintenance and care of animals of the Institutional Committee for the Protection of Animals (CIPA) at the Centre Hospitalier de l'Université de Montréal, in alignment with the Canadian Counsel on Animal Care (CCAC) guidelines.
2. Animals can be group housed in technoplast cages (Allentown Edge Cage # 410822-1, Allentown Cage cover # EBT 224423-1, Allentown Feeding Grid # EBT 224423-1) with neslet (Anacare) and Enviro-dri bedding/litter (Shepherd), enrichment (plastic house NexGen Mouse 500 Safe Harbor Mouse Retreat), and custom-made clear polycarbonate cylinder tubes 5 cm ID × 13 cm, with ad libitum access to food (Teklad Irradiated Global 18% Protein Rodent Diet) and water.
3. Heat pad (water circulating system Heating Pump Androit HTP-1500) to provide supplemental heat in order to maintain adequate environmental temperature.

4. Caloric supplements (ClearH2O DietGel® Boost 72-04-5022 2 oz. (56 g) cups).

2.2 Blood Collection

1. Anesthesia chamber (Small Induction Chamber by Dispomed).
2. Isoflurane (Forane DIN 02188856, Baxter Corporation).
3. Bone scissors.
4. Heparin-coated blood collection tubes.
5. Centrifuge (cooled to 4 degrees C).
6. Ice bucket.
7. Corticosterone ELISA kit (e.g., ENZO, Cat. No. ADI-900-097).

3 Methods

3.1 Early-Life Stress Protocol

3.1.1 Breeding Strategy

1. Pair two female mice with one male mouse of breeding age (P42–60) in a cage (*see* **Note 1**).
2. Monitor females for pregnancy as soon as 1 week post-pairing (*see* **Note 2**). The mouse can be held by the tail to assess pregnancies visually; they are mostly noticed by E7–10. When the mouse is held by the tail, a rounded belly with bulging sides can be observed. Females are not evaluated for plugs nor palpated to determine pregnancies to avoid unnecessary stress.
3. Separate the female mice as soon as pregnancies are confirmed. Each female is separated in their respective cage. A water bottle is added to the cage to supplement the automatic valve delivery system to ensure easy access for optimal hydration during pregnancy and lactation.
4. After single housing the female mice, check for births twice a day, starting at 17 days after first possible mating (*see* **Note 3**).
5. At P7, pups are counted, and sexes are determined. Sexing neonatal pups is performed by determining the anogenital distance. The distance between the genitalia and the anus is considerably larger in males as compared to females.

3.2 Maternal Separation (MS) and Limited Bedding Model

1. At P10, reduce home cage bedding by ¾ for early-life stress conditions. This will be maintained until P17 when conditions return to normal and daily separation bouts are terminated (*see* **Note 4**).
2. From P10 to P16, separate mother from her pups for 4 h per day (*see* **Note 4**), move the litter to a clean cage, and place half of the cage containing litter on a heat pad on the lowest setting.

3. At P17, the early-life stress protocol ends with termination of separation bouts and return of standard amounts of bedding and nesting materials.
4. Pups remain with the mother until P19–21 when weaning can occur. At this time, mouse numbers and sex are confirmed. Female and male progeny are housed separately. Wet food is added to the bottom of the cage, and a water bottle is provided for easy access to ensure survival of the pups. If pups are smaller than usual, caloric and nutritional supplements can be added as well. Supplements are palatable and easily digestible, consequently enhancing survival rate.

3.3 Hormonal Validation

Validating the hormonal response to stress is paramount for a robust investigation of the influence of stress (acute or chronic) on rodent physiology and behavior. In addition, time of day of stress implementation as well as the time between stress and experiment are critical variables that must be considered as these can strongly influence many physiological and behavioral outcomes.

3.3.1 Time of Day

Glucocorticoids (primarily corticosterone in rodents) fluctuate on circadian rhythm (Fig. 1). As such, the zeitgeber time (ZT) at which the stress paradigm is started may be an important factor to consider. In principle, protocols conducted at (or close to) glucocorticoid circadian nadirs allow for larger magnitude increases in blood glucocorticoids. In contrast, protocols conducted at circadian peaks could result in ceiling effects, potentially inducing more discrete changes and therefore false-negative interpretation of the short-term effects of stress. Robust stressors involving footshock, swim, etc., should, nevertheless, increase corticosterone levels far above the circadian peak.

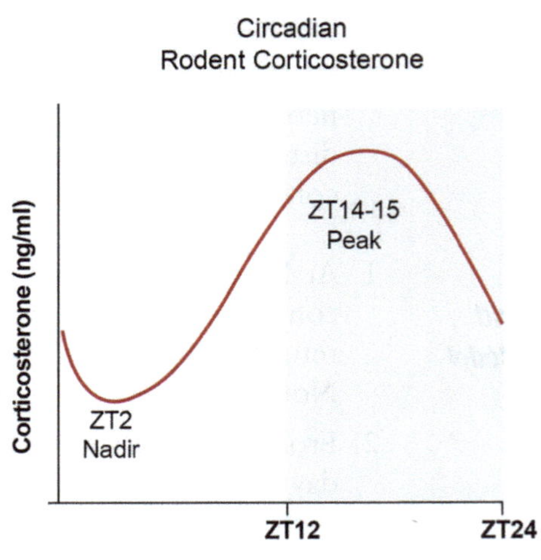

Fig. 1 Circadian fluctuations in rodent corticosterone

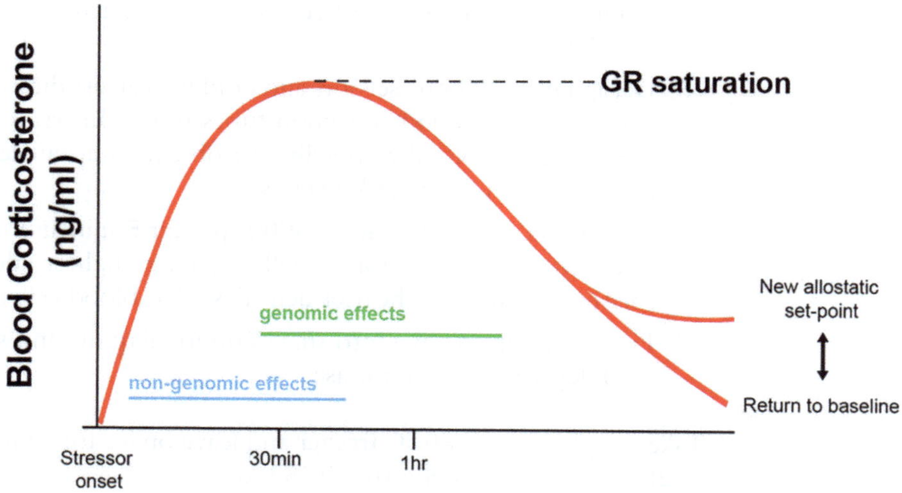

Fig. 2 Hormonal response to acute stress. Blood corticosterone increases quickly after stressor onset and subsides over the course of hours. GRs reach peak saturation after 30 min–1 h post-stress onset

3.3.2 Post-stress Interval

Timing post-stress is another key variable that will influence the genomic or non-genomic effects of stress hormones on desired outcome variables. To allow for maximal influence of glucocorticoids on the brain, consider a post-wait time of at least 30 min (Fig. 2). To investigate genomic signalling effects of glucocorticoids such as direct suppression/augmentation/tethering of genes with glucocorticoid response elements (GREs), a minimum wait time of 1 h is recommended to allow for saturation of glucocorticoid receptor signalling [34].

3.4 Blood Corticosterone Collection and Analysis

The following protocol for obtaining rodent corticosterone measurements has been optimized to minimize the effects of handling induced elevations in corticosterone.

3.4.1 Blood Collection

1. During blood collection grouped/single-housed rodents should be kept outside of the room to ensure rodents do not smell previous blood collection or odors that may induce a stress response.
2. Turn on the anesthesia chamber at maximum (5% gaseous isoflurane) for 2 min.
3. Unplug anesthesia chamber, and take it out of the necropsy room next to the rodent cage.
4. Place one rodent in anesthesia camber and quickly close to minimize escape of isoflurane gas. Check the time at which the rodent is placed in an anesthesia chamber. Rodents must be sacrificed maximum 2 min after first initial handling to

minimize elevations in corticosterone due to acute handling/anesthesia stress.

5. Decapitate with bone scissors and hold the carcass directly over heparin-coated blood collection tubes to obtain trunk blood. Take as much blood as possible to obtain sufficient levels of plasma/serum for ELISA analysis.
6. Spin blood in a centrifuge at 5000 rpm for 5 min at 4 degrees celcius. Centrifugal force will separate lighter densities (plasma/serum) and heavier densities (Red blood cells).
7. Pipette serum/plasma into small 200 μL aliquots and store at −80 degrees celsius until use.

3.4.2 Corticosterone ELISA Kit

Take samples from −80 °C freezer and leave on ice for a minimum of 20 min before running the ELISA kit.

Follow instructions with ELISA kit (*see* **Notes 5–7**).

4 Notes

1. It is important to add the females to the male cage for breeding success and for social territorial dominance purposes. A sufficient amount of nesting material and a hut should be provided for environmental enrichment to reduce stress and to offer protection and privacy. It also promotes nesting and burrowing behaviors.
2. Plan months in advance and be flexible around the dates. Males and females are placed in the cage for around a week to 10 days and pregnancy is roughly estimated.
3. Do not disturb the female and pups for at least 7 days after birth.
4. ELS lasts 7 days starting at P10, meaning there are seven rounds of MS (one per day). The last round of MS is therefore at P16. At P17, there is no MS; you simply restore bedding to end the ELS.
5. All manipulations should occur around the same time every day: cage cleaning, maternal separation, restoring and removing nesting material, weaning, etc.
6. ELISA calculates corticosterone values based on correct inputs of non-specific binding solution, blanks, and corticosterone standards. Be very precise with dilution of assay, and wash buffers ensure to accurate readings.
7. Run triplicate samples to obtain higher confidence of normalized values.

8. If running two plates at the same time, ensure to spread control and stress samples across plates to allow for error variations in standards and pipetting.

Acknowledgments

This work was supported by a Fondation CHUM, Fondation Courtois, Canadian Institutes of Health Research Project Grant (478629), an NSERC Discovery Grant (RGPIN-2021-03211), Fonds de Recherche du Québec—Santé (296562 and 309889), and the Brain & Behavior Research Foundation (NARSAD Young Investigator Award 28589).

Author Contributions All authors contributed to chapter along lines of expertise. All authors read and approved the final version of the chapter.

References

1. Nagai J, Yu X, Papouin T, Cheong E, Freeman MR, Monk KR et al (2021) Behaviorally consequential astrocytic regulation of neural circuits. Neuron 109(4):576–596
2. Jimenez-Blasco D, Busquets-Garcia A, Hebert-Chatelain E, Serrat R, Vicente-Gutierrez C, Ioannidou C et al (2020) Glucose metabolism links astroglial mitochondria to cannabinoid effects. Nature 583(7817): 603–608
3. Papouin T, Dunphy JM, Tolman M, Dineley KT, Haydon PG (2017) Septal cholinergic neuromodulation tunes the astrocyte-dependent gating of hippocampal NMDA receptors to wakefulness. Neuron 94(4): 840–854.e7
4. García-Cáceres C, Quarta C, Varela L, Gao Y, Gruber T, Legutko B et al (2016) Astrocytic insulin signaling couples brain glucose uptake with nutrient availability. Cell 166(4):867–880
5. MacDonald AJ, Holmes FE, Beall C, Pickering AE, Ellacott KLJ (2020) Regulation of food intake by astrocytes in the brainstem dorsal vagal complex. Glia 68(6):1241–1254
6. Adedipe II, Depaauw-Holt LR, Latraverse-Arquilla J, Vaugeois J, Duquenne M, Rogers B et al (2022) Astrocyte glucocorticoid signaling mediates cognitive impairment induced by early-life stress
7. Murphy-Royal C, Johnston AD, Boyce AKJ, Diaz-Castro B, Institoris A, Peringod G et al (2020) Stress gates an astrocytic energy reservoir to impair synaptic plasticity. Nat Commun 11(1):2014
8. Oliveira JF, Araque A (2022) Astrocyte regulation of neural circuit activity and network states. Glia 70(8):1455–1466
9. Murphy-Royal C, Ching SN, Papouin T (2023) A conceptual framework for astrocyte function. Nat Neurosci 26(11):1848–1856
10. Murphy-Royal C, Gordon GR, Bains JS (2019) Stress-induced structural and functional modifications of astrocytes – further implicating glia in the central response to stress. Glia 67(10): 1806
11. Tertil M, Skupio U, Barut J, Dubovyk V, Wawrzczak-Bargiela A, Soltys Z et al (2018) Glucocorticoid receptor signaling in astrocytes is required for aversive memory formation. Transl Psychiatry 8(1):255
12. Di Benedetto B, Malik VA, Begum S, Jablonowski L, Gómez-González GB, Neumann ID et al (2016) Fluoxetine requires the endfeet protein Aquaporin-4 to enhance plasticity of astrocyte processes. Front Cell Neurosci 2(10):8
13. Banasr M, Chowdhury GMI, Terwilliger R, Newton SS, Duman RS, Behar KL et al (2010) Glial pathology in an animal model of depression: reversal of stress-induced cellular, metabolic and behavioral deficits by the glutamate-modulating drug riluzole. Mol Psychiatry 15(5):501–511

14. Elsayed M, Banasr M, Duric V, Fournier NM, Licznerski P, Duman RS (2012) Antidepressant effects of fibroblast growth factor-2 in behavioral and cellular models of depression. Biol Psychiatry 72(4):258–265
15. Skupio U, Tertil M, Bilecki W, Barut J, Korostynski M, Golda S et al (2019) Astrocytes determine conditioned response to morphine via glucocorticoid receptor-dependent regulation of lactate release. Neuropsychopharmacology 29:1–12
16. Bohn MC, Howard E, Vielkind U, Krozowski Z (1991) Glial cells express both mineralocorticoid and glucocorticoid receptors. J Steroid Biochem Mol Biol 40(1–3):105–111
17. Di Benedetto B, Rupprecht R (2013) Targeting glia cells: novel perspectives for the treatment of neuropsychiatric diseases. Curr Neuropharmacol 11(2):171–185
18. Lu CL, Ren J, Mo JW, Fan J, Guo F, Chen LY et al (2022) Glucocorticoid receptor–dependent astrocytes mediate stress vulnerability. Biol Psychiatry 92(3):204–215
19. Lopez J, Bagot RC (2021) Defining valid chronic stress models for depression with female rodents. Biol Psychiatry 90:226
20. Hammels C, Pishva E, De Vry J, van den Hove DLA, Prickaerts J, van Winkel R et al (2015) Defeat stress in rodents: from behavior to molecules. Neurosci Biobehav Rev 1(59): 111–140
21. Peña CJ, Kronman HG, Walker DM, Cates HM, Bagot RC, Purushothaman I et al (2017) Early life stress confers lifelong stress susceptibility in mice via ventral tegmental area OTX2. Science 356(6343):1185–1188
22. Walker CD, Bath KG, Joels M, Korosi A, Larauche M, Lucassen PJ et al (2017) Chronic early life stress induced by limited bedding and nesting (LBN) material in rodents: critical considerations of methodology, outcomes and translational potential. Stress 20:421–448
23. Zhang X, Alnafisah RS, Hamoud ARA, Shukla R, Wen Z, McCullumsmith RE et al (2021) Role of astrocytes in major neuropsychiatric disorders. Neurochem Res 46(10): 2715–2730
24. Rajkowska G, Stockmeier C (2013) Astrocyte pathology in major depressive disorder: insights from human postmortem brain tissue. Curr Drug Targets 14(11):1225–1236
25. O'Leary LA, Mechawar N (2021) Implication of cerebral astrocytes in major depression: a review of fine neuroanatomical evidence in humans. Glia 69(9):2077–2099
26. Vivi E, Di Benedetto B (2024) Brain stars take the lead during critical periods of early postnatal brain development: relevance of astrocytes in health and mental disorders. Mol Psychiatry 2024:1–13
27. Hösli L, Zuend M, Bredell G, Zanker HS, Porto de Oliveira CE, Saab AS et al (2022) Direct vascular contact is a hallmark of cerebral astrocytes. Cell Rep 39(1)
28. Dudek K, Paton S, Lebel M, Lavoie O, Bouchard J, Kaufmann FN et al (2023) Astrocytic cannabinoid receptor 1 promotes resilience by dampening stress-induced blood-brain barrier alterations
29. Dudek KA, Dion-Albert L, Lebel M, LeClair K, Labrecque S, Tuck E et al (2020) Molecular adaptations of the blood–brain barrier promote stress resilience vs. depression. Proc Natl Acad Sci USA 117(6):3326–3336
30. Rajkowska G, Hughes J, Stockmeier CA, Javier Miguel-Hidalgo J, Maciag D (2013) Coverage of blood vessels by astrocytic endfeet is reduced in major depressive disorder. Biol Psychiatry 73(7):613–621
31. Walker CD, Bath KG, Joels M, Korosi A, Larauche M, Lucassen PJ et al (2017) Chronic early life stress induced by limited bedding and nesting (LBN) material in rodents: critical considerations of methodology, outcomes and translational potential. Stress (Amsterdam, Netherlands) 20(5):421–448
32. Farhy-Tselnicker I, Allen NJ (2018) Astrocytes, neurons, synapses: a tripartite view on cortical circuit development. Neural Dev 13: 1–12
33. Rurak GM, Simard S, Freitas-Andrade M, Lacoste B, Charih F, Van Geel A et al (2022) Sex differences in developmental patterns of neocortical astroglia: a mouse translatome database. Cell Rep 38(5)
34. Gray JD, Kogan JF, Marrocco J, McEwen BS (2017) Genomic and epigenomic mechanisms of glucocorticoids in the brain. Nature Publishing Group, p 13

Chapter 17

Cellular Investigation of Astrocytic Neuroendocrine Responses to Stress

Lewis Depaauw-Holt, Sarah Peyrard, Anthony Bosson, and Ciaran Murphy-Royal

Abstract

Canonical stress signaling pathways involve adaptive neuroendocrine responses. Typically, stress activation of the hypothalamic–pituitary–adrenal axis will result in an increase in glucocorticoids that act on almost all tissues including the brain. Recent work has investigated the impact of glucocorticoid signaling in astrocytes, highlighting sensitivity of these cells to stress hormones. Further understanding whether specific glucocorticoid signaling changes exist in astrocytes and whether regional heterogeneity exists remains unknown. Here, we outline a method to investigate glucocorticoid signaling in astrocytes in fixed tissue.

Key words Astrocytes, Early-life stress, Neuroendocrinology, Glucocorticoids, Immunohistochemistry, Cellular imaging

1 Introduction

Stressors can affect biological systems, brain circuits, and cells through genomic and non-genomic mechanisms [1]. Our previous work looking at the effects of acute swim stress on hippocampal and cortical synaptic plasticity found transcriptional changes in astrocytes thought to partially underlie the astrocytic metabolic syncytium disruption [2]. The influence of glucocorticoid signaling in astrocytes has garnered widespread attention in recent years, leading many research teams to question the impact of these signaling pathways on astrocyte function and the downstream consequences on stress responses [1–9].

Using IHC, the following method allows for a rapid qualitative investigation of glucocorticoid receptor localization in astrocytes in fixed brain slices from naive and stress mice. Localization of glucocorticoid receptors can be used as a proxy for receptor activity in fixed brain tissue, as the mechanism of action requires translocation to the nucleus to induce gene expression changes [3–5]. Here, we

Barbara Di Benedetto (ed.), *Astrocytes: Methods and Protocols*, Methods in Molecular Biology, vol. 2896,
https://doi.org/10.1007/978-1-0716-4366-2_17,

describe a method using a masking approach to isolate DAPI-associated anti-glucocorticoid receptor (GR) fluorescence and determine whether stress influences neuroendocrine signaling pathways in astrocytes.

2 Materials

2.1 Animals

1. Jax laboratory C57BL/6 J mice are used for all experiments (P45–P90). All experiments were approved and performed in accordance with guidelines for maintenance and care of animals of the Institutional Committee for the Protection of Animals (CIPA) at the Centre Hospitalier de l'Université de Montréal.

2.2 Preparation of Brains for Immunohistochemistry (IHC)

1. Mini-peristaltic pump for transcardial perfusion.
2. Phosphate-buffered saline (PBS, 10×): 137 mM NaCl, 27 mM KCl, 8 mM Na_2HPO_4, and 20 mM KH_2PO_4 diluted to 1× with Milli-Q H_2O. Sterilize by autoclaving.
3. 4% Paraformaldehyde (PFA) in 1× PBS.
4. Butterfly needles (21 gauge).
5. 33 gauge needles.
6. 30% sucrose in PBS 1× for cryoprotection.
7. Dissection tools for brain extraction.

2.3 Preparation of Slices for Free-Floating IHC

1. Cryostat.
2. 12-well plates.
3. Optimal cutting temperature (OCT) compound to fix brains for cryostat.
4. Cryoprotective solution (for storing slices at −20 degrees Celsius): 30% ethylene glycol and 30% glycerol in PBS 1×.

2.4 Immunohistochemistry

1. 24-well plates.
2. Wash buffer: PBS 1×.
3. Blocking/permeabilization buffer: PBS 1×, 0.5% Triton ×-100, and 3% bovine serum albumin.
4. Antibody buffer solution: PBS 1×, 0.5% Triton ×-100, and 3% bovine serum albumin.
5. Primary antibodies:
   (a) Anti-GFAP (chicken) (1:500, e.g., PA1-10004).
   (b) Anti-S100β (rabbit) (1:500, e.g., ab52642).
   (c) Anti-glucocorticoid receptor (mouse) (1:500, e.g., MA1-510).

6. Secondary antibodies.
   (a) Anti-chicken (goat) (1:1000, e.g., A11041).
   (b) Anti-rabbit (goat) (1:1000, e.g., 111-545-144).
   (c) Anti-mouse (goat) (1:1000, e.g., A32728).
7. Mounting medium: ProLong Gold.
8. Plastic pipette dish to mount slices on slides.
9. Pre-cleaned microscope slides (e.g., Superfrost/Plus, Fisher Scientific).
10. Microscope cover glass (22 × 50 mm, e.g., Fisherbrand, Fisher Scientific).
11. Pipettes and other general laboratory equipment.
12. Horizontal shaker.
13. Confocal microscope.
14. ImageJ software (http://www.imagescience.org).

3 Methods

3.1 Perfusion

1. Anesthetize mouse with 5% gaseous isoflurane. Wait until breathing has slowed and there is an absence of toe pinch response.
2. Set up peristaltic pump, and run both 1× PBS and 4% PFA through their respective tubes connecting to the 21 gauge butterfly needles.
3. With continuous isoflurane, stabilize mouse on the back by pinning forepaws to a polystyrene foam with 33 gauge needles.
4. With forceps, hold up the cartilage of the xiphoid process of the animal, and open a small hole below sternum and carefully. Cut laterally at the base of the ventral ribcage.
5. Cut up the two sides of the chest to expose the expose thoracic field.
6. Clamp the sternum with large locking hemostats, and lift it up to expose the heart and lungs.
7. Insert the perfusion needle at the apex of the heart directly into the left ventricle, and hold it there with a pair of hemostats. Perfuse the mouse with PBS 1× flow at a steady drip of 30 mL/min. A good perfusion and flow will result in the liver turning pale (approx. 3–5 min).
8. Switch to PFA 4%, and perfuse body for 5 min or until extremities become turgid.
9. Decapitate and extract the brain with small scissors and forceps.

10. Place each brain in a 15 mL Falcon tube containing 4% PFA at 4 °C.
11. After 24 h, switch brains into a 30% sucrose solution until they sink to the bottom of the 15 mL Falcon tube.
12. Once sunk, wrap brains in aluminum foil, and store at −80 degrees celsius until ready for cryosectioning.

3.2 Preparation of Slices for IHC–Cryosectioning

1. Remove brains from −80 °C and place into a polystyrene foam box sitting on dry ice.
2. Place aluminum-covered brains into the cryostat to equilibrate with desired cutting temperature (−22 °C).
3. Prepare a 12-well plate with 1 mL/well of cryoprotector solution for brain slices.
4. Cover the brain holder of cryostat with thin layer of OCT compound, and balance brain on the cryostat holder in the vertical position.
5. Once base of the brain is stuck to holder, proceed to cover sides and front of brain with OCT compound.
6. Place brain (on holder) into cryostat, and wait until OCT compound has completely frozen.
7. Place block into cryostat and slice at 30 μm.
8. Continue cutting until reaching desired brain region.
9. Collect slices with a thin paint brush previously dampened with cryoprotector solution. Place slices into wells of the 12-well plate (5 slices per well).
10. Store slices at −20 °C until IHC.

3.3 Immunostaining Protocol

The following free-floating slice IHC protocol is optimized for 4% PFA fixed tissue obtained via transcardial perfusion.

1. Wash (shake) free-floating slices three times in 1× PBS (500 μL per well) for 15 min to remove the cryostat mounting medium (total time = 45 min).
2. Permeabilize slices (one slice per well) in a Block-Perm solution (250 μL per well) for 1 h (Block-Perm Solution: 1× PBS with 3% bovine serum albumin, 0.5% of 10% Triton ×-100 stock).
3. Incubate slices in primary antibodies (250 μL per well) for 24 h at +4 degrees Celsius.
4. Wash slices three times in 1× PBS for 15 min to remove nonspecific binding of primary antibodies.
5. Incubate slices in secondary antibodies (250 μL per well) for 1 h at room temperature.
6. Wash slices three times in 1× PBS for 15 min to remove nonspecific binding of secondary antibodies.

7. Mount slices onto slides using ProLong Gold mounting medium. Gently add coverslip, leave slides at room temperature overnight to dry, and then, store at +4 degrees Celsius until imaged.

3.4 Glucocorticoid Receptor Localization

3.4.1 GR Fluorescence in response to 5 h of Sleep Deprivation

A 5-h sleep deprivation paradigm (usually commencing at zeitgeber time (ZT)0–ZT5) is an easy and effective method of elevating endogenous blood corticosterone [6] without any metabolic side effects (i.e., rodents have ad libitum access to food and water). Prolonged wakefulness at the beginning of the rodent light cycle (sleeping period) prevents the drop to circadian nadir levels of blood corticosterone. *See* Fig. 1 for representative images and data obtained using the described protocol with Fiji.

Analysis of confocal Z-stack of S100β, GR, and DAPI IHC staining (*see* **Note 1**).

1. Draw ROI around astrocytes of interest.
2. Image > Color > split channels.
3. Isolate S100β channel to identify astrocytes pups remain with the mother until P19–21 when weaning can occur. At this time, numbers and genders are confirmed.
   (a) Threshold.
   (b) Edit > Selection > create selection.
   (c) Add to ROI manager (Ctrl + T, then add).
4. Apply S100β saved selection to DAPI channel.
   (a) Edit > clear outside.
5. Threshold DAPI channel.

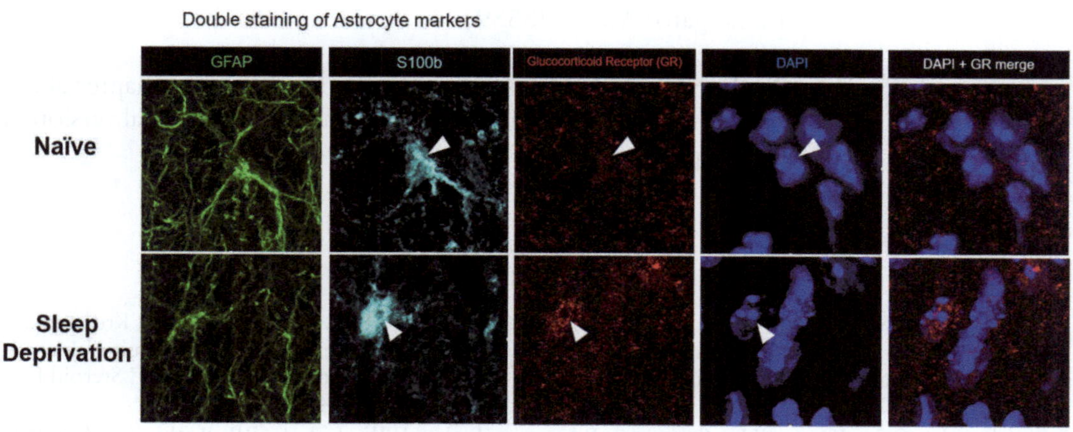

Fig. 1 Confocal images of astrocytes immunostained for GFAP, S100β, GR, and DAPI. *Top panel*: Naïve non-stressed mice with minimal nuclear localized GR. *Bottom row*: Mice subjected to a 5 h sleep deprivation prior to sacrifice. DAPI + GR merge: notable increased GR signal inside the astrocyte nucleus, indicative of GR activation

   (a) Edit > Selection > create selection.
   (b) Add to the ROI manager.
6. Apply DAPI saved selection to GR channel.
   (a) Edit > Clear outside.
7. Threshold remaining S100β+ + DAPI+ GR fluorescence to obtain values for astrocytic nuclear GR fluorescence.
   (b) Analyze > Measure.
8. Measured data can be normalized to DAPI or S100β astrocyte size. Analyze a minimum of five cells per brain slice to account for heterogeneity of GR signaling in astrocytes.
9. Plot individual astrocyte intensities as observational repeats with a minimum of five animals ($N = 5$, $n = 25$).

4 Note

1. This method of analysis with Fiji can be leveraged for looking at nuclear glucocorticoid receptor composition for any cell type (replace the S100β marker with a somatic marker of your cell of interest).

Acknowledgments

This work was supported by a Fondation CHUM, Fondation Courtois, Canadian Institutes of Health Research Project Grant (478629), an NSERC Discovery Grant (RGPIN-2021-03211), Fonds de Recherche du Québec—Santé (296562 and 309889), and the Brain & Behavior Research Foundation (NARSAD Young Investigator Award 28589).

Author Contributions All authors contributed to chapter along lines of expertise. All authors read and approved the final version of the chapter.

References

1. Lu CL, Ren J, Mo JW, Fan J, Guo F, Chen LY et al (2022) Glucocorticoid receptor–dependent astrocytes mediate stress vulnerability. Biol Psychiatry 92(3):204–215
2. Tombaugh GC, Yang SH, Swanson RA, Sapolsky RM (1992) Glucocorticoids exacerbate hypoxic and hypoglycemic hippocampal injury in vitro: biochemical correlates and a role for astrocytes. J Neurochem 59(1):137–146
3. Bohn MC, Howard E, Vielkind U, Krozowski Z (1991) Glial cells express both mineralocorticoid and glucocorticoid receptors. J Steroid Biochem Mol Biol 40(1–3):105–111
4. Taylor WW, Imhoff BR, Sathi ZS, Liu WY, Garza KM, Dias BG (2021) Contributions of glucocorticoid receptors in cortical astrocytes to memory recall. Learn Mem 28(4):126–133

5. Vielkind U, Walencewicz A, Levine JM, Bohn MC (1990) Type II glucocorticoid receptors are expressed in oligodendrocytes and astrocytes. J Neurosci Res 27(3):360–373
6. Tertil M, Skupio U, Barut J, Dubovyk V, Wawrzczak-Bargiela A, Soltys Z et al (2018) Glucocorticoid receptor signaling in astrocytes is required for aversive memory formation. Transl Psychiatry 8(1):255
7. O'Callaghan JP, Brinton RE, McEwen BS (1991) Glucocorticoids regulate the synthesis of glial fibrillary acidic protein in intact and adrenalectomized rats but do not affect its expression following brain injury. J Neurochem 57(3): 860–869
8. Murphy-Royal C, Gordon GRGR, Bains JSJS (2019) Stress-induced structural and functional modifications of astrocytes – further implicating glia in the central response to stress. Glia 67: 1806
9. Guayasamin M, Depaauw-Holt LR, Adedipe II, Ghenissa O, Vaugeois J, Duquenne M et al (2022) Early-life stress induces persistent astrocyte dysfunction resulting in fear generalisation. https://doi.org/10.1101/2022.12.11.519598

Chapter 18

Morphological Investigation of Astrocytic Responses to Stress

Lewis Depaauw-Holt, Sarah Peyrard, Anthony Bosson, and Ciaran Murphy-Royal

Abstract

There are many ways in which astrocytes are likely to respond to stress, but one of the most reliable phenotypes has been glial fibrillary acidic protein (GFAP) and morphological changes. GFAP is usually a reliable indicator for morphological reorganization but cannot be used alone for detailed morphological reconstruction and analysis. Sparse labeling of astrocytes with a fluorescent indicator, e.g., eGFP, is a robust way to determine discrete morphological changes in these cells. Here, we outline both methods to study stress-induced changes in astrocyte morphology.

Key words Astrocytes, Early-life stress, Stress, Immunohistochemistry, Cellular imaging, GFAP, Morphology, Single-cell morphology

1 Introduction

Some of the most robust astrocytic adaptations following stress are changes in glial fibrillary acidic protein (GFAP) expression and morphological reorganization [1–3].

GFAP is an intermediate filament-III protein found in the main branching structures of astrocytes and one of the most reliable phenotypes for morphological change. Changes in GFAP can be indicative of "reactivity" states in a number of pathological conditions and neurodegenerative models [4–8]. While GFAP is a reliable indicator for morphological reorganization, it cannot be used alone for detailed morphological reconstruction as GFAP is not present throughout the entire cytosol and is absent from the fine processes. As such, further validation of discrete morphological changes with injected dyes [9, 10], genetic labeling of astrocytes, or reconstruction using electron microscopy [11] is therefore recommended. Nevertheless, raw or normalized intensity measures

Barbara Di Benedetto (ed.), *Astrocytes: Methods and Protocols*, Methods in Molecular Biology, vol. 2896,
https://doi.org/10.1007/978-1-0716-4366-2_18,

of GFAP can be a simple yet effective starting point for integrating astrocytic responses across stress models.

While GFAP gives a general indication of any changes in major branching processes in response to stressors, it remains a limited marker for cell volumes and process complexity. Below, we use a stereotaxically injected viral approach to isolate astrocytes for morphological reconstruction with IMARIS. Using reduced viral titers and adjusted stereotaxic coordinates, we obtain sparse labeling of astrocytes at the viral injection perimeter while maintaining brain region specificity.

Finally, GFAP only labels a subset of astrocytes and should not be considered as a pan-astrocyte marker. Changes in astrocyte markers are common and well-documented phenotypes associated with stress [3, 12–15]. In the context of early-life stress (ELS) commencing during critical developmental windows, it remains important to assess changes in morphology alongside potential alterations in cell density or population sizes (neuronal and glial). For this reason, we recommend cell counting using a pan-astrocyte marker such as S100β instead of GFAP. S100β is a protein containing 2EF-hand calcium-binding motifs with significantly higher localization with astrocytic cell somas compared to processes (Fig. 2). Colocalization of S100β with DAPI nuclei serves as a robust method to measure of astrocyte population sizes.

2 Materials

2.1 Animals

1. Jax laboratory C57BL/6 J mice are used for all experiments (P45–P90). All experiments were approved and performed in accordance with guidelines for maintenance and care of animals of the Institutional Committee for the Protection of Animals (CIPA) at the Centre Hospitalier de l'Université de Montréal, in alignment with the Canadian Counsel on Animal Care (CCAC) guidelines.

2.2 Preparation of Brains for Immunohistochemistry (IHC)

1. Mini-peristaltic pump for transcardial perfusion.
2. Phosphate-buffered saline (PBS, 10×): 137 mM NaCl, 27 mM KCl, 8 mM Na_2HPO_4, and 20 mM KH_2PO_4 diluted to 1× with Milli-Q H_2O. Sterilize by autoclaving.
3. 4% Paraformaldehyde (PFA) in 1× PBS.
4. Butterfly needles (21 gauge).
5. 33 gauge needles.
6. 30% sucrose in PBS 1× for cryoprotection.
7. Dissection tools for brain extraction.

2.3 Preparation of Slices for Free-Floating IHC

1. Cryostat.
2. 12-well plates.
3. Optimal cutting temperature (OCT) compound to fix brains for cryostat.
4. Cryoprotective solution (for storing slices at −20 °C): 30% ethylene glycol and 30% glycerol in PBS 1×.

2.4 Immunohistochemistry

1. 24-well plates.
2. Wash buffer: PBS 1×.
3. Blocking/permeabilization buffer: PBS 1× 0.5% Triton ×-100, and 3% bovine serum albumin.
4. Antibody buffer solution: PBS 1×, 0.5% Triton ×-100, and 3% bovine serum albumin.
5. Primary antibodies:
   (a) Anti-GFAP (chicken) (1:500, e.g., PA1-10004).
   (b) Anti-S100β (rabbit) (1:500), e.g., ab52642).
   (c) Anti-eGFP (1:500, e.g., TA100022).
6. Secondary antibodies.
   (a) Anti-chicken (goat) (1:1000, e.g., A11041).
   (b) Anti-rabbit (goat) (1:1000, e.g., 111-545-144).
   (c) Anti-mouse (goat) (1:1000, e.g., A32728).
7. Mounting medium: ProLong Gold.
8. Plastic pipette dish to mount slices on slides.
9. Pre-cleaned microscope slides (e.g., Superfrost/Plus, Fisher Scientific).
10. Microscope cover glass (22 × 50 mm, e.g., Fisherbrand, Fisher Scientific).
11. Pipettes and other general laboratory equipment.
12. Horizontal shaker.
13. Confocal microscope.

2.5 Stereotaxic Surgeries

1. Stereotaxic frame.
2. Sterile surgeries tools: scissors, clamps, tweezers, cotton tips, and sterile compresses.
3. Sterile drapes.
4. Heating pad.
5. Needles and syringes.
6. Eye lubricant.
7. Saline 0.9%.
8. Lidocaine and carprofen.

9. Drill.
10. Hamilton syringe.
11. Sterile sutures.

2.6 Software

1. Fiji plugins.
   (a) https://imagej.net/plugins/sholl-analysis: Sholl analysis.
   (b) https://imagej.net/plugins/volume-calculator: volume calculator.
2. Reference.
   (a) GliaMorph toolkit: https://currentprotocols.onlinelibrary.wiley.com/doi/full/10.1002/cpz1.654.
3. IMARIS.
   (a) Volume and 3D complexity: https://imaris.oxinst.com/learning/view/article/first-steps-in-imaris-image-analysis-software.
   (b) Toolkit for process lengths. https://imaris.oxinst.com/learning/view/article/shortest-distance-overlapped-volume-where-to-find-the-measurements-microscopy-image-analysis
   (c) IMARIS reference manual. https://www.bitplane.com/download/manuals/ReferenceManual6_1_0.pdf

3 Methods

3.1 Perfusion

1. Anesthetize mouse with 5% gaseous isoflurane. Wait until breathing has slowed and there is an absence of toe pinch response.
2. Set up peristaltic pump, and run both 1× PBS and 4% PFA through their respective tubes connecting to the 21 gauge butterfly needles.
3. With continuous isoflurane, stabilize mouse on the back by pinning forepaws to a polystyrene foam with 33 gauge needles.
4. With forceps, hold up the cartilage of the xiphoid process of the animal, and open a small hole below sternum and carefully. Cut laterally at the base of the ventral ribcage.
5. Cut up the two sides of the chest to expose the expose thoracic field.
6. Clamp the sternum with large locking hemostats, and lift it up to expose the heart and lungs.
7. Insert the perfusion needle at the apex of the heart directly into the left ventricle, and hold it there with a pair of hemostats. Perfuse the mouse with PBS 1× flow at a steady drip of 30 mL/

min. A good perfusion and flow will result in the liver turning pale (approx. 3–5 min).

8. Switch to PFA 4%, and perfuse body for 5 min or until extremities become turgid.
9. Decapitate and extract the brain with small scissors and forceps.
10. Place each brain in a 15 mL Falcon tube containing 4% PFA at 4 °C.
11. After 24 h, switch brains into a 30% sucrose solution until they sink to the bottom of the 15 mL Falcon tube.
12. Once sunk, wrap brains in aluminum foil, and store at −80 °C until ready for cryosectioning.

3.2 Preparation of Slices for IHC–Cryosectioning

1. Remove brains from −80 °C and place into a polystyrene foam box sitting on dry ice.
2. Place aluminum-covered brains into the cryostat to equilibrate with desired cutting temperature (−22 °C).
3. Prepare a 12-well plate with 1 mL of cryoprotector solution for brain slices.
4. Cover the brain holder of cryostat with thin layer of OCT compound, and balance brain on the cryostat holder in the vertical position.
5. Once base of brain is stuck to holder, proceed to cover sides and front of the brain with OCT compound.
6. Place brain (on holder) into cryostat and wait until OCT compound has completely frozen.
7. Place block into cryostat and slice at 30 μm.
8. Continue cutting until reaching desired brain region.
9. Collect slices with a thin paint brush previously dampened with cryoprotector solution. Place slices into wells of the 12-well plate (5 slices per well).
10. Store slices at −20 °C until IHC.

3.3 Immunostaining Protocol for Canonical Structural Markers of Astrocytes

The following free-floating slice IHC protocol is optimized for 4% PFA-fixed tissue obtained via transcardial perfusion.

1. Wash (shake) free-floating slices three times in 1× PBS (500 μL per well) for 15 min to remove the cryostat mounting medium (total time = 45 min).
2. Permeabilize slices (one slice per well) in a Block-Perm solution (250 μL per well) for 1 h (Block-Perm Solution: 1× PBS with 3% bovine serum albumin, 0.5% of 10% Triton ×-100 stock).
3. Incubate slices in primary antibodies (250 μL per well) for 24 h at +4 degrees Celsius (Fig. 1; *see* **Notes 1** and **2**).

Fig. 1 Astrocytic GFAP expression across hippocampal layers. Note remarkable heterogeneity across three subregions of the hippocampus (left to right—CA2, stratum pyramidale, and stratum oriens). Using canonical landmarks such as neuronal cell layers in the cortex and hippocampus allows for robust comparisons between naïve and stress groups

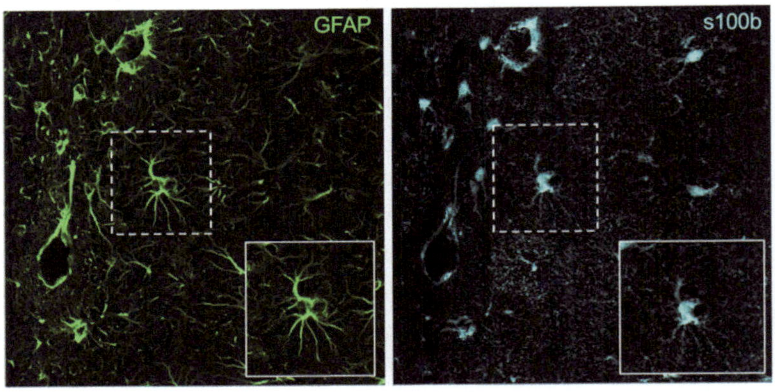

Fig. 2 Comparison of GFAP staining (left) and pan-astrocyte marker S100β (right). Note higher density of nuclear expression with S100β staining compared to main branching process of GFAP compatible for cell counting protocols

4. Wash slices three times in 1× PBS for 15 min to remove nonspecific binding of primary antibodies (Fig. 2).
5. Incubate slices in secondary antibodies (250 μL per well) for 1 h at room temperature.
6. Wash slices three times in 1× PBS for 15 min to remove nonspecific binding of secondary antibodies.
7. Mount slices onto slides using ProLong Gold mounting medium. Gently add coverslip, leave slides at room

temperature overnight to dry, and then, store at +4 degrees Celsius until imaged.

3.4 Sparse Labeling of Astrocytes for Detailed Morphological Analysis

3.4.1 Viral Injection for Sparse Labeling of Astrocytes (See Notes 3 and 4)

1. Place rodents onto stereotax, with ear bars immobilizing the animal, before taking Bregma and Lambda coordinates for the brain region of interest.
2. Calculate dorsal–ventral (DV) coordinate for region of interest, but reduce subtraction from bregma by 20%.
   (a) For example, for targeting CA3 astrocytes, replace original DV coordinate of −1.8 with −1.6 (Fig. 3).
3. Inject virus of interest paying specific attention to volume injected and flow rate; in this example, we injected AAV2/5-GfaABC1D-eGFP.
4. Wait 7–10 min after injection of virus before removing the syringe.
5. Close the skull with sterile sutures.
6. Wait 3 weeks for viral expression and then perfuse mice with 4% paraformaldehyde.
7. Dissect brains, fix with 4% PFA for 24 h, and leave in a 30% sucrose solution until brains sink.
8. Embed brains in OCT solutions and cut at 40 μm thickness.
9. Stain brains with anti-eGFP primary antibodies to amplify signal of AAV2/5-GfaABC1D-eGFP.
10. Incubate with secondary and DAPI to stain the nucleus of cells.
11. Image on confocal microscope to obtain z-stacks of isolated eGFP+ astrocytes (*see* Fig. 4).

3.4.2 Image Analysis with IMARIS (See Note 5)

1. Open IMARIS.
   (a) Convert files from .CZI to native IMARIS files.
2. IMARIS tab > Surface.
   (a) Algorithm selection parameters.
      (i) Segment only ROI.
      (ii) Object–object statistics.
3. Isolate astrocyte of interest by entering exact coordinates in the IMARIS ROI 1 manager.
   (a) Source > channel > AF488 (or associated AlexaFluor from IHC).
4. Isolate channel amplifying eGFP signal.
5. Select surface detail 0.118 μm.
6. Threshold with absolute intensity.

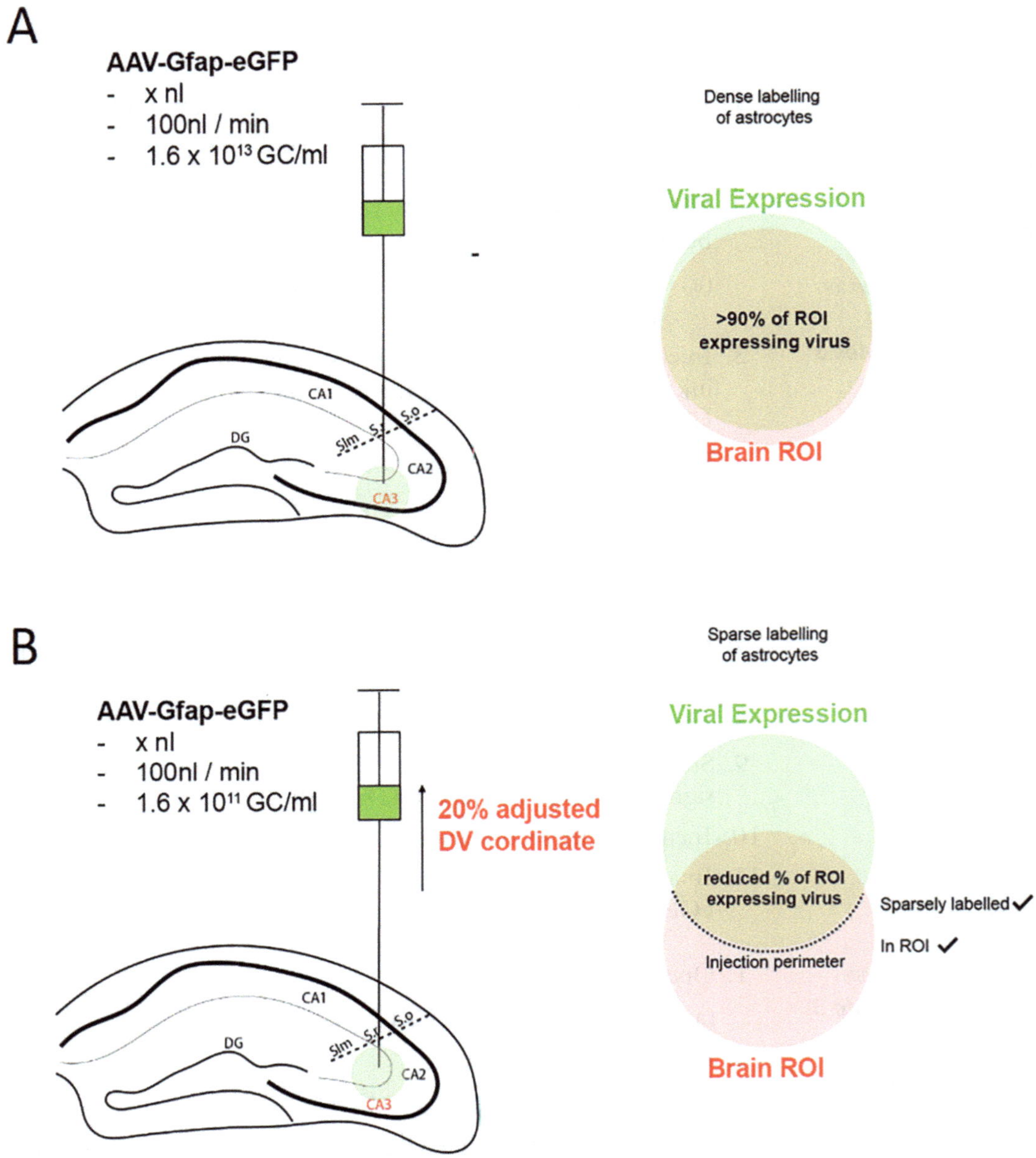

Fig. 3 Schematic representation of sparse labeling approach using viral vectors to express fluorescent proteins in astrocytes. (**a**) Standard method for targeting astrocytes with dense expression, optimized for 90% expression in brain ROI. (**b**) Adjusted DV coordinate for sparse expression of virus at expression perimeter with concomitant expression in brain ROI

(a) Consider background subtraction (local contrast) if staining is sub-optimal.

(b) Enable split touching object (region growing).

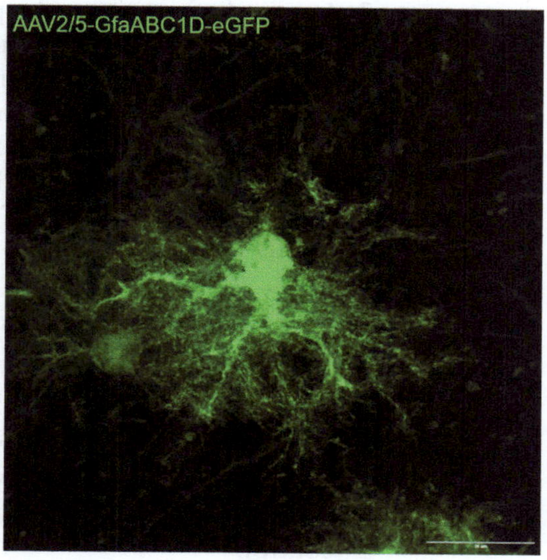

Fig. 4 Representative max intensity z-stack projection of AAV2/5-GfaABC$_1$D-eGFP-expressing astrocytes isolated from neighboring astrocytes using sparse labeling method of stereotaxic injection. Image acquisition: slice = 30 μm thickness. Imaged with *ZEISS LSM 980 with Airyscan 2* scanning confocal microscope with oil immersion Plan-Apochromat 63× objective 1.4 NA. Sixteen-bit images of 246 × 246 μm areas at 400 Hz (frame size (x*y); 1024 × 1024, pixel size; 250 nm). A total of 25–30 μm z-stacks were acquired with a step size of 0.5 μm

   (i) This will allow for post hoc editing of surfaces if software oversamples an astrocyte process.

   (ii) Seed points diameter: 0.300 μm.

7. Filter surfaces: to delete small objects not considered an astrocyte process.

   (a) Using circle select mode: delete sampled background noise or neighboring astrocytes (*see* **Note 6**).

8. Finish: in the selection tab under statistics, there is a volume value (μm^3), which gives the 3D volume of the isolates' eGFP-expressing astrocyte.

4 Notes

1. Minimize exposure to light at all times after **step 4** to prevent degradation of fluorescent markers. ELS lasts 7 days starting at P10, meaning there are seven rounds of maternal separation (MS) (one per day). The last round of MS is therefore at P16. At P17, there is no MS, you simply restore bedding to end the ELS.

2. Confocal microscopy of hippocampal astrocytes gives a detailed example of the large branching events visible with GFAP staining at a 1:500 concentration. Image acquisition parameters are critical: Slice = 30 μm thickness. Primary antibody: anti-GFAP. Imaged using a Leica TCS SP5 laser scanning confocal microscope with oil immersion Plan-Apochromat 63× objective 1.4 NA. Sixteen-bit images of 246 × 246 μm areas at 400 Hz (frame size (x*y); 1024 × 1024, pixel size; 250 nm). A total of 25–30 μm z-stacks were acquired with a step size of 0.5 μm (maximum intensity z-projections).
3. Using reduced viral titers and adjusted stereotaxic coordinates, we obtain sparse labeling of astrocytes at the viral injection perimeter with maintaining brain region specificity.
4. Astrocyte morphology can vary across and within brain regions [16]; therefore, it is critical to have a brain region-matched control.
5. The IMARIS platform has a high throughput with a user-friendly interface for obtaining astrocyte volume and branching complexity (see IMARIS user manuals in Materials). *See* Materials for alternative Fiji plug-ins and referenced toolkits optimized for analyzing astrocyte morphology.
6. The low titer of virus injection plays a big part in isolating astrocytes and makes post hoc editing of processes quicker. With low-titer injections, individual astrocytes will express eGFP with an absence of expression in neighboring astrocytes.

Acknowledgments

This work was supported by a Fondation CHUM, Fondation Courtois, Canadian Institutes of Health Research Project Grant (478629), an NSERC Discovery Grant (RGPIN-2021-03211), Fonds de Recherche du Québec—Santé (296562 and 309889), and the Brain & Behavior Research Foundation (NARSAD Young Investigator Award 28589).

Author Contributions All authors contributed to chapter along lines of expertise. All authors read and approved the final version of the chapter.

References

1. Hol EM, Pekny M (2015) Glial fibrillary acidic protein (GFAP) and the astrocyte intermediate filament system in diseases of the central nervous system. Curr Opin Cell Biol 32:121–130
2. Banasr M, Chowdhury GMI, Terwilliger R, Newton SS, Duman RS, Behar KL et al (2010) Glial pathology in an animal model of depression: reversal of stress-induced cellular,

metabolic and behavioral deficits by the glutamate-modulating drug riluzole. Mol Psychiatry 15(5):501–511

3. Murphy-Royal C, Gordon GR, Bains JS (2019) Stress-induced structural and functional modifications of astrocytes – further implicating glia in the central response to stress. Glia 67:1806–1820
4. van Ballegoij WJ, van de Stadt SI, Huffnagel IC, Kemp S, Willemse EA, Teunissen CE, Engelen M (2020) Plasma NfL and GFAP as biomarkers of spinal cord degeneration in adrenoleukodystrophy. Ann Clin Transl Neurol 7(11):2127–2136
5. Liedtke W, Edelmann W, Bieri PL, Chiu FC, Cowan NJ, Kucherlapati R et al (1996) GFAP is necessary for the integrity of CNS white matter architecture and long-term maintenance of myelination. Neuron 17(4):607–615
6. Abdelhak A, Foschi M, Abu-Rumeileh S, Yue JK, D'Anna L, Huss A et al (2022) Blood GFAP as an emerging biomarker in brain and spinal cord disorders. Nat Rev Neurol 18(3): 158–172
7. Zhang X, Alnafisah RS, Hamoud ARA, Shukla R, Wen Z, McCullumsmith RE et al (2021) Role of astrocytes in major neuropsychiatric disorders. Neurochem Res 46(10): 2715–2730
8. Pekny M, Pekna M (2014) Astrocyte reactivity and reactive astrogliosis: Costs and benefits. Physiol Rev 94(4):1077–1098
9. Murphy-Royal C, Johnston AD, Boyce AKJ, Diaz-Castro B, Institoris A, Peringod G et al (2020) Stress gates an astrocytic energy reservoir to impair synaptic plasticity. Nat Commun 11(1)
10. Bushong EA, Martone ME, Ellisman MH (2004) Maturation of astrocyte morphology and the establishment of astrocyte domains during postnatal hippocampal development. Int J Dev Neurosci 22(2):73–86
11. Salmon CK, Syed TA, Kacerovsky JB, Alivodej N, Schober AL, Sloan TFW et al (2023) Organizing principles of astrocytic nanoarchitecture in the mouse cerebral cortex. Curr Biol 33(5):957–972
12. Di Benedetto B, Malik VA, Begum S, Jablonowski L, Gómez-González GB, Neumann ID et al (2016) Fluoxetine requires the endfeet protein aquaporin-4 to enhance plasticity of astrocyte processes. Front Cell Neurosci 10:8
13. Nagy C, Suderman M, Yang J, Szyf M, Mechawar N, Ernst C et al (2015) Astrocytic abnormalities and global DNA methylation patterns in depression and suicide. Mol Psychiatry 20(3):320–328
14. Torres-Platas SG, Nagy C, Wakid M, Turecki G, Mechawar N (2016) Glial fibrillary acidic protein is differentially expressed across cortical and subcortical regions in healthy brains and downregulated in the thalamus and caudate nucleus of depressed suicides. Mol Psychiatry 21(4):509–515
15. Ernst C, Nagy C, Kim S, Yang JP, Deng X, Hellstrom IC et al (2011) Dysfunction of astrocyte connexins 30 and 43 in dorsal lateral prefrontal cortex of suicide completers. Biol Psychiatry 70(4):312–319
16. Viana JF, Machado JL, Abreu DS, Veiga A, Barsanti S, Tavares G et al (2023) Astrocyte structural heterogeneity in the mouse hippocampus. Glia 71(7):1667–1682

Chapter 19

Functional Evaluation of Astrocytic Responses to Stress

Lewis Depaauw-Holt, Sarah Peyrard, Anthony Bosson, and Ciaran Murphy-Royal

Abstract

In order to fully understand the contribution of astrocytes to stress-related psychiatric disorders, functional assays are required. There are numerous approaches that can be used to study astrocyte function including live-cell imaging of surface protein trafficking, patch-clamp recording of glutamate transporter currents, or calcium imaging using inorganic dyes and genetically encoded calcium indicators. Here, we will outline a method to record and analyze calcium activity in astrocytes using virus-mediated expression of the genetically encoded calcium indicator GCaMP. This approach gives insight into specific astrocyte changes in response to stress which could include hyper- or hypo-activity in these cells.

Key words Astrocytes, Early-life stress, Stress, Live-cell imaging, Calcium imaging

1 Introduction

The direct implication of astrocytes as active partners in neural communication gained momentum following the work of Ann Cornell-Bell in 1990 identifying a particular form of excitability-based fluctuations of intracellular Ca^{2+} concentrations in astrocytes [1]. Several distinct calcium events have now been identified including microdomain events, whole-cell activities, and calcium waves that propagate between cells which are believed to be critical for information processing [1–5].

Astrocytic Ca^{2+} activities reveal the activation state of the cell and potential modulation of synaptic circuits they're involved in [6]. As such, proper quantification of astrocyte calcium activity and how this goes awry after stress remains crucial to understanding the contribution of these cells to neural circuit dysfunction.

While various approaches can be used to measure astrocyte calcium activity, here, we outline a viral overexpression of the genetically encoded calcium indicator GCaMP (Fig. 1).

Barbara Di Benedetto (ed.), *Astrocytes: Methods and Protocols*, Methods in Molecular Biology, vol. 2896, https://doi.org/10.1007/978-1-0716-4366-2_19,

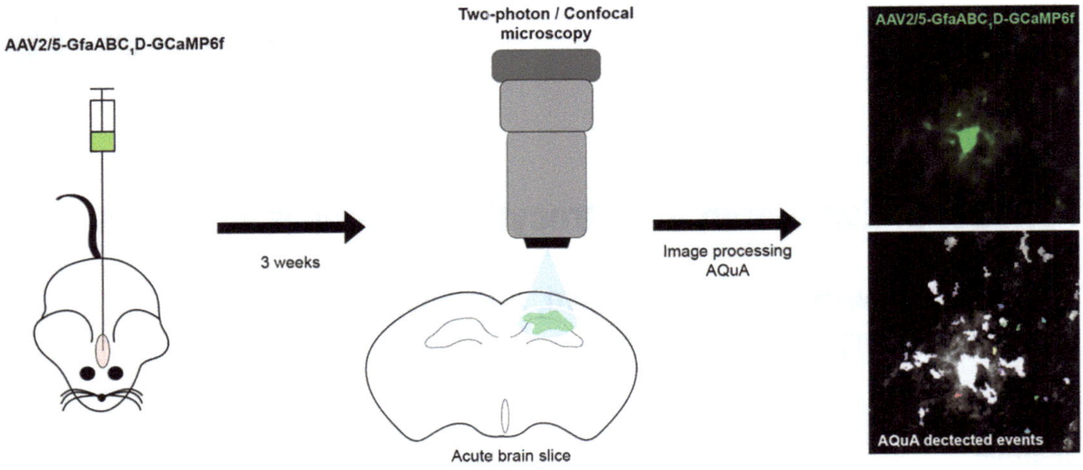

Fig. 1 Timeline for GCaMP imaging experiments. Viral injection of AAV2/5-GfaABC$_1$D-GCaMP6f into a region of interest in the brain. Wait 3 weeks post-injection to allow for viral expression in astrocytes and reduced brain inflammation. Image under confocal or two-photon microscopy, and analyze calcium events with AQuA software

2 Materials

2.1 Animals

1. Jax laboratory C57BL/6 J mice are used for all experiments (P45–P90). All experiments were approved and performed in accordance with guidelines for maintenance and care of animals of the Institutional Committee for the Protection of Animals (CIPA) at the Centre Hospitalier de l'Université de Montréal, in alignment with the Canadian Counsel on Animal Care (CCAC) guidelines.

2.2 Stereotaxic Surgeries

1. Stereotaxic frame.
2. Sterile surgeries tools: Scissors, clamps, tweezers, cotton tips, and sterile compresses.
3. Sterile drapes.
4. Heating pad.
5. Needles and syringes.
6. Eye lubricant.
7. Saline 0.9%.
8. Lidocaine and carprofen.
9. Drill.
10. Hamilton syringe.
11. Sterile sutures.

2.3 Acute Brain Slices

1. Slicing solution can be sucrose-based [7] or NMDG [8] depending on your research paradigm. Some use choline chloride-based slicing solution for older animals [9].
2. Recording solution, artificial cerebrospinal fluid (ACSF) [7–9]
3. Vibratome Leica VT1100, 1200, or 1200S.
4. Dissection tools.
5. Slice anchor.
6. AAV-gfaABC1D-GCaMP6f.
7. Confocal or two-photon microscope.
8. AquA analysis tool [10]

3 Methods

3.1 Astrocyte Calcium Imaging Using Viral GCaMP Expression (See Note 1)

Carry out stereotaxic surgery following local ethics committee guidelines.

1. In brief, mount rodents onto stereotax and take Bregma and Lambda coordinates for the brain region of interest.
2. Take anterior–posterior (AP), medial–lateral (ML), and dorsal–ventral (DV) coordinates.
3. Inject AAV-gfaABC1D-GCaMP6f (or any relevant variants) and 800 ηL (or lower for sparse labeling, see sections above) at 100–200 ηL/min to label the entire astrocyte network of the targeted brain region (*see* **Note 2**).
4. Wait for 7–10 min after injection of AAV-gfaABC1D-GCaMP6f and remove the syringe (*see* **Note 3**).
5. Close the wound with sterile sutures, and hydrate the animal with subcutaneous injection of saline 0.9%.
6. Wait 3 weeks prior imaging to allow good viral expression and most inflammatory processes to go back to physiological states.
7. Prepare acute brain slices (300–400 μm) of your region of interest.
8. Acquisition of Ca^{2+} time series (*see* **Note 4**):
   (a) Imaging chamber should be perfused with ACFS at 1–2 mL/min.
   (b) Use a slice anchor to prevent any z-plane movements.
   (c) Imaging wavelengths: confocal, 488 nm; two-photon, 920 nm.
   (d) Typical acquisition speed 0.8–2 Hz.
   (e) 16 bits or higher, as most of the post processing methods to increase spatial resolutions you might want to use require 16-bit images.

(f) Take note of the pixel size (within the imaging software or later in metadata files) as it will be required for AQuA analysis.

(g) Obtain the best signal-to-noise ratio using the minimal laser power to prevent beaching and allow for longer acquisitions.

9. Imaging field selection, quality over quantity: sharp cell edges and absence of varicosity or any "blob" inside the cells. Try to image a focal plane where you can see the soma and processes (of one or more astrocytes) as it will help monitor any drifts and overall cell viability during acquisition. If an astrocyte is in physiological conditions, you will observe spontaneous activities (focal and expanded, as well as some somatic activities) in most brain regions (*see* **Note 5**).
10. Once acquisition is done, you can process images in Fiji and use plug-ins of your choice to correct for orthogonal drifts.

3.2 Data Analysis (See Note 6)

1. Open AQuA in Matlab (more stable than the ImageJ version) and load your file.
2. Adjust the parameters based on your experimental setup and research question you are addressing. Note that you can save your configuration and used masks within AQuA, which will allow you to reanalyze each dataset with the same parameters if needed.
3. Once the analysis is completed, you can save the movies and ROI overlays for future analyses and visualization purposes.
4. Non-exhaustive list of the data you can extract using AQuA for astrocyte Ca^{2+} dynamics characterization:

(a) Area, perimeter, max Df (peak of the dF/F_0 curve for current event), duration, frequency, rising and decay time of events, scores for propagating events (orientation, size, and speeds can be inferred), temporal density (number of events that share spatial footprint with current event), spatial density (number of events co-occurred with current event), and more detailed features are available.

4 Notes

1. Imaging single astrocyte Ca^{2+} activities in gene targeted free environment (soma and processes) can be achieved by delivering chemical calcium probes such as fluo-4 via a patch pipette using patch-clamp technique [7, 9, 11].

2. These concentrations and delivery rates work for most brain regions (hippocampus, cortex, striatum, hypothalamus, amygdala) in our hands but might require specific adjustment for other brain regions.
3. Remove it slowly; it should take approximately 2 min to remove the syringe entirely.
4. A 5 min acquisition time will allow to sample most of calcium dynamics exhibited by astrocytes in physiological conditions; this may need adjustments to fit your research goals.
5. Astrocyte calcium activities are more frequent in the processes than in the somatic area and could be highly localized, restricted to 1–2 μm, or propagate to neighboring domains.
6. Due to the complexity of Ca^{2+} spatiotemporal dynamics in astrocytes, it is important to use tools that are non-ROI-based approaches such as AQuA [10]. Utilization of such tools will also help with data reproducibility across research teams. Based on astrocyte physiology knowledge, AQuA has been developed for the quantification of complex calcium and neurotransmitter dynamics using event-based approaches built around machine learning principles. This software then allows the user free classification of astrocyte functional domains as they evolve in space and time rather than a single location (see [10] for complete AQuA description and user guide).

Acknowledgments

This work was supported by a Fondation CHUM, Fondation Courtois, Canadian Institutes of Health Research Project Grant (478629), an NSERC Discovery Grant (RGPIN-2021-03211), Fonds de Recherche du Québec—Santé (296562 & 309889), and the Brain & Behavior Research Foundation (NARSAD Young Investigator Award 28589).

Author Contributions All authors contributed to chapter along lines of expertise. All authors read and approved the final version of the chapter.

References

1. Cornell-Bell AH, Finkbeiner SM, Cooper MS, Smith SJ (1990) Glutamate induces calcium waves in cultured astrocytes: long-range glial signaling. Science 247(4941):470–473
2. Shigetomi E, Bushong EA, Haustein MD, Tong X, Jackson-Weaver O, Kracun S et al (2013) Imaging calcium microdomains within entire astrocyte territories and endfeet with GCaMPs expressed using adeno-associated viruses. J Gen Physiol 141(5):633–647

3. Yu X, Nagai J, Khakh BS (2020) Improved tools to study astrocytes. Nat Rev Neurosci 21(3):121–138
4. Yu X, Taylor AM, Nagai J, Golshani P, Evans CJ, Coppola G et al (2018) Reducing astrocyte calcium signaling in vivo alters striatal microcircuits and causes repetitive behavior. Neuron 99:1170–1187
5. Rupprecht P, Duss SN, Becker D, Lewis CM, Bohacek J, Helmchen F (2024) Centripetal integration of past events in hippocampal astrocytes regulated by locus coeruleus. Nat Neurosci 27:927
6. Bazargani N, Attwell D (2016) Astrocyte calcium signaling: the third wave. Nat Neurosci 19(2):182–189
7. Bosson A, Paumier A, Boisseau S, Jacquier-Sarlin M, Buisson A, Albrieux M (2017) TRPA1 channels promote astrocytic Ca2+ hyperactivity and synaptic dysfunction mediated by oligomeric forms of amyloid-β peptide. Mol Neurodegener 12(1):53
8. Murphy-Royal C, Johnston AD, Boyce AKJ, Diaz-Castro B, Institoris A, Peringod G et al (2020) Stress gates an astrocytic energy reservoir to impair synaptic plasticity. Nat Commun 11(1):2014
9. Matos M, Bosson A, Riebe I, Reynell C, Vallée J, Laplante I et al (2018) Astrocytes detect and upregulate transmission at inhibitory synapses of somatostatin interneurons onto pyramidal cells. Nat Commun 9(1):4254
10. Wang Y, DelRosso NV, Vaidyanathan TV, Cahill MK, Reitman ME, Pittolo S et al (2019) Accurate quantification of astrocyte and neurotransmitter fluorescence dynamics for single-cell and population-level physiology. Nat Neurosci 22(11):1936–1944
11. Mehina EMF, Murphy-Royal C, Gordon GR (2017) Steady-state free Ca2+ in astrocytes is decreased by experience and impacts arteriole tone. J Neurosci 37(34):8150

Chapter 20

Unveiling Ketamine's Influence on Astrocytic Kir4.1 Channels Through Multimodal Analysis: Confocal Microscopy, Immunocytochemistry, Fluorescence Analysis, and Electrophysiology

Samo Pirnat, Katja Fink, Matjaž Stenovec, Marko Kreft, and Robert Zorec

Abstract

Understanding the elusive mechanisms responsible for the therapeutic efficacy of ketamine in major depressive disorder (MDD) is crucial. Astrocytes play a vital role in regulating extracellular potassium concentration ($[K^+]_o$), which is essential for maintaining proper neuronal excitability and overall brain function. Dysregulation of $[K^+]_o$ can lead to significant changes in neuronal activity, potentially contributing to the pathophysiology of various neurological and psychiatric conditions, including depression. To this end, we conducted a multifaceted investigation to elucidate the effects of ketamine on the inwardly rectifying K^+ channel Kir4.1, which is critical for neuronal excitability and K^+ homeostasis. Using cultured rat cortical astrocytes expressing fluorescently labeled Kir4.1 (Kir4.1-EGFP), we followed the dynamics of Kir4.1-EGFP vesicles after ketamine exposure. In addition, using live cell immunolabeling and patch-clamp assays in cultured mouse astrocytes, we investigated the effects of ketamine on Kir4.1 surface density and voltage-activated currents, similar to Ba^{2+} inhibition. This comprehensive methodological approach sheds light on the modulation of Kir4.1 dynamics by ketamine and thus provides valuable insights into its therapeutic mechanisms in MDD.

Key words Confocal microscopy, Vesicle mobility, Electrophysiology, Astrocytes, Kir4.1, Ketamine, Major depressive disorder

1 Introduction

Astrocytes, prevalent neuroglial cells in the central nervous system (CNS), perform a variety of essential functions including maintaining homeostasis, supporting neuronal survival and function [1, 2], secreting gliosignaling molecules [3, 4], and regulating synaptic activity critical for learning and memory [5, 6]. One notable function is their role in spatial buffering of extracellular potassium

The Authors Samo Pirnat and Katja Fink have equally contributed to this work.

Barbara Di Benedetto (ed.), *Astrocytes: Methods and Protocols*, Methods in Molecular Biology, vol. 2896, https://doi.org/10.1007/978-1-0716-4366-2_20,

concentration ($[K^+]_o$), predominantly facilitated by inwardly rectifying K^+ channels such as Kir4.1 [7–10]. Dysregulation of astrocytic K^+ buffering has been identified in psychiatric disorders, including major depressive disorder (MDD), in which altered neuronal excitability contributes to pathogenesis.

In particular, studies have revealed impaired astroglial Kir4.1 expression in models of depression, which is associated with altered $[K^+]_o$ and aberrant neuronal firing patterns in regions such as the lateral habenula (LHb) [11, 12]. Ketamine, a dissociative anesthetic with rapid antidepressant effects, has shown promise as a treatment for MDD [13, 14], distinct from conventional antidepressants that target monoamine systems. Understanding the mechanisms of ketamine, particularly its effects on astrocytes, is crucial.

Previous research suggests that ketamine affects various astrocytic functions, including vesicular secretion of peptides [15, 16] and plasmalemmal cholesterol composition [17, 18], likely via cAMP signaling pathways. These alterations may contribute to its rapid antidepressant effect, which contrasts with the delayed onset of conventional antidepressants [13, 14, 19]. However, the precise effects of ketamine on astroglial Kir4.1 dynamics and membrane conductance remain unexplored.

In this context, we present methods aimed at investigating the effects of ketamine on astrocytic Kir4.1 channels. We used confocal microscopy and performed fixed and live immunocytochemistry and fluorescence analysis in combination with commercial and self-developed software to investigate Kir4.1 vesicle trafficking and channel surface density. In addition, we used electrophysiological methods to measure uncompensated macrocurrents in whole-cell configuration and to observe changes in membrane conductance after ketamine exposure. The methods used help to expand the understanding of the intricate interplay between ketamine, astrocytes, and Kir4.1 channels and provide further insight into the rapid antidepressant mechanisms underlying the therapeutic efficacy of ketamine in MDD [20].

2 Materials

2.1 Cell Cultures

1. DMEM (e.g., Sigma-Aldrich, cat. no. 5671).
2. Penicillin–streptomycin (e.g., Sigma-Aldrich, cat. no. P0781).
3. Bovine serum albumin/BSA (e.g., Sigma-Aldrich, cat. no. A2153).
4. L-15 medium (e.g., Sigma-Aldrich, cat. no. L5520).
5. Fetal bovine serum (e.g., Sigma-Aldrich, cat. no. F6178).

6. Sodium pyruvate solution (e.g., Sigma-Aldrich, cat. no. S8636).
7. L-Glutamine (e.g., Sigma-Aldrich, cat. no. 63126).
8. Trypsin/EDTA (e.g., Sigma-Aldrich, cat. no. T3924).
9. Microdissection scissors.
10. Dissecting scissors.
11. Small forceps.
12. Small curved forceps.
13. 2 mL and 5 mL syringes.
14. 1.1, 0.8, and 0.6 mm needles.
15. 60 mm Petri dishes.
16. 35 mm Petri dishes.
17. 25 cm^2 tissue culture flasks.
18. 10 cm^2 tissue culture flat tubes.
19. 15 mL conical tubes.
20. 10 mL sterile serological pipettes.
21. 75 μm nylon/polyester cell strainer.
22. Parafilm.
23. Centrifuge.
24. Benchtop platform shaker.
25. Incubator (37 °C, 5% CO_2),
26. Poly-L-lysine (e.g., Sigma-Aldrich, cat. no. P1524).
27. 22 mm coverslips.
28. Sterile tips.

2.2 Plasmid Isolation and Cell Transfection

1. LB (Luria-Bertani) medium.
   (a) Tryptone (e.g., Biolife Italiana, 4122902).
   (b) Yeast extract (e.g., Biolife Italiana, 412220).
   (c) Potassium chloride (e.g., Carlo Erba, 479687).
2. PureYield™ Plasmid Midiprep System (e.g., Promega, Madison, ZDA).
3. pKir4.1-EGFP encoded plasmid (a gift from Drs. M. Eaton and S. Skatchkov).
4. Transfection medium.
   (a) Sodium pyruvate solution (e.g., Sigma-Aldrich S8636).
   (b) L-Glutamine (Sigma-Aldrich 63126).
   (c) DMEM (Sigma-Aldrich 5671).

5. Transfection mixture.
   (a) FuGENE 6 (e.g., Promega Corporation, Madison, WI, USA).
   (b) pKir4.1-EGFP plasmid.
   (c) Transfection medium.
6. Centrifuge.
7. Orbital shaker.

2.3 LysoTracker Labeling

1. 60 mm Petri dishes.
2. LysoTracker Red DND-99 (e.g., LyTR; Thermo Fisher Scientific, Waltham, MA, USA) (*see* **Note 1**).
3. Mounting medium: Extracellular solution A (ECS-A).
   (a) 130 mM NaCl (e.g., Sigma-Aldrich, cat. no. S9888).
   (b) 5 mM KCl (e.g., Sigma-Aldrich, cat. no. P3911).
   (c) 2 mM $CaCl_2$ (e.g., Sigma-Aldrich, cat. no. 21115).
   (d) 1 mM $MgCl_2$ (e.g., Sigma-Aldrich, cat. no. 63609).
   (e) 10 mM HEPES (e.g., Sigma-Aldrich, cat. no. H3375).
   (f) 10 mM D-(+)-Glucose (e.g., Sigma-Aldrich, cat. no. G8270).

   Titrate with 1 M HCl or 1 M NaOH to achieve pH 7.2–7.4.
4. Recording chamber.
5. Pipettes and other general laboratory equipment.
6. Vortex mixer.
7. Confocal microscope (LSM 780, Zeiss, Germany).

2.4 Fixed and Live Cell Immunochemistry and Fluorescence Analysis

2.4.1 Fixed Cell Immunochemistry/Fluorescence Analysis

1. 60 mm Petri dishes.
2. Washing buffer: PBS 1×.
3. Permeabilization/fixation buffer: 0.1% Triton ×-100 in 4% paraformaldehyde (PFA) in PBS 1×.
4. Blocking buffer: 10% (v/v) goat serum and 3% (w/v) bovine serum albumin (BSA) in PBS 1×.
5. Antibody buffer solution: 3% (w/v) BSA in PBS 1×.
6. Primary antibodies:
   (a) Rabbit polyclonal anti-Kir4.1 (1:500; e.g., APC-165, Alomone, Jerusalem, Israel) (*see* **Note 2**).
   (b) Rabbit polyclonal anti-aquaporin 4 (AQP4; 1:400, e.g., sc20812, Santa Cruz Biotechnology, Dallas, TX, USA) (*see* **Note 3**).

(c) Mouse polyclonal anti-synaptotagmin IV (SytIV; 1:200, e.g., sc30095, Santa Cruz Biotechnology) (*see* **Note 4**).

(d) Mouse monoclonal anti-synaptobrevin 2 (Syb2/VAMP2; 1:500, e.g., 104211, Synaptic Systems, Goettingen, Germany) (*see* **Note 4**).

(e) Rabbit polyclonal anti-VAMP3 (1:1000, e.g., Abcam, Cambridge, UK) (*see* **Note 4**).

(f) Mouse monoclonal anti-microtubule-associated protein 1 light chain 3 (LC3; 1:100, e.g., M152-3, MBL, Woburn, MA, USA) (*see* **Note 5**).

(g) Rabbit monoclonal anti-Rab7 (1:400; e.g., ab137029, Abcam) (*see* **Note 5**).

(h) Mouse monoclonal anti-α-tubulin (1:100–1:200; e.g., T-6199, Merck) (*see* **Note 6**).

(i) Mouse monoclonal anti-β-actin (1:500–1:1000; e.g., Clone AC-15, A5441, Merck) (*see* **Note 6**).

7. Secondary antibodies:

   (a) Goat anti-mouse IgG (H + L) Cross-Adsorbed Secondary Antibody, Alexa Fluor™ 546 (1:800; Thermo Fisher Scientific) (*see* **Note** 7).

   (b) Goat anti-rabbit IgG (H + L) Cross-Adsorbed Secondary Antibody, Alexa Fluor™ 546 (1:800; Thermo Fisher Scientific) (*see* **Note** 7).

8. Mounting medium: SlowFade Gold with or without DAPI (4′,6-diamidino-2-phenylindole) (Thermo Fisher Scientific).

9. Recording chamber.

10. Pipettes and other general laboratory equipment.

11. Vortex mixer.

12. Confocal microscope (e.g., LSM 780, Zeiss).

2.4.2 Live Cell Immunochemistry/ Fluorescence Analysis

1. 60 mm Petri dishes.

2. Washing buffer: PBS 1×.

3. Antibody buffer solution: 3% (w/v) BSA in PBS 1×.

4. Primary antibody:

   (a) Rat polyclonal anti-Kir4.1 (clone 20F9, 1:500; a gift from Prof. Dr. Bernhard Hemmer, Technische Universität München, Munich, Germany) (*see* **Note 8**).

5. Secondary antibody:

   (a) Goat anti-rat IgG (H + L) Cross-Adsorbed Secondary Antibody, Alexa Fluor™ 546 (1:800; Thermo Fisher Scientific).

6. Mounting medium: ECS-A.
7. Recording chamber.
8. Pipettes and other general laboratory equipment.
9. Vortex mixer.
10. Confocal microscope (e.g., LSM 780, Zeiss).

2.5 Analysis of Kir4.1 Vesicle Mobility

1. Mounting medium: ECS-A.
2. Recording chamber.
3. Pipettes and other general laboratory equipment.
4. Vortex mixer.
5. Confocal microscope (e.g., LSM 780, Zeiss).

2.6 Electrophysiology

1. Intracellular solution (ICS).
   (a) 130 mM KCl (Sigma-Aldrich, cat. no. P3911).
   (b) 1 mM $CaCl_2$ (Sigma-Aldrich, cat. no. C3881).
   (c) 1 mM $MgCl_2$ (Sigma-Aldrich, cat. no. 208337).
   (d) 10 mM HEPES (Sigma-Aldrich, cat. no. H3375).
   (e) 2 mM Na_2ATP (Sigma-Aldrich, cat. no. A2383).
   (f) 10 mM EGTA (Sigma-Aldrich, cat. no. E4378).
   (g) 0.25 mM spermine/HCl (Merck, cat. no. 306-67-2).
2. Extracellular solution B (ECS-B).
   (a) 138 mM NaCl (Sigma-Aldrich, cat. no. S9888).
   (b) 5 mM KCl (Sigma-Aldrich, cat. no. P3911).
   (c) 2 mM $CaCl_2$ (Sigma-Aldrich, cat. no. C3881).
   (d) 1.9 mM $MgCl_2$ (Sigma-Aldrich, cat. no. 208337).
   (e) 10 mM HEPES (Sigma-Aldrich, cat. no. H3375).
   (f) 3 mM Tetraethylammonium chloride (Sigma-Aldrich, cat. no. T2265).
   (g) 6 mM 4-Aminopyridine/HCl (Sigma-Aldrich, cat. no. A0152).
3. Freezing-point osmometer (Osmomat 030, Gonotec, Berlin, Germany).
4. pH meter (Mettler Toledo, MP220).
5. NaOH (Sigma-Aldrich, cat. no. S8045) (*see* **Note 9**).
6. KOH (Sigma-Aldrich, cat. no. P5958) (*see* **Note 9**).
7. Ketamine hydrochloride (Tocris Bioscience RD3131).
8. Barium chloride (Sigma-Aldrich B0750).
9. Borosilicate glass capillaries, outer diameter 1.5 mm, inner diameter 0.86 mm (Harvard Apparatus, cat. no. 30-0058).

10. Flaming/Brown micropipette puller (Shutter Instrument, any model).
11. Inverted microscope 40×/air objective (Axio Observer.A1, Zeiss).
12. Lock-in amplifier coupled to a standard patch-clamp amplifier (any manufacturer, e.g., Celica SWAM IIc) with acquisition and analysis software (usually supplied by manufacturer).
13. Micromanipulator (e.g., InjectMan NI 2; Eppendorf).
14. Digital oscilloscope, two to four channels, with a roll option mode (any manufacturer, e.g., EZ).
15. Software-generated square wave stimulation WinWCP version 5.7.8 (Strathclyde Electrophysiology Software (2011)).
16. Low-pass filter, four channels (any manufacturer, e.g., WPI, Harvard Apparatus).
17. Minimum four-port A/D converter with Bayonet Neill–Concelman (BNC) connectors (any manufacturer, e.g., National Instruments; can be already built into the amplifier).

3 Methods

3.1 Cell Cultures

3.1.1 Preparation of Primary Rat or Mouse Astrocyte Cultures

Preparation of primary rat astrocytic culture is based upon a previously described procedure [21].

Experiments involving rodents must conform to all relevant institutional and governmental regulations.

1. Decapitate 2- or 3-day-old rat or mouse pups (P2–P3) with scissors according to institutional guidelines, and place their heads into a 60 mm Petri dish (*see* **Notes 10** and **11**).
2. Hold the snout with fine forceps or fingers, and gently insert small scissors between the scalp and the skull. Cut the skin from the base of the skull to the mid-eye area, then toward the ears, and fold back the skin flaps.
3. Cut the skull along the midline of the brain and toward the ears without damaging the brain.
4. Using forceps, separate and open the raised skullcap from the brain surface..
5. Extract the brain from the skull cavity, from the olfactory bulbs to the cerebellum, using curved forceps or spatula. Place the cortices into a 60 mm Petri dish with isolation medium (*see* **Note 12**).
6. While holding the brain with forceps, separate the cerebrum from the cerebellum and brain stem with a scalpel or forceps (*see* **Note 13**).

7. Divide the cerebral hemispheres along the midline with a scalpel or forceps, ensuring not to cut completely through. Remove the cortex (grey matter containing cell bodies) from both hemispheres. Discard the midbrain, preserving only the cerebral hemispheres.
8. Dissect the hippocampus from the cortex with curved forceps (it is visible as a moon-shape tissue of brighter color) on both hemispheres. Discard the prefrontal cortex.
9. Transfer the cortices into a 60 mm Petri dish with 3 mL of isolation medium.
10. Peel away the meninges from the cortex surface and inner side using fine forceps and a magnifier (*see* **Note 14**).
11. Transfer the cortices into a new 60 mm Petri dish with 3 mL of isolation medium. Repeat twice.
12. Transfer the tissue suspension into a 15 mL conical tube.
13. Dissociate the tissue by pipetting up and down (three strokes) with a 10 mL serological pipette (*see* **Note 15**).
14. Centrifuge the cells in 15 mL conical tube 1200 rpm for 4 min, resuspend the pellet with 5 mL of isolation medium, and repeat the dissociation **step 13** twice.
15. Place the dissociated tissue into a 35 mm Petri dish, and triturate it through 1.1, 0.8, and 0.6 mm needles, three times each needle, changing the dish each time.
16. Aspirate the culture medium to remove cell debris, and feed the cultures with 5 mL of fresh astrocyte culture medium (*see* **Note 16**).
17. After 5–9 days, when cells reach 70% confluency, aspirate medium, rinse cultures with culture medium to remove floating cells, and add fresh astrocyte culture medium (*see* **Note 17**).
18. Place the flask back into the incubator, and allow it to equilibrate until the evening.
19. Tighten flask caps, wrap each flask in parafilm or plastic, and then attach to the shaker platform with adhesive tape.
20. Shake flasks at 225 rpm for 12 h (overnight) at room temperature (RT).
21. In the morning, remove the medium, rinse cells with culture medium, and add fresh astrocyte culture medium. Repeat **steps 17–20** three times (*see* **Note 18**).
22. Remove the medium and rinse cultures with 4 mL DMEM.
23. Trypsinize cells with 4 mL trypsin/EDTA solution; incubate at 37 °C for 5 min.

24. Centrifuge the cells in 15 mL conical tube at 900 rpm for 5 min, resuspend the pellet with 4 mL astrocyte culture medium, and gently dissociate cells by pipetting up and down (three to five strokes) with a 10 mL serological pipette.
25. Plate cells at desired density in the 10 cm^2 tissue culture flat tubes (add 2 mL of astrocyte culture medium to the tube, and then, add 0.5 mL of cell suspension). Subculture cells onto the poly-L-lysine-coated coverslips at least 1 day before experiments.

3.2 Plasmid Isolation and Cell Transfection

3.2.1 Plasmid Isolation from Escherichia coli with Encoded Plasmid pKir4.1-EGFP

Plasmid isolation from *Escherichia coli* with encoded plasmid pKir4.1-EGFP is aimed at obtaining plasmid DNA from bacterial culture for experimental use. The procedure is adapted from the PureYield™ Plasmid Midiprep System kit (Promega, A2492) plasmid isolation manual.

Only trained personnel should conduct work involving genetically modified organisms (GMOs) to ensure proper handling and safety. Use protective equipment (lab coat, gloves). Dispose of waste via autoclaving. Clean surfaces with 70% alcohol. Deactivate bacteria on glass/plastic with a biocidal agent and then autoclave. Accurately record all activities in the lab logbook. Provide clear documentation of the destruction/inactivation of GMO-containing materials.

The isolation procedure consists of three parts: preparation of LB (Luria-Bertani) liquid medium for bacterial cell inoculation, inoculation of transformed bacterial cells, and the plasmid isolation procedure using the Midiprep kit.

3.2.2 Preparation of LB (Luria-Bertani) Medium for Bacterial Cell Inoculation

1. Weight tryptone, yeast extract, and KCl in deionized water (Milli-Q), and adjust the pH to 7.0 with NaOH.
2. Pour the prepared medium into Schott bottles, and autoclave using the liquid cycle.
3. Use LB media at RT after cooling.

3.2.3 Inoculation of Transformed (Plasmid-Containing) Bacterial Cells

This procedure is conducted aseptically. In a sterile Erlenmeyer flask, add 60 mL of LB medium. Incorporate the appropriate antibiotic into the medium (e.g., 100 μg/mL of ampicillin or 25 μg/mL of kanamycin). Bacterial cells with introduced plasmids can be inoculated into LB medium in two ways:

1. Remove the bacterial plate from the refrigerator. Sterilize forceps with flame, cool, and pick up a sterile tip. Touch a well-separated colony, and release it into the Erlenmeyer flask with LB medium and an antibiotic. Shake the culture at 37 °C and 260 rpm for 16 h.
2. Remove the vial with bacteria from the −80 °C freezer and place it on ice. Sterilize forceps with flame, cool, and pick up a

sterile tip. Touch the frozen bacterial culture, and release it into the Erlenmeyer flask with LB/antibiotic medium. Return the vial with bacteria to freezer (−80 °C). Shake the culture at 37 ° C and 260 rpm for 16 h.

3.2.4 Plasmid Isolation Procedure with Midiprep

1. Before isolation, spectrophotometrically check the density of grown bacterial cells by measuring the culture absorbance at a wavelength of 600 nm. Optimal optical density (OD) range is 2–4. Use LB/antibiotic medium for control. Isolate plasmid DNA (pDNA) according to the manufacturer's instructions using the vacuum DNA isolation kit (PureYield™ Plasmid Midiprep System; Promega, A2492).
2. After shaking, transfer the bacterial culture into a 50 mL centrifuge tube ("Falcon"), and centrifuge at room temperature at 5000 g for 10–15 min. Rinse the Erlenmeyer flask with residual bacterial material with 15–20 mL of 70% ethanol, and pour it into a glass container for liquid waste.
3. After centrifugation, pour off the supernatant; add 3 mL of cell resuspension solution to the pellet, and vortex until a homogeneous solution is obtained.
4. Add 3 mL of cell lysis solution, and gently invert the Falcon tube three to five times. Incubate for 3 min at RT.
5. Add 5 mL of neutralization solution, and invert the Falcon tube five to ten times to mix the solution. Wait for 2–3 min for the liquid to clarify, and centrifuge with a fixed-angle rotor at RT at 15,000 g for 15–25 min. While centrifuging, assemble a vacuum pump and columns (Erlenmeyer flask, tube, rubber bands, adapter, 2× adapter plug), and then, place the blue cleaning column over the white collecting column on the adapter on the vacuum pump.
6. After centrifugation, transfer the supernatant to the blue column, wait for 2–3 min for larger particles to move to the top, and open the water on the tap to create a vacuum. Wait for all the liquid to pass through both columns before closing the water tap. Remove the blue column, but leave the white column on the adapter.
7. Add 5 mL of endotoxin removal wash to the white column, and run it through the column with a vacuum. Then, add 20 mL of column wash solution, and also, run it through the column with a vacuum. After all the solution has passed through the column, let the vacuum run for 30–60 s or until you no longer smell ethanol to allow the membrane at the bottom of the white column to dry completely.
8. Remove the white column from the pump adapter, and place it in a fresh 50 mL centrifuge tube.

9. Pipette 300–600 μL of nuclease-free water over the entire surface of the white column membrane, and wait for 1 min. Centrifuge with a swing-out rotor at RT at 1500 g for 5 min. During this process, Falcon tubes should not be covered with lids. Check the purity and concentration of isolated pDNA spectrophotometrically: dilute 7 μL of plasmid DNA sample in 693 μL of dH_2O (Milli-Q) in a test tube, mix well with a pipette, and transfer to a glass cuvette. Measure the DNA concentration (see spectrophotometer instructions) in two replicates. Rinse the cuvette well with Milli-Q between measurements.
10. Determine purity by calculating the ratio of absorbance at 260 nm (A260) to 280 nm (A280). A ratio of 1.7–2.0 is considered to contain DNA with relatively few protein contaminants.
11. Calculate the concentration of isolated pDNA from the absolute value of A260 using the equation:

$$c\_pDNA\ ((\mu g)/(mL)) = A260 \times DF \times 50\ ((\mu g)/(mL))$$

where DF represents the dilution factor.

12. Store 40 μL of isolated pDNA at −20 °C until use.

3.2.5 Cell Transfection with Plasmid pKir4.1-EGFP

Transfection of cells with FuGENE 6 Transfection Reagent (Promega, Cat. No. E2691) allows the entry of nucleic acids into mammalian cells for studies of gene expression. Here's the procedure using FuGENE 6 Transfection Reagent (*see* **Note 19**):

1. Ensure that cell culture on a coverslip is sufficiently dense for a large number of transfected cells.
2. Dilute 3 μL of FuGENE reagent in 100 μL of serum-free transfection medium (use the same medium as for cell culture preparation but without serum and antibiotics). Add FuGENE carefully, avoiding contact with the walls of the tube. Incubate at RT for 5 min.
3. Add 1–2 mg of DNA in the diluted FuGENE tube. Mix well and incubate at RT for 15 min.
4. Meanwhile, examine the cells, remove the culture medium, and add fresh 900 μL of culture medium. Then, add the medium with DNA to the cells, gently mix, and incubate overnight at 37 °C.
5. After 24 h, replace the transfection medium with culture medium.

3.3 LysoTracker Labeling

1. Incubate transfected cells in culture medium containing 200 nM LysoTracker Red DND-99 for 5 min at 37 °C and 5% CO_2 to label acidic late endo-/lysosomes.

2. Wash LysoTracker-labeled cells once with ECS-A.
3. Mount the cells onto the recording chamber.
4. Supply the mounted cells with ECS-A.
5. Observe the cells under a confocal microscope (LSM 780, Zeiss, Jena, Germany) to visualize the labeled acidic late endo-/lysosomes.

3.4 Fixed and Live Cell Immunochemistry and Fluorescence Analysis

3.4.1 Fixed Cell Immunochemistry

1. Fixation and Permeabilization:
   (a) Fix Kir4.1-EGFP transfected rat astrocytes in 4% PFA for 15 min at RT.
   (b) Permeabilize the fixed cells with 0.1% Triton ×-100 in 4% PFA for 10 min at RT.
   (c) Wash the cells four times with PBS at RT.
2. Blocking and Primary Antibody Incubation:
   (a) Reduce nonspecific background staining by incubating the cells in blocking buffer with 10% (v/v) goat serum in 3% (w/v) bovine serum albumin (BSA) in PBS for 1 h at 37 °C (*see* **Note 20**).
   (b) Wash the cells with PBS once.
   (c) Incubate the cells with primary antibodies diluted in 3% BSA in PBS overnight at 4 °C (*see* **Note 21**).
3. Secondary Antibody Staining:
   (a) Wash the cells four times in PBS to remove unbound primary antibodies.
   (b) Stain the cells with secondary anti-rabbit or anti-mouse antibodies conjugated to Alexa Fluor 546 (1:800; Thermo Fisher Scientific) for 45 min at 37 °C.
   (c) Wash the cells four times in PBS.
4. Mounting:
   (a) Mount the coverslips onto glass slides using the SlowFade Gold antifade mountant with or without DAPI to preserve fluorescence and provide nuclear counterstaining.
5. Analysis: Quantification of Co-localization:
   (a) Export tiff files (Zeiss ZEN software or equivalent) of recordings containing green-emitting fluorophores (Kir4.1-EGFP) and red-emitting fluorophores (e.g., anti-aquaporin 4 Ab and others).
   (b) Analyze the exported tiff files using ColocAna (Celica Biomedical) (or similar application) for automated high-throughput co-localization analysis. Count all green, red, and co-localized (green and red) pixels in each image.

(c) Determine fluorescence co-localization (%) with reference to green pixels identifying Kir4.1 vesicles.

(d) Set the threshold for the co-localized pixel count to 20% of the maximal fluorescence to reduce the apparent overlap from closely positioned fluorescent structures.

3.4.2 Live Cell Immunochemistry

1. Antibody Incubation and Staining:
   (a) Wash live mouse astrocytes once with 3% BSA in PBS for 2 min.
   (b) Incubate the cells with rat polyclonal anti-Kir4.1 primary antibody diluted in 3% BSA in PBS for 45 min at 37 °C (*see* **Note 22**).
2. Secondary Antibody Staining:
   (a) Wash the cell-loaded coverslips three times with 3% BSA in PBS for 2 min.
   (b) Incubate the cells with corresponding secondary antibodies for 45 min at 37 °C (*see* **Note 22**).
3. Imaging and Analysis:
   (a) Wash the cell-loaded coverslips three times with 3% BSA in PBS for 2 min.
   (b) Transfer the coverslips to the recording chamber. Add 400 μL of ECS-A.
   (c) Observe immunofluorescent cells with a confocal microscope equipped with a plan apochromatic oil immersion objective 63×/NA 1.4 (or similar).
   (d) Obtain confocal images (single planes or z-stacked) with 561-nm diode-pumped solid-state laser excitation.
   (e) Bandpass filter the fluorescence emission at 565–615 nm.
   (f) Use ImageJ software to analyze confocal images of live cells, measuring the apparent size (surface area) and cumulative (integrated; ε) fluorescence intensity of immunofluorescent Kir4.1 at the surface of the astrocytes. Determine the number of immunofluorescent Kir4.1-positive puncta. Compare treated groups with the control group using ANOVA on ranks followed by Dunn's post hoc test to assess the surface area or cumulative intensity of Kir4.1 fluorescence relative to controls (Fig. 1).

3.5 Analysis of Kir4.1 Vesicle Mobility

1. Mount coverslips with astrocytes into the recording chamber.
2. Supply the mounted cells with ECS-A.
3. Transfer the recording chamber to a confocal microscope (LSM 780; Zeiss) equipped with a plan apochromatic oil immersion objective 63×/NA 1.4.

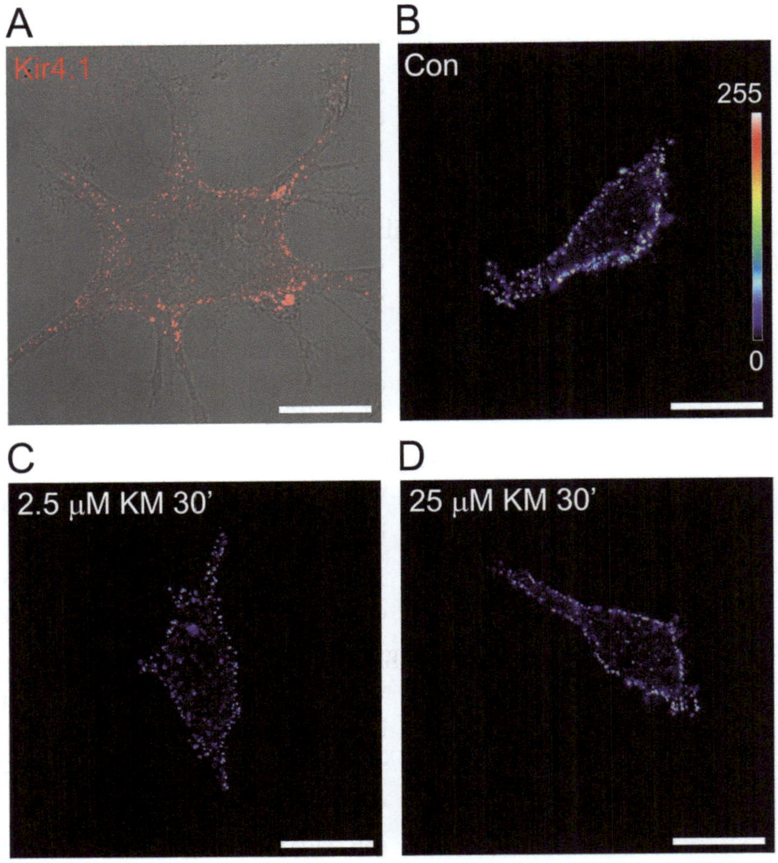

Fig. 1 Impact of short-term ketamine treatment on Kir4.1 surface density in cultured mouse astroglia. (**a**) Superimposed transmitted light and confocal image depicting immunofluorescent Kir4.1, labeled with anti-Kir4.1 and Alexa546-conjugated secondary antibody (red), observed at the surface of live mouse astrocytes. (**b**–**d**) Pseudocolored representations illustrate immunofluorescent Kir4.1 at the surface of a non-treated control cell (Con; **b**) and astrocytes treated for 30 min with 2.5 μM (**c**) or 25 μM ketamine (KM; **d**). The pseudocolored intensity scale (0–255 intensity levels) indicates the Kir4.1 immunolabeling intensity. Scale bars (**a**–**d**): 20 μm

4. Excite Kir4.1-EGFP using a 488 nm argon laser line and filter emission fluorescence with a bandpass filter (495–545 nm).
5. Analyze vesicle mobility in exported tiff files using the ParticleTR 2 software (Celica Biomedical) or functionally equivalent software package (*see* **Note 23**).
6. Select 20 mobile vesicles per transfected astrocyte and track their movement automatically, ensuring they remain in focus in each image plane (*see* **Note 24**).
7. Halt automatic tracking when a tracked vesicle moves close to another vesicle, and manually determine its position in subsequent frames until it separates enough from the neighboring vesicle to resume automated tracking.

8. Obtain the following mobility parameters for 15 s epochs: TL (track length; total length of the traveled pathway), MD (maximal displacement; the farthest translocation of a vesicle), DI (directionality index; the ratio of MD/TL), and speed.

3.6 Electrophysiology

The whole-cell patch-lamp setup was equipped with CELL software (Celica Biomedical), and current measurements were conducted using open-source software WinWCP V5.7.8.

1. Connect the setup: Swam IIC, filters, both oscilloscopes, micromanipulator, and computer, and run software programs (*see* **Note 25**).
2. Before recordings, adjust settings in both programs as follows (*see* **Note 26**):
   (a) Settings in CELL (and adjust settings on the amplifier accordingly):
      1. Lock-in frequency: 800 Hz.
      2. Vrev: −50 Hz.
      3. Im: 50 pA/V.
      4. Greal: 100 pS/V.
      5. Gimag: 10 pS/V.
      6. Vm: 100 mV/V.
      7. Pulse duration: 50 ms.
      8. Amplitude: 1 mV (low) and 5 mV (high).
      9. Calculated resistance: 1.
      10. Signal input: 50 pA.
   (b) Settings in WinWCP V5.7.8:
      1. Stimulus:
         (i) Delay: 50 ms.
         (ii) Duration: 100 ms.
         (iii) Amplitude increments: 10 mV (or proportionally higher in case of stimulus input attenuation, e.g., 100 mV in case of tenfold attenuation).
      2. Recording:
         (i) Recording duration: 204.79 ms.
         (ii) Sample interval: 0.2 ms.
         (iii) Stimulus repeat period: 400 ms.
         (iv) Number of records: 15.
         (v) Number of repetitions per increment: 1.
         (vi) Number of output channels: 2.

(vii) Number of samples (per channel): 1024.

(viii) External stimulus: N.

3. Place the coverslip with seeded cells into the holder, fill it with the chosen volume of ECS-B, attach the reference electrode, submerge it into the bath solution, and plug it into the head-stage reference connector.
4. Fill the patch pipette and holder with ICS, and connect to the connector, and attach the suction tube to the side of the pipette holder.
5. Ensure that the ICS in the pipette and holder are free of air bubbles, as they can interfere with measurements by acting as additional resistors.
6. Use the micromanipulator to position the pipette with the electrode into the solution.
7. Nullify the junction potentials in tracking mode (offset the potentiometer or the appropriate software function), and select the voltage clamp mode. Apply a square voltage pulse (1 mV).
8. Read the resistance of the pipette on the oscilloscope function in CELL (e.g., 1–3.5 MΩ at gain 50 pA/V).
9. Locate the tip of the pipette: use the micromanipulator to move across the visual field until you see the shadow. Then, turn the coarse screw of the focus three-quarters of a turn to raise the focus, and move the pipette in all directions, slightly downward to find the tip (the meniscus is very bright and just before the tip). Bring it into focus by moving downward slowly. When the pipette is in focus, return the focus to top of the selected cell.
10. Prepare suction and slowly lower the pipette onto the cell, by the side of the nucleus. When contact is made, the current on the oscilloscope decreases. Apply suction and observe current steps that should straighten completely at a successful patch (*see* **Note 27**). Switch to VC mode. Gradually add negative potential with Vm to −70 mV.
11. Compensate transient currents at the beginning and end of the stimulus with Cfast (≈3 pF) and µs (≈8), turn off the oscilloscope mode in CELL, and start the measurement.
12. Switch to measure (ME) mode on amplifier.
13. Perforate with fast suction. If the membrane is successfully ruptured into the whole-cell configuration, continue measurement of capacitance. Save the file.
14. Open WinWCP and change settings on the amplifier: switch measure mode to tracking mode (TR), stimulus gain from 0.001 to 0.1, and main gain (A/V) from 50 pA/V to 0.2 nA/V.

15. Before step recording, switch Vhold on zero and after the measurement back on −70 mV (*see* **Note 28**).
16. Record voltage steps before the stimuli and then after in desired time points (e.g., after 30 s, 1 min, and 3 min). Save the recording.
17. Analyze the recording using Matlab (MathWorks Inc., Natick, MA, USA) to extract the data.

4 Notes

1. To assess Kir4.1-EGFP accumulation in acidic organelles like lysosomes, analyze co-localization with LyTR-laden acidified late endo-lysosomes, and compare their mobility properties.
2. Analyze co-localization of transfected Kir4.1-EGFP with endogenous Kir4.1, labeled with anti-Kir4.1 antibody, to determine their subcellular location and ratio.
3. Expect high co-localization of Kir4.1 with aquaporin 4, a water channel protein found near astrocytic Kir4.1.
4. Investigate co-localization of Kir4.1 with synaptotagmin IV, VAMP2, and VAMP3, indicating potential involvement in SNARE- and Ca^{2+}-dependent exocytosis.
5. Determine co-localization of Kir4.1-EGFP with LC3 to assess its presence in autophagosomes or late endosomes and multi-vesicular bodies (Rab7).
6. To optophysiologically examine whether Kir4.1-EGFP-positive vesicles structurally associate with cytoskeletal elements, immunofluorescently label microtubules (anti-α-tubulin Ab) or actin filaments (anti-β-actin Ab) in Kir4.1-EGFP-transfected astrocytes.
7. Alexa Fluor™ 555 or alternative fluorophore/secondary antibody combination can also be used. Use appropriate fluorophore/secondary antibody combinations, avoiding spectral overlap, especially for double labeled cells.
8. Ketamine-evoked impact on the mobility of Kir4.1-positive vesicles may in turn influence the flux of vesicles toward the plasmalemma and affect the incorporation of Kir4.1 by regulated exocytosis. Live immunocytochemistry using anti-Kir4.1 antibody that targets the extracellular part of Kir4.1 can be used to examine whether short-term ketamine treatment affects the surface density of native Kir4.1 in live astrocytes.
9. Handle NaOH and KOH cautiously to prevent severe burns or eye damage. Use in a well-ventilated area with proper protective gear.

10. Use 2- or 3-day-old rat pups to minimize neuronal contamination in cell suspension during tissue dissection.
11. Minimize the time between brain collection and cortex dissection to prevent tissue degradation due to blood supply deprivation. It is recommended that the pups are sacrificed one at a time.
12. Completely cover brains with isolation buffer during all dissection procedures to prevent tissue drying.
13. Use a scalpel for faster and cleaner cuts. Using fine forceps may be more precise, but it is more time-consuming and damages the tissue more easily.
14. Ensure complete removal of meninges to avoid fibroblast contamination in astrocyte culture.
15. Cortical tissue pieces break apart, resulting in an opaque solution.
16. Change culture medium every 2–3 days until the final astrocyte purification.
17. Culture flasks contain a monolayer of astrocytes, which are flat, fibroblast-like cells, with a few oligodendrocytes scattered on the surface. To achieve adequate stratification of astrocytes and oligodendrocytes, maintain the initial culture period between 5 and 9 days. Extending this period results in clustering of astrocytes above the monolayer.
18. If more than 50% of cells detach do not remove the medium in the morning, but place the flask in the incubator and allow it to equilibrate until the next morning.
19. The described protocol is suitable for small coverslips and for work in small Petri dishes. Adjust reagent quantities for larger coverslips and containers. Transfection media can be left overnight depending on cell sensitivity. The expression time after transfection depends on the type of promoter in the plasmid, the cell type, etc. After 24–48 h after transfection, the plasmid pKir4.1-EGFP expression was high enough for recordings.
20. Ideally, the blocking serum should come from the same species as the one in which the secondary antibody is raised. In our lab, we have had good results using normal goat serum for secondary antibodies from different species.
21. We always used one primary antibody per coverslip. You can also use multiple antibodies per coverslip, but keep in mind they will have to originate from different species and require appropriate secondary antibody pairs to minimize the chance of cross-reactivity. Additionally, each secondary antibody must be labeled with a different fluorophore to minimize spectral overlap. In general, it's not feasible to use more than two or

three primary antibodies on the same coverslip. It's essential to validate the specificity and performance of each antibody individually and in combination to ensure reliable results.

22. Shorten incubation time to 15 min on ice for both primary and secondary antibodies if internalization rate is high. Fill a small Styrofoam box with crushed ice, carefully put the Petri dish containing the coverslip on ice, cover the box, and then put the box in the refrigerator. The wash cycles after primary and secondary antibody incubations remain the same (3× 2 min).
23. The ParticleTR 2 software employs two-dimensional (2D) Gaussian curve fitting to analyze vesicle intensity profiles [22, 23]. Utilizing a simplex algorithm [24] with a least squares estimator, the software fits a 2D Gaussian curve to each vesicle intensity profile, recording the peak of the curve as the vesicle's coordinates in each frame.
24. Considering that cultured astrocytes are relatively thin (1–2 μm), comparable to the depth of field of our imaging system, any Z-axis vesicle mobility is unlikely to significantly affect the measurements of vesicle tracks.
25. Electrophysiological equipment should be shielded from electromagnetic events and vibrations.
26. Set and verify all settings with a model cell for individual systems, as Vrev varies for different cell types.
27. Ensure gigaseal (gigaOhm resistance) formation. If leakage occurs, change the electrode and select another cell.
28. If you want to keep the voltage to −70 mV and proceed with steps from there, modify the settings in WinWCP protocol.

Acknowledgments

We are grateful to B. Hemmer for the donation of rat polyclonal anti-Kir4.1 (clone 20F9) and M. Eaton and S. Skatchkov for the donation of the pKir4.1-EGFP. This work was supported by the Slovenian Research Agency (research core funding P3-310 and projects J3 6790 and J3 9266 to R.Z.).

References

1. Seri B, Garcia-Verdugo JM, McEwen BS, Alvarez-Buylla A (2001) Astrocytes give rise to new neurons in the adult mammalian hippocampus. J Neurosci 21(18):7153–7160. https://doi.org/10.1523/JNEUROSCI.21-18-07153.2001
2. Verkhratsky A, Parpura V, Vardjan N, Zorec R (2019) Physiology of Astroglia. Adv Exp Med Biol 1175:45–91. https://doi.org/10.1007/978-981-13-9913-8_3
3. Verkhratsky A, Matteoli M, Parpura V, Mothet JP, Zorec R (2016) Astrocytes as secretory cells of the central nervous system: idiosyncrasies of vesicular secretion. EMBO J 35(3):239–257. https://doi.org/10.15252/embj.201592705

4. Zorec R, Verkhratsky A, Rodriguez JJ, Parpura V (2016) Astrocytic vesicles and gliotransmitters: slowness of vesicular release and synaptobrevin2-laden vesicle nanoarchitecture. Neuroscience 323:67–75. https://doi.org/10.1016/j.neuroscience.2015.02.033
5. Clarke LE, Barres BA (2013) Emerging roles of astrocytes in neural circuit development. Nat Rev Neurosci 14(5):311–321. https://doi.org/10.1038/nrn3484
6. Zorec R, Horvat A, Vardjan N, Verkhratsky A (2015) Memory formation shaped by Astroglia. Front Integr Neurosci 9:56. https://doi.org/10.3389/fnint.2015.00056
7. Seifert G, Huttmann K, Binder DK, Hartmann C, Wyczynski A, Neusch C, Steinhauser C (2009) Analysis of astroglial K+ channel expression in the developing hippocampus reveals a predominant role of the Kir4.1 subunit. J Neurosci 29(23):7474–7488. https://doi.org/10.1523/JNEUROSCI.3790-08.2009
8. Tang X, Taniguchi K, Kofuji P (2009) Heterogeneity of Kir4.1 channel expression in glia revealed by mouse transgenesis. Glia 57(16):1706–1715. https://doi.org/10.1002/glia.20882
9. Nwaobi SE, Cuddapah VA, Patterson KC, Randolph AC, Olsen ML (2016) The role of glial-specific Kir4.1 in normal and pathological states of the CNS. Acta Neuropathol 132(1):1–21. https://doi.org/10.1007/s00401-016-1553-1
10. Chever O, Djukic B, McCarthy KD, Amzica F (2010) Implication of Kir4.1 channel in excess potassium clearance: an in vivo study on anesthetized glial-conditional Kir4.1 knockout mice. J Neurosci 30(47):15769–15777. https://doi.org/10.1523/JNEUROSCI.2078-10.2010
11. Cui Y, Yang Y, Ni Z, Dong Y, Cai G, Foncelle A, Ma S, Sang K, Tang S, Li Y, Shen Y, Berry H, Wu S, Hu H (2018) Astroglial Kir4.1 in the lateral habenula drives neuronal bursts in depression. Nature 554(7692):323–327. https://doi.org/10.1038/nature25752
12. Yang Y, Cui Y, Sang K, Dong Y, Ni Z, Ma S, Hu H (2018) Ketamine blocks bursting in the lateral habenula to rapidly relieve depression. Nature 554(7692):317–322. https://doi.org/10.1038/nature25509
13. Newport DJ, Carpenter LL, McDonald WM, Potash JB, Tohen M, Nemeroff CB, Biomarkers APACoRTFoN, Treatments (2015) Ketamine and other NMDA antagonists: early clinical trials and possible mechanisms in depression. Am J Psychiatry 172(10):950–966. https://doi.org/10.1176/appi.ajp.2015.15040465
14. Zarate CA Jr, Singh JB, Carlson PJ, Brutsche NE, Ameli R, Luckenbaugh DA, Charney DS, Manji HK (2006) A randomized trial of an N-methyl-D-aspartate antagonist in treatment-resistant major depression. Arch Gen Psychiatry 63(8):856–864. https://doi.org/10.1001/archpsyc.63.8.856
15. Stenovec M, Lasic E, Bozic M, Bobnar ST, Stout RF Jr, Grubisic V, Parpura V, Zorec R (2016) Ketamine inhibits ATP-evoked exocytotic release of brain-derived neurotrophic factor from vesicles in cultured rat astrocytes. Mol Neurobiol 53(10):6882–6896. https://doi.org/10.1007/s12035-015-9562-y
16. Lasic E, Rituper B, Jorgacevski J, Kreft M, Stenovec M, Zorec R (2016) Subanesthetic doses of ketamine stabilize the fusion pore in a narrow flickering state in astrocytes. J Neurochem 138(6):909–917. https://doi.org/10.1111/jnc.13715
17. Lasic E, Lisjak M, Horvat A, Bozic M, Sakanovic A, Anderluh G, Verkhratsky A, Vardjan N, Jorgacevski J, Stenovec M, Zorec R (2019) Astrocyte specific remodeling of Plasmalemmal cholesterol composition by ketamine indicates a new mechanism of antidepressant action. Sci Rep 9(1):10957. https://doi.org/10.1038/s41598-019-47459-z
18. Senese NB, Rasenick MM (2021) Antidepressants produce persistent Galpha (s)-associated signaling changes in lipid rafts after drug withdrawal. Mol Pharmacol 100(2):66–81. https://doi.org/10.1124/molpharm.120.000226
19. Berman RM, Cappiello A, Anand A, Oren DA, Heninger GR, Charney DS, Krystal JH (2000) Antidepressant effects of ketamine in depressed patients. Biol Psychiatry 47(4):351–354. https://doi.org/10.1016/s0006-3223(99)00230-9
20. Bozic M, Pirnat S, Fink K, Potokar M, Kreft M, Zorec R, Stenovec M (2023) Ketamine reduces the surface density of the Astroglial Kir4.1 channel and inhibits voltage-activated currents in a manner similar to the action of Ba(2+) on K(+) currents. Cells 12(10). https://doi.org/10.3390/cells12101360
21. Schwartz JP, Wilson DJ (1992) Preparation and characterization of type 1 astrocytes cultured from adult rat cortex, cerebellum, and striatum. Glia 5(1):75–80. https://doi.org/10.1002/glia.440050111
22. Anderson CM, Georgiou GN, Morrison IE, Stevenson GV, Cherry RJ (1992) Tracking of cell surface receptors by fluorescence digital

imaging microscopy using a charge-coupled device camera. Low-density lipoprotein and influenza virus receptor mobility at 4 degrees C. J Cell Sci 101(Pt 2):415–425. https://doi.org/10.1242/jcs.101.2.415

23. Schütz GJ, Axmann M, Freudenthaler S, Schindler H, Kandror K, Roder JC, Jeromin A (2004) Visualization of vesicle transport along and between distinct pathways in neurites of living cells. Microsc Res Techniq 63(3):159–167. https://doi.org/10.1002/jemt.20016

24. Lagarias JC, Reeds JA, Wright MH, Wright PE (1998) Convergence properties of the Nelder-Mead simplex method in low dimensions. Siam J Optimiz 9(1):112–147. https://doi.org/10.1137/S1052623496303470

Chapter 21

Revealing and Characterizing Astrocytic Atrophy in an Animal Model of Depression

Si-Si Lin, Bin Zhou, Yong Tang, and Alexei Verkhratsky

Abstract

Astrocyte atrophy is the main histopathological hallmark of major depressive disorder (MDD) in humans and in animal models of depression. In particular, depression and depressive-like behaviors are associated with a substantial decrease in size and complexity of astrocytes in the prefrontal cortex. This results in the reduced homeostatic support of synaptic transmission, which arguably translates into abnormal activity of neuronal networks and pathological changes in mood and behavior. Treatment of experimental animals with antidepressants as well as with acupuncture in the specific acupoints alleviates depressive-like behaviors and rescues astrocytic atrophy. Here, we describe the methodology for inducing and monitoring the depressive-like behaviors in mice using chronic unpredictable mild stress (CUMS), with subsequent in-depth analysis of astrocytic morphology using confocal microscopy in conjunction with astrocyte-specific labeling by virally transfected genetically encoded fluorescent probe mCherry and intracellular injection of Lucifer yellow. We also describe immunocytochemical visualization of astrocyte-specific cytoskeletal linker ezrin, which is involved in controlling perisynaptic astrocytic leaflets.

Key words Astrocytes, Chronic unpredictable stress, Depression, Confocal microscopy, Viral transfection, Lucifer yellow, Imaris–Fiji software

1 Introduction

Astrocytes are principal homeostatic cells of the central nervous system (CNS) supporting nervous tissue at all levels of organization, from molecular to organ-wide [1, 2]. Astrocytes present a complex spongiform morphology defined by numerous tiny peripheral processes known as leaflets emanating from primary and higher order processes named branches [3, 4]. Astrocytic leaflets establish close associations with synapses and form synaptic cradle to foster, regulate, and sustain synaptic connectivity [5, 6]. Plasma membrane of leaflets is densely packed with Na^+-K^+ pumps and Na^+-dependent solute carrier (SLC) transporters responsible for removal of neurotransmitters, supply of neurotransmitter precursors, ion homeostasis, and provision of reactive

Barbara Di Benedetto (ed.), *Astrocytes: Methods and Protocols*, Methods in Molecular Biology, vol. 2896,
https://doi.org/10.1007/978-1-0716-4366-2_21,

oxygen species scavengers [7–9]. In addition, astrocytes secrete numerous molecules for modulating and regulating synaptogenesis, synaptic elimination, and synaptic function [10]. Astrocytic leaflets are endowed with high degree of morphological plasticity, which, through modulating geometry of leaflet–synapse associations and the distance between leaflet membranes and active synaptic zones, regulate neurotransmitter removal and spillover, thus affecting synaptic transmission [11].

Astrocytes contribute to all neurological and neuropsychiatric diseases; astrogliopathology is complex, mutable, age, context, and disease-specific [12]. Pathological changes in astrocytes range from reactive astrogliosis (an evolutionary conserved defensive program [13]) to astroglial death (clamstodendrosis, [14]). In aging and various forms of neuropathology, shrinkage of leaflets leads to reduced homeostatic support, aberrant neurotransmitter spillover, and impaired synaptic plasticity [15–17]. Astrocytic atrophy with reduced leaflet presence is prominent in brain aging [18, 19] and is implicated in pathophysiology of a wide range of neurological diseases [12, 20], including neurodegenerative disorders [21, 22], epilepsy [23], addiction [24], and various psychiatric disorders [25, 26]. Astrocytic atrophy is the leading histopathological signature of mood disorders in response to stress, in particular in the major depressive disorder (MDD) [27–29] and post-traumatic stress disorder [30]. Exposure of animals to the chronic unpredictable mild stress (CUMS) is one of the most widespread techniques to model depressive-like behaviors (Fig. 1), [27, 31]; these models are used for characterizing pathophysiological mechanisms and refining possible treatments [32]. Treatment with CUMS induces depressive-like behaviors in mice, including decrease in sucrose preference (reflecting anhedonia) and prolongation of the immobility time (indicative of depressive-like changes) in forced swimming test (FST) and tail suspension test (TST), *see* Fig. 2 and [32]. At the cellular level, treatment with CUMS resulted in prominent morphological atrophy of astrocytes, manifested in the decrease of their territorial domains, complexity, and length of processes (Fig. 3, [32]). Treatment with antidepressants (such as fluoxetine) or electroacupuncture at the specific acupoint (Zusanli, ST36) alleviated depressive-like behaviors and restored astrocytic size and morphological complexity [32]. Astrocytic atrophy following exposure to CUMS was accompanied with a decrease in the expression of ezrin [32], a member of ezrin-radixin-moesin (ERM) family, which tethers actin filaments to the plasma membrane and is critical for formation and structural plasticity of astrocytic leaflets and astrocyte–synaptic interactions [33, 34]. Pharmacological treatment with fluoxetine or electroacupuncture restored astrocyte expression of ezrin [32]. Pharmacological management of MDD remains a challenge, with many patients being resistant to traditional antidepressants. Therefore, developing reliable detection

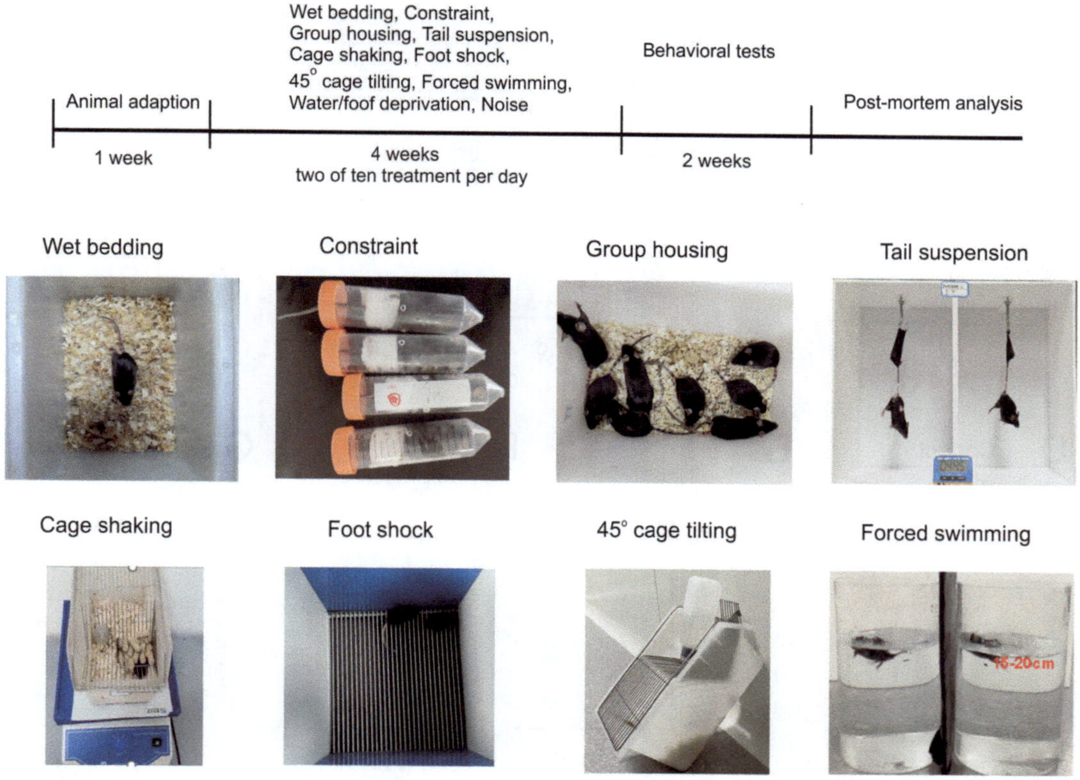

Fig. 1 Chronic unpredictably mild stress regimen: timeline and stressors

methods to reveal deficits of astrocytic morphology might help to identify astrocyte as cellular target for treatment of depressive disorders. In this chapter, we describe the establishment of the CUMS protocol, quantifying depressive-like behaviors with specific tests, and in-depth morphological examination of astrocytes. We describe the technique of labeling astrocytes by targeted expression of fluorescent probe mCherry or intercellular injection of Lucifer yellow, the latter technique being superior to visualize complex arborization of protoplasmic astrocytes (Fig. 4).

2 Materials

2.1 Animals

1. All experiments were performed on C57BL/6 mice (obtained from Chengdu Dossy Experimental Animal Co., Chengdu, China); the mice were 6 weeks old at the beginning of the experimental protocol, which lasted 10 weeks. All mice were adapted to the standard laboratory conditions (24 ± 2 °C room temperature and 65 ± 5% humidity on 12/12 h light–dark cycles) with drinking water and food available ad libitum. The experimental procedures were made in accordance with the National Institute of Health Guidelines for the Care and Use

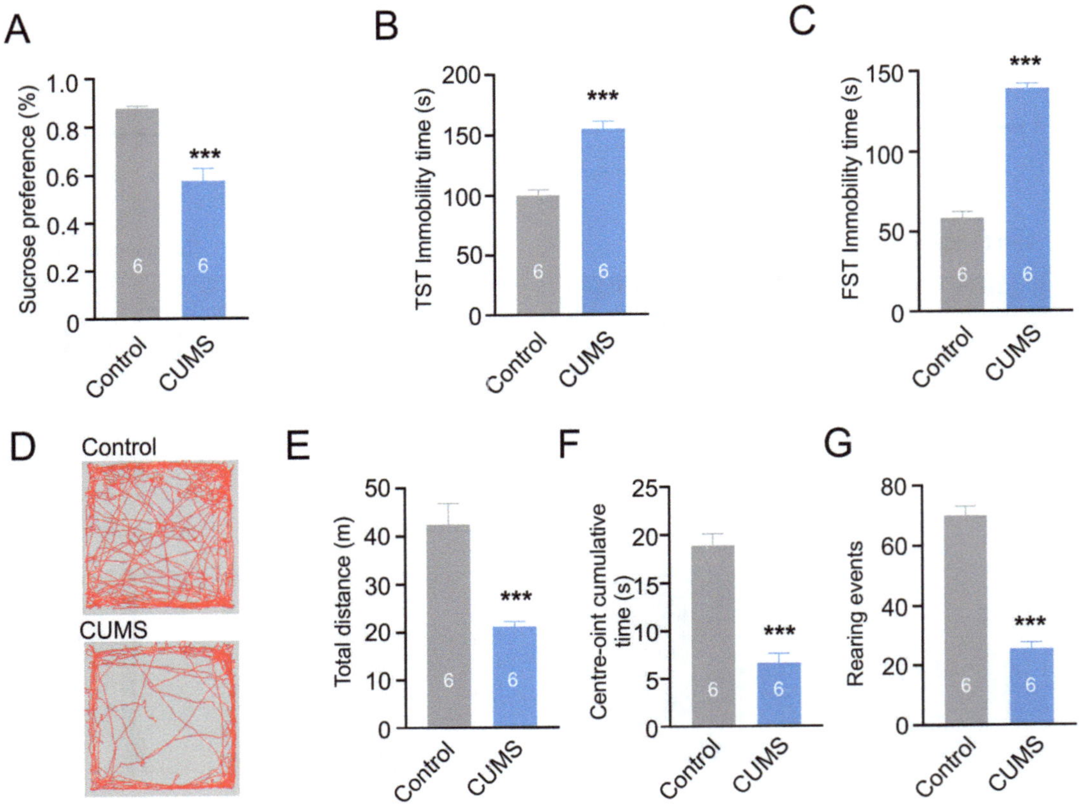

Fig. 2 Exposure to CUMS regimen triggers depressive-like phenotypes in mice. (**a–c**) CUMS reduces sucrose intake and prolongs the TST and FST immobility time. (**d**) CUMS affects exploratory behavior in an open field test; the panel shows representative running traces in open field test; the observation time was 10 min. (**e–g**) Total running distance, center-point cumulative duration, and number of rearing events. All data are presented as mean ± sem. *$p < 0.05$, **$p < 0.01$, ***$p < 0.001$. The number of experiments is indicated on each column. (Reproduced from [32])

of Laboratory Animals and approved by the Animal Ethics Committee of Chengdu University of Traditional Chinese Medicine (protocol code, AM3520, 8 May 2019).

2.2 Preparation of CUMS Regimen and Behavioral Tests (see Notes 1–4)

1. Learned helplessness experiment system.
2. 50 mL tube with hole.
3. The cylindrical tanks (30 cm height × 20 cm diameters).
4. Laboratory shaker.
5. Tail suspension box.
6. EthoVision XT video tracking system.
7. Special water bottle with tick marks.
8. Thermometer.
9. 75% ethyl alcohol.

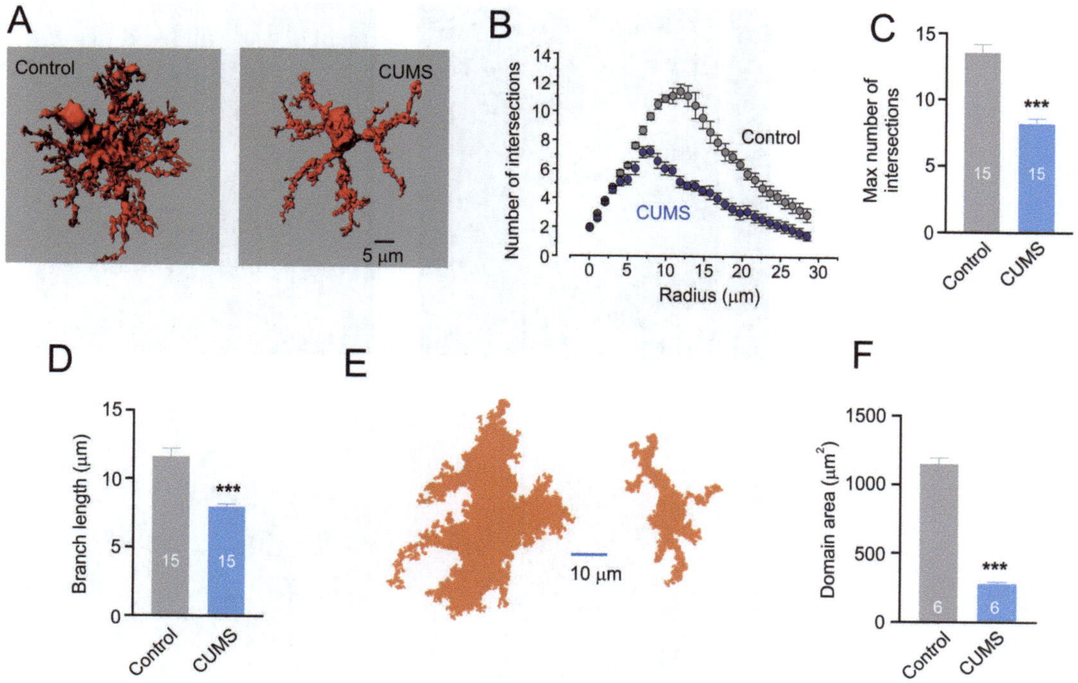

Fig. 3 CUMS induces morphological atrophy in astrocytes in the prefrontal cortex. (**a**) Representative 3D reconstruction of astrocyte in control and CUMS groups. (**b**) Sholl analysis of astrocytic morphology for control and CUMS groups shows the number of intersections of astrocytic branches with concentric spheres centered in the middle of cell soma. (**c**) Maximal number of intersections for astrocytes in control and CUMS groups. (**d**) Average length of astrocytic processes in control and CUMS groups. (**b–d**) $n = 15$ for each group. (**e**) Representative examples of astrocytic territorial domains obtained as a projection of astrocytes along the z-axis projection for control and CUMS animals. (**f**) Average astrocytic domain for control and CUMS group. All data are presented as mean ± sem. $^*p < 0.05$, $^{**}p < 0.01$, $^{***}p < 0.001$. The number of experiments is indicated in each column

10. Double distilled water (dd water).
11. Hand towel.
12. Sucrose.

2.3 Preparation of Virus Injection

1. Stereotaxic system.
2. 75% ethanol, carprofen 20 mg/kg (50 mg/mL injection solution), sterile saline, and PBS with glycerol small volumes.
3. Virus: AAV2/5-gfaABC1D-mCherry (*see* **Notes 5–6**).
4. 1 µL microsyringe and medical sutures.
5. General laboratory tools.

2.4 Preparation of Brain for Immunofluorescence and Lucifer Yellow Iontophoretic Injection

1. Mini-peristaltic pump for intracardial perfusion for small animals, water bath.
2. Laboratory dissection tools for brain preparation and extraction.

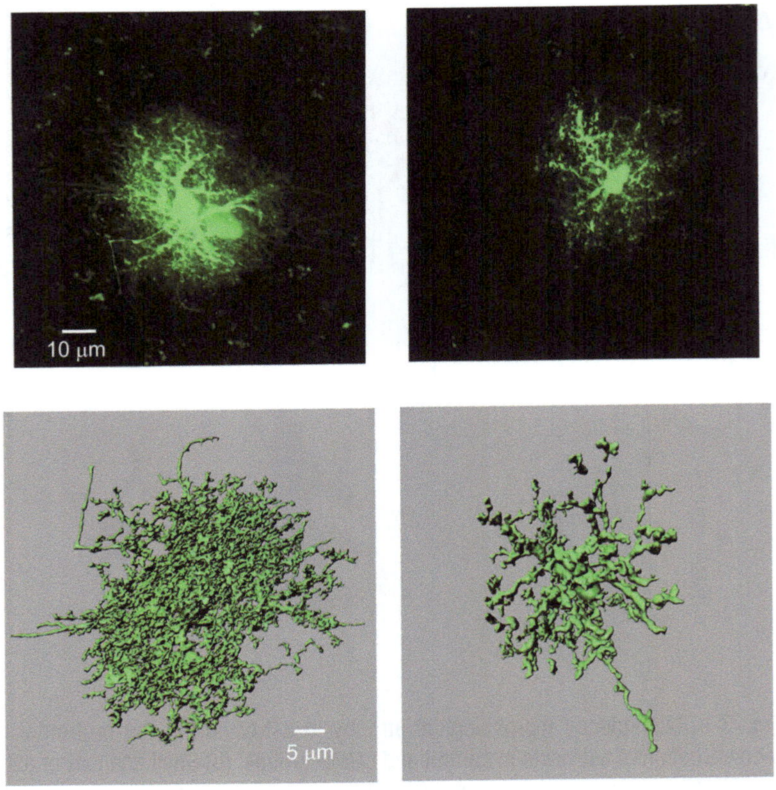

Fig. 4 Representative confocal images of Lucifer yellow–labeled astrocyte in mPFC. (**a**) Representative confocal images of Lucifer yellow–labeled astrocyte in mPFC in control and CUMS groups. (**b**) Representative 3D reconstruction of astrocyte in control and CUMS groups

3. Perfusion fluid: saline, 4% paraformaldehyde (PFA), pH 7.5, 10% formalin solution, and PBS with heparin sodium and lidocaine.
4. Cryoprotective solution (for brains after perfusion): 4% PFA and 30% sucrose in PBS 1×.

2.5 Preparation of Brain Slices for Immunofluorescence Recordings (*see* Note 7)

1. Cryostat.
2. Optimal cutting temperature compound, to fix brains for cutting in the cryostat.
3. Microtome blades, adhesion microscope slides, and slice box.

2.6 Preparation of Brain Slices for Lucifer Yellow Iontophoretic Injection (*see* Note 7)

1. Vibratome, microtome blades, and round plates.
2. Washing buffer: PBS 1× solution containing 8 mM Na_2HPO_4, 137 mM NaCl, and 2 mM NaH_2PO_4.

2.7 Immunofluorescence Staining

1. Washing buffer: PBST 1× solution containing 135 mM NaCl, 2.7 mM KCl, 3.2 mM Na_2HPO_4, 0.5 mM KH_2PO_4, and 0.05% Tween 20.
2. Blocking/permeabilization buffer: 10% serum and 5% Triton ×-100.
3. Antibody buffer: PBST 1×, 2% serum, and 0.2% Triton ×-100.
4. Primary antibody: rabbit monoclonal anti-ezrin antibody (1: 100, Cell Signalling Technology).
5. Secondary antibody: Alexa Fluor 488 SuperBoost kit (Invitrogen, *see* **Note 8**).
6. Working buffer for secondary antibody:1 μL H_2O_2 and 100 μL reaction buffer (component from SuperBoost kit) (*see* **Note 9**).
7. Stopping buffer for secondary antibody: 1 μL stop buffer (component D from SuperBoost kit):110 μL PBST 1× (*see* **Note 9**).
8. Antifade mounting medium.
9. Microscope cover glass.
10. Nature balsam.
11. Confocal microscope.

2.8 Lucifer Yellow Iontophoretic Injection

1. Patch clamp system.
2. Borosilicate glass (BF100-58-10).
3. Microelectrode puller.
4. Electrical circuit for iontophoresis.
5. 1–1.5% Lucifer yellow solution (dissolve in dd water).
6. Adhesion microscope slides and microscope cover glass.
7. Antifade mounting medium.
8. Nature balsam.
9. Confocal microscope.

3 Methods

3.1 CUMS Protocol (Fig. 1)

1. Forced swimming: prepare 4 °C water and tank, and put mice into the tank for 5 min.
2. Tail suspension: prepare tail suspension system, and put mice onto it for 5 min.
3. Cage shaking: prepare lab shaker, put mice cage on it for 5 min, and the velocity of the shaker is 45–60 r/min.
4. Wet bedding: use water to wet mice bedding, and let mice stay with it for 12 h.
5. Constraint: put mice into lab 50 mL tube for 1 h.

6. Foot shock: put mice into learned helplessness box, and set parameter as 1 mA for 5 min.
7. Group housing: group housing mice as six or seven mice per cage.
8. 45° cage tilting: put mice cage in a 45° cage tilt for 12 h.
9. Water and (or) food deprivation: water and (or) food deprivation for 12 h.
10. Noise: use learned helplessness system, and set parameter noise as 120db for 3 h.

3.2 Behavioral Tests (*see* Notes 1–4)

1. Open filed test: prepare for EthoVision XT video tracking system, and set all parameters. Track the mouse in the box for 10 min. The box should be cleaned with dd water after each session to get rid of the smell.
2. Tail suspension test: prepare EthoVision XT video tracking system, tail suspension box, and tape. Tape mice tail, and the approximate distance between the mouse's nose and the apparatus floor is 20–25 cm. Track the mouse in the box for 6 min. The box should be cleaned with dd water after each session to get rid of the smell.
3. Forced swimming test: prepare EthoVision XT video tracking system, cylindrical tanks, and thermometer. The tanks should be filled with tap water about 20 cm high and make sure the temperature is around 23–25 °C. Track the mouse in the tanks for 6 min. The water should be changed after each session.
4. Sucrose preference test: prepare water bottle with tick marks, sucrose. Provide mice with two bottles of water in their home cages, leave the mice with these two bottles for 3 days, then change two bottles with 1% sucrose solution for the following 2 days, and then treat mice with water and food deprivation for 24 h. On the test day, provide mice two bottles with one containing 1% sucrose solution and one containing water for 12 h.

3.3 Virus Injection (*see* Notes 5–6)

1. Anesthetize mice with isoflurane (0.3–0.5 min/L).
2. Shave mice hair, and cut the scalp to make an anterior–posterior incision.
3. Fully expose and dry the skull, and using a sterile cotton swab, clean the surface of the skull to make sure the bregma and the lambda are visible.
4. Adjust bregma and lambda and maintain a difference around ±0.05 mm.
5. Find the coordinates of mPFC based on the atlas of the mouse brain: 1.9 mm posterior to the bregma, ± 0.3 mm lateral to the midline, and −1.3 mm ventral to the surface of the skull, and label it with a marker.

6. Using a fine dental drill (0.5 mm drill bit) carefully drill the skull at the target site.
7. Clean the microsyringe with sterile PBS, and fill it with 0.7 μL of the concentrated virus solution.
8. Place the microsyringe over the hole and slowly lower it vertically until it reaches the surface of the brain. Set the dorsal/ventral (D/V) coordinate (depth) to 0.
9. Lower the syringe to reach the mPFC, and set D/V to 1.3 mm.
10. Set the digital pump to 0.035 μL/min and start infusion.
11. Slowly remove the syringe out of the brain.
12. Suture the wound with absorbable thread.

3.4 Immunofluorescence

1. Anesthetize mice with isoflurane first, and then inject pentobarbital sodium (1%, 0.2 mL). Quickly place mice on the perfusion table and fix limbs and tail.
2. With the help of big forceps and scissors, hold the skin of the animals above the chest, open it by cutting a small hole below the sternum, and carefully cut laterally just under the diaphragm. Then, carefully cut the diaphragm along the ribcage. Cut up the two sides of the chest, clamp the sternum with a pair of hemostats, and lift it up to expose the chest cavity.
3. Take the apex of the heart with your thumb and forefinger or a good pair of tissue forceps. Insert the perfusion needle into the left ventricle and hold it there with a pair of hemostats. Let the cold (4 °C) saline flow at a steady drip of 14 mL/min until the liver turns to a pale color, which indicates a good flush of arteries/veins from blood with saline.
4. Change the saline to cold PFA 4%, and continue perfusing the animal for 5–10 min until all paws and liver become whitish and blenched, respectively.
5. After perfusion, open the skull and remove the brain.
6. Place each brain in a 15 mL Falcon tube for post-fixation overnight in 4% PFA at 4 °C in the freezer.
7. After 1 day, wash brains with saline three times, and substitute PFA 4% with 30% sucrose solution (30% sucrose in PBS 1×). Let the brains in this solution at 4 °C in the freezer until they sink.
8. After sinking of the brains, place them at −20 or −80 °C until ready for cutting.

3.5 Immunostaining

1. Take out brains from the freezer, wash them shortly in saline, and then, wipe them from excess of liquid.
2. Prepare the brain holder of the cryostat with a thin layer of OCT compound, and immerge the brain into it in a vertical

position (olfactory bulbs toward you), surround it at the base with additional OCT, and keep it in this position until the OCT is completely frozen.

3. Place the block in the correct position in the cryostat and start cutting at 40 μm (*see* **Note 7**).
4. Use adhesion microscope slides to stick brain slice and put them into slice box.
5. Check virus expression in the slices by standard fluorescence microscopy.
6. Put slice box at −20 or −80 °C until ready for staining.
7. Draw a circle around the brain slice with immunity staining guard pen.
8. Wash brain slice with PBST 1× three times for 5 min.
9. Incubate with 5% Triton ×-100 in a 37 °C incubator, 30 min.
10. Incubate with 10% goat serum in a 37 °C incubator, 1 h.
11. Remove the goat serum, and incubate slices with primary antibody for 48 h at 4 °C in the freezer.
12. Wash brain slice with PBST 1× in a special box on a horizontal shaker three times for 10 min each.
13. Incubate with the second antibody for 1 h at the room temperature. From step 13 onward, the brain slice should be protected from light.
14. Wash the brain slice with PBST 1× three times for 10 min in a special box on a horizontal shaker.
15. Incubate with working buffer for 8–9 min at room temperature.
16. Remove the working buffer, and incubate with stop solution for 8–10 min at room temperature.
17. Wash the brain slice with PBST 1× three times for 10 min in a special box on a horizontal shaker.
18. Remove the buffer from the brain slice, and wipe the slice from excess of liquid using a dust-free mirror wiping paper.
19. Put few drops of mounting medium on slides and cover them with a microscope cover glass.
20. Glue cover glass with nature balsam, and keep sections at 4 °C in the freezer until imaging.

3.6 Lucifer Yellow Iontophoretic Injection

1. Take brains from the freezer, wash them shortly in PBS and then wipe them from excess of liquid.
2. Glue the brain to the stage (olfactory bulbs toward you), fix the holder to the vibratome, and fix microtome blades.
3. Set the top and the bottom on the vibratome, and start cutting at 150 μm (*see* **Note 7**).

4. Collect brain slices, which contain mPFC, and place them into PBS 1× on ice.
5. Use microelectrode puller to make a sharp microelectrode for injection.
6. Fill microelectrode with Lucifer yellow solution, and fix to the drug delivery system on self-made patch clamp holder.
7. Identify single astrocyte by the small, round, or oval cell body in the mPFC under the microscope.
8. Inject Lucifer yellow with the help of iontophoretic electrical circuit into single astrocyte slowly, until the arborization of the astrocyte can be clearly seen.
9. Put slices from PBS, and mount them on adhesion slide with the help of a thin brush to gently drift them.
10. Wipe brain slices from the excess of liquid using dust-free mirror wiping paper.
11. Put few drops of mounting medium on slides and cover them with a microscope cover glass.
12. Glue cover glass with nature balsam, and keep sections at 4 °C in the freezer until imaging.

3.7 Confocal Microscopy

1. Use 4× objective to scan the sections to localize the region of interest.
2. Use 60× or 100× oil immersion objective to find single astrocyte, find the surface of astrocyte, set zero confocal plane, and then find the deepest surface to set the end confocal plane.
3. Start z-stack imaging (0.25 μm z-step size for each image) with 100× oil immersion objective and the image frame size of 1024 × 1024 μm.
4. Save files.

3.8 Behavioral Test Analysis

1. Analyze immobility time of tail suspension test and forced swimming test only the last 4 min.
2. Calculate sucrose and water intake separately. Calculate sucrose preference using the following formula: preference = (sucrose intake/total intake) × 100%.

3.9 Sholl Analysis

1. Import z-stack images files into Fiji software.
2. Use the "Z projection" to generate a 2D image.
3. Use "Process," "Binary," and "Make Binary" to generate a binary image for Sholl analysis.
4. Draw a straight line between astrocytic soma and the longest branch.

5. Use "Analysis" and "Sholl Analysis" to set the parameter: starting radius, ending radius, and radius step size.
6. The "Sholl analysis" results will show the degree of astrocyte complexity.

3.10 3D Reconstruction

1. Convert images from the original *.oib format into *.ims format by Imaris Viewer.
2. Import *.ims images into Imaris software.
3. Use the "Slice" function of Imaris to measure soma and "Filament function" of Imaris to measure branch distance.
4. Use "Surface" and "Add new surface" to generate different surface channels.
5. Import distances into "Diameter of largest Sphere which fits into the Object," and import 0.1 times of soma or branch distance into "Surface Area Detail Level."
6. Adjust the "Threshold" until the 3D surface covered the fluorescence image.
7. Choose "Surface" to show the 3D reconstruction image.
8. If there are different channels, repeat **steps 1–7** to generate a 3D reconstruction image.
9. Choose "Tools" and "Surface-Surface coloc" plug-in to generate the surface–surface colocalization.
10. Choose "Statistics," and find "Total number of surface," which shows the surface–surface colocalization data.
11. Choose "Snapshot" to export the 3D reconstruction image.

3.11 Statistical Analysis

1. All statistical analyses were performed by GraphPad Prism 8. Data with more than two groups were tested for significance using one-way ANOVA test followed by the Holm–Sidak test. Multiple comparisons between the data were performed in case of their non-normal distribution, using the Kruskal–Wallis ANOVA on ranks, followed by Tukey's test. Significance was defined as $P < 0.05$.
2. For each brain: ~10 images of the single astrocyte were obtained, and 3D reconstructions were made.

4 Notes

1. There should be enough time for mice to have a break between two stressors during CUMS protocol.
2. Before behavioral tests, mice should be at behavioral room for at least 1 h to adapt to light, smell, and sound.

3. The sequence of behavioral tests is recommended to be as follows: open field test, tail suspension test, force swimming test, and sucrose preference test. Every two tests should be separated by at least 1 day.
4. It is very important to change two bottle positions every 12 h in sucrose preference test.
5. We prefer to inject virus with 10^{11}–10^{12} gc/mL titer to label astrocyte.
6. We prefer to inject virus in cortex with the velocity around 0.02–0.025 μL/min.
7. The thickness of the brain slice for immunostaining is 30–45 μm, for Lucifer yellow injection 150–200 μm.
8. For ezrin immunostaining, we prefer goat anti-rabbit Alexa 488 (Invitrogen, Carlsbad, California, USA) as the second antibody.
9. It is important to control the working and stopping buffer time for ezrin secondary antibody.

Acknowledgments

This work was supported by grants from NSFC-RSF (82261138557 and 23-44-00103), Sichuan Provincial Administration of Traditional Chinese Medicine (2023zd024), NSFC (82274668, 82230127), the Innovation Team and Talents Cultivation Program of the National Administration of Traditional Chinese Medicine (ZYYCXTD-D-202003), and the Sichuan Science and Technology Program (2022YFH0006).

References

1. Verkhratsky A, Nedergaard M (2016) The homeostatic astroglia emerges from evolutionary specialization of neural cells. Philos Trans R Soc Lond Ser B Biol Sci 371(1700):20150428
2. Verkhratsky A, Butt AM (2023) Neuroglia: function and pathology. Elsevier, New York
3. Semyanov A, Verkhratsky A (2021) Astrocytic processes: from tripartite synapses to the active milieu. Trends Neurosci 44(10):781–792
4. Khakh BS, Sofroniew MV (2015) Diversity of astrocyte functions and phenotypes in neural circuits. Nat Neurosci 18(7):942–952
5. Verkhratsky A, Nedergaard M (2014) Astroglial cradle in the life of the synapse. Philos Trans R Soc Lond Ser B Biol Sci 369(1654): 20130595
6. Nedergaard M, Verkhratsky A (2012) Artifact versus reality – how astrocytes contribute to synaptic events. Glia 60(7):1013–1023
7. Rose CR, Verkhratsky A (2016) Principles of sodium homeostasis and sodium signalling in astroglia. Glia 64(10):1611–1627
8. Verkhratsky A, Rose CR (2020) Na^+-dependent transporters: the backbone of astroglial homeostatic function. Cell Calcium 85: 102136
9. Rose CR, Verkhratsky A (2024) Sodium homeostasis and signalling: the core and the hub of astrocyte function. Cell Calcium 117: 102817
10. Augusto-Oliveira M, Arrifano GP, Takeda PY, Lopes-Araújo A, Santos-Sacramento L,

Anthony DC, Verkhratsky A, Crespo-Lopez ME (2020) Astroglia-specific contributions to the regulation of synapses, cognition and behaviour. Neurosci Biobehav Rev 118:331–357

11. Henneberger C, Bard L, Panatier A, Reynolds JP, Kopach O, Medvedev NI, Minge D, Herde MK, Anders S, Kraev I, Heller JP, Rama S, Zheng K, Jensen TP, Sanchez-Romero I, Jackson CJ, Janovjak H, Ottersen OP, Nagelhus EA, Oliet SHR, Stewart MG, Nägerl UV, Rusakov DA (2020) LTP induction boosts glutamate spillover by driving withdrawal of perisynaptic astroglia. Neuron 108(5):919
12. Verkhratsky A, Butt A, Li B, Illes P, Zorec R, Semyanov A, Tang Y, Sofroniew MV (2023) Astrocytes in human central nervous system diseases: a frontier for new therapies. Signal Transduct Target Ther 8(1):396
13. Escartin C, Galea E, Lakatos A, O'Callaghan JP, Petzold GC, Serrano-Pozo A, Steinhauser C, Volterra A, Carmignoto G, Agarwal A, Allen NJ, Araque A, Barbeito L, Barzilai A, Bergles DE, Bonvento G, Butt AM, Chen WT, Cohen-Salmon M, Cunningham C, Deneen B, De Strooper B, Diaz-Castro B, Farina C, Freeman M, Gallo V, Goldman JE, Goldman SA, Gotz M, Gutierrez A, Haydon PG, Heiland DH, Hol EM, Holt MG, Iino M, Kastanenka KV, Kettenmann H, Khakh BS, Koizumi S, Lee CJ, Liddelow SA, MacVicar BA, Magistretti P, Messing A, Mishra A, Molofsky AV, Murai KK, Norris CM, Okada S, Oliet SHR, Oliveira JF, Panatier A, Parpura V, Pekna M, Pekny M, Pellerin L, Perea G, Perez-Nievas BG, Pfrieger FW, Poskanzer KE, Quintana FJ, Ransohoff RM, Riquelme-Perez M, Robel S, Rose CR, Rothstein JD, Rouach N, Rowitch DH, Semyanov A, Sirko S, Sontheimer H, Swanson RA, Vitorica J, Wanner IB, Wood LB, Wu J, Zheng B, Zimmer ER, Zorec R, Sofroniew MV, Verkhratsky A (2021) Reactive astrocyte nomenclature, definitions, and future directions. Nat Neurosci 24(3):312–325
14. Balaban D, Miyawaki EK, Bhattacharyya S, Torre M (2021) The phenomenon of clasmatodendrosis. Heliyon 7(7):e07605
15. Verkhratsky A, Augusto-Oliveira M, Pivoriunas A, Popov A, Brazhe A, Semyanov A (2021) Astroglial asthenia and loss of function, rather than reactivity, contribute to the ageing of the brain. Pflugers Arch 473(5):753–774
16. Verkhratsky A, Rodrigues JJ, Pivoriunas A, Zorec R, Semyanov A (2019) Astroglial atrophy in Alzheimer's disease. Pflugers Arch 471(10):1247–1261
17. Yang Z, Gong M, Jian T, Li J, Yang C, Ma Q, Deng P, Wang Y, Huang M, Wang H, Yang S, Chen X, Yu Z, Wang M, Chen C, Zhang K (2022) Engrafted glial progenitor cells yield long-term integration and sensory improvement in aged mice. Stem Cell Res Ther 13(1):285
18. Popov A, Brazhe A, Denisov P, Sutyagina O, Li L, Lazareva N, Verkhratsky A, Semyanov A (2021) Astrocyte dystrophy in ageing brain parallels impaired synaptic plasticity. Aging Cell 20(3):e13334
19. Popov A, Brazhe N, Morozova K, Yashin K, Bychkov M, Nosova O, Sutyagina O, Brazhe A, Parshina E, Li L, Medyanik I, Korzhevskii DE, Shenkarev Z, Lyukmanova E, Verkhratsky A, Semyanov A (2023) Mitochondrial malfunction and atrophy of astrocytes in the aged human cerebral cortex. Nat Commun 14(1):8380
20. Verkhratsky A, Zorec R, Parpura V (2017) Stratification of astrocytes in healthy and diseased brain. Brain Pathology (Zurich, Switzerland) 27(5):629–644
21. Olabarria M, Noristani HN, Verkhratsky A, Rodríguez JJ (2010) Concomitant astroglial atrophy and astrogliosis in a triple transgenic animal model of Alzheimer's disease. Glia 58(7):831–838
22. Yeh C-Y, Vadhwana B, Verkhratsky A, Rodríguez JJ (2011) Early astrocytic atrophy in the entorhinal cortex of a triple transgenic animal model of Alzheimer's disease. ASN Neuro 3(5):271–279
23. Plata A, Lebedeva A, Denisov P, Nosova O, Postnikova TY, Pimashkin A, Brazhe A, Zaitsev AV, Rusakov DA, Semyanov A (2018) Astrocytic atrophy following status epilepticus parallels reduced Ca2+ activity and impaired synaptic plasticity in the rat hippocampus. Front Mol Neurosci 11:215
24. Scofield MD, Li H, Siemsen BM, Healey KL, Tran PK, Woronoff N, Boger HA, Kalivas PW, Reissner KJ (2016) Cocaine self-administration and extinction leads to reduced glial fibrillary acidic protein expression and morphometric features of astrocytes in the nucleus accumbens core. Biol Psychiatry 80(3):207–215
25. Kim R, Healey KL, Sepulveda-Orengo MT, Reissner KJ (2018) Astroglial correlates of neuropsychiatric disease: From astrocytopathy to astrogliosis. Prog Neuro-Psychopharmacol Biol Psychiatry 87(Pt A):126–146
26. Verkhratsky A, Parpura V (2016) Astrogliopathology in neurological, neurodevelopmental and psychiatric disorders. Neurobiol Dis 85:254–261

27. Aten S, Du Y, Taylor O, Dye C, Collins K, Thomas M, Kiyoshi C, Zhou M (2023) Chronic stress impairs the structure and function of astrocyte networks in an animal model of depression. Neurochem Res 48(4): 1191–1210
28. Rajkowska G, Hughes J, Stockmeier CA, Javier Miguel-Hidalgo J, Maciag D (2013) Coverage of blood vessels by astrocytic endfeet is reduced in major depressive disorder. Biol Psychiatry 73(7):613–621
29. Rajkowska G, Stockmeier CA (2013) Astrocyte pathology in major depressive disorder: insights from human postmortem brain tissue. Curr Drug Targets 14(11):1225–1236
30. Li B, Zhang D, Verkhratsky A (2022) Astrocytes in post-traumatic stress disorder. Neurosci Bull 38(8):953–965
31. Li X, Chen B, Zhang D, Wang S, Feng Y, Wu X, Cui L, Ji M, Gong W, Verkhratsky A, Xia M, Li B (2023) A novel murine model of mania. Mol Psychiatry 28(7):3044–3054
32. Lin S-S, Zhou B, Chen B-J, Jiang R-T, Li B, Illes P, Semyanov A, Tang Y, Verkhratsky A (2023) Electroacupuncture prevents astrocyte atrophy to alleviate depression. Cell Death Dis 14(5):343
33. Badia-Soteras A, Heistek TS, Kater MSJ, Mak A, Negrean A, van den Oever MC, Mansvelder HD, Khakh BS, Min R, Smit AB, Verheijen MHG (2022) Retraction of astrocyte leaflets from the synapse enhances fear memory. Biol Psychiatry 94:226
34. Derouiche A, Frotscher M (2001) Peripheral astrocyte processes: monitoring by selective immunostaining for the actin-binding ERM proteins. Glia 36(3):330–341

Correction to: Astrocytes

Barbara Di Benedetto

Correction to:
Barbara Di Benedetto (ed.), ***Astrocytes: Methods and Protocols,***
Methods in Molecular Biology, vol. 2896,
https://doi.org/10.1007/978-1-0716-4366-2

The below acknowledgement text from FM has been moved to chapter 4

"The work presented here was supported by the Deutsche Forschungsgemeinschaft (DFG, German Research Foundation; Projekt# 461542557) and the Federal Ministry of Education and Research (BMBF), Germany (Project SynGluCross to CRR)."

The updated version of this book can be found at
https://doi.org/10.1007/978-1-0716-4366-2
https://doi.org/10.1007/978-1-0716-4366-2_4

Barbara Di Benedetto (ed.), *Astrocytes: Methods and Protocols*, Methods in Molecular Biology, vol. 2896,
https://doi.org/10.1007/978-1-0716-4366-2_22,

INDEX

Barbara Di Benedetto (ed.), *Astrocytes: Methods and Protocols*, Methods in Molecular Biology, vol. 2896, https://doi.org/10.1007/978-1-0716-4366-2,

MIX
Papier aus verantwortungsvollen Quellen
Paper from responsible sources
FSC® C105338

If you have any concerns about our products,
you can contact us on
ProductSafety@springernature.com

In case Publisher is established outside the EU,
the EU authorized representative is:
Springer Nature Customer Service Center GmbH
Europaplatz 3, 69115 Heidelberg, Germany

Printed by Libri Plureos GmbH
in Hamburg, Germany